JAMAICA

Primary Mathematics

Second Edition

NSC Edition

Grade 6

Michelle Holgarth

hachette LEARNING

The Publishers would like to thank the following for permission to reproduce copyright material.

Photo credits

t = top, *b* = below, *l* = left, *r* = right, *c* = centre, *b/g* = background

p.8: (t) © cosmic_pony/stock.adobe.com, (c) © Dejan Jovanovic/stock.adobe.com, (b) © Jut/stock.adobe.com; p.19: © sabelskaya/stock.adobe.com; p.24: © Thomas Bredenfeld/stock.adobe.com; p.27: © Keitma/stock.adobe.com; p.32: © Victority/stock.adobe.com; p.35: © lukszczepanski/stock.adobe.com; p.39: © ASEF/stock.adobe.com; p.41: © Primastock/stock.adobe.com; p.43: © Prostock-studio/stock.adobe.com; p.44: © absolut/stock.adobe.com; p.45: © chachamp/stock.adobe.com; p.49: (l) © Marko/stock.adobe.com, (c) © Ildar/stock.adobe.com, (r) © Chaosamran_Studio/stock.adobe.com; p.50: © Peter Hermes Furian/stock.adobe.com; p.56: © Sergei Sizkov/stock.adobe.com; p.61: © iofoto/stock.adobe.com; p.76: (l) © Silverpics/stock.adobe.com, (r) © Fotokon/stock.adobe.com; p.77: (t) © Marina Marr/stock.adobe.com, (b) © hary_cz/stock.adobe.com; p.79: © Natis/stock.adobe.com; p.80: (tl) © cloud7days/stock.adobe.com, (tc) © Vasily Merkushev/stock.adobe.com, (tr) © bjphotographs/stock.adobe.com, (cl) © bigmouse108/stock.adobe.com, (c) © New Africa/stock.adobe.com, (cr) © Dmitri Stalnuhhin/stock.adobe.com, (b) © Gudellaphoto/stock.adobe.com; p.81: © YummyBuum/stock.adobe.com; p.82: © _nastassia/stock.adobe.com; p.83: © yarm_sasha/stock.adobe.com; p.84: © Peter Hermes Furian/stock.adobe.com; p.85: © tashka2000/stock.adobe.com; p.87: © deke/stock.adobe.com; p.100: © photka/stock.adobe.com; p.102: © Alexe/stock.adobe.com; p.109: © Tony Baggett/stock.adobe.com; p.111: © tadamichi/stock.adobe.com; p.116: © irinagoleva/stock.adobe.com; p.118: © StopperOhana/Shutterstock.com; p.121: © Pixel-Shot/stock.adobe.com; p.122: (l) © ImageFlow/stock.adobe.com, (r) © Koonsiri/stock.adobe.com; p.128: © photoeverywhere/stock.adobe.com; p.133: © Viktorija Reuta/Shutterstock.com; p.134: (tl) © Andy Dean/stock.adobe.com, (tc) © Hogan Imaging/stock.adobe.com, (tr) © pololia/stock.adobe.com, (bl) © pololia/stock.adobe.com, (c) © Samuel B./stock.adobe.com, (br) © pololia/stock.adobe.com; p.135: © pavel siamionov/stock.adobe.com; p.137: © Yuri Bizgaimer/stock.adobe.com; p.138: (t) © Jiri Hera/stock.adobe.com, (c) © WavebreakMediaMicro/stock.adobe.com; p.145: © Andriy Bezuglov/stock.adobe.com; p.146: © Nata Bene/stock.adobe.com; p.151: © Joke Phatrapong/stock.adobe.com; p.154: © vectorfusionart/stock.adobe.com; p.156: © Christoph/stock.adobe.com; p.159: © Olga Miltsova/stock.adobe.com; p.161: © muratart/stock.adobe.com; p.172: © dimazel/stock.adobe.com; p.174: (l) © nexusby/stock.adobe.com, (r) © Maor_Winetrob/Shutterstock.com; p.175: © tacna/stock.adobe.com; p.182: © DmitryStock/stock.adobe.com; p.184: © Wayhome Studio/stock.adobe.com; p.189: © Gedion van Eiken/Shutterstock.com; p.191: © JackF/stock.adobe.com; p.195: © magdal3na/stock.adobe.com; p.196: © domnitsky/stock.adobe.com; p.197: © lilkin/stock.adobe.com; p.199: © mavoimages/stock.adobe.com; p.200: © pololia/stock.adobe.com; p.201: © Dariia/stock.adobe.com; p.205: © Joshhh/stock.adobe.com; p.207: © microvictor/Shutterstock.com; p.208: (l) © skyNext/stock.adobe.com, (c) © DanielKoell/stock.adobe.com, (r) © Dariusz Jarzabek/stock.adobe.com; p.210: (tl) © Alekss/stock.adobe.com, (tr) © Richard Blaker/stock.adobe.com, (bl) © Alexstar/stock.adobe.com, (br) © butus/stock.adobe.com; p.226: © AKaiser/Shutterstock.com; p.228: © Gianfranco Vivi/Shutterstock.com; p.231: © jethita/stock.adobe.com; p.233: (c) © matiasdelcarmine/stock.adobe.com, (b) © corbacserdar/stock.adobe.com; p.237: © Prostock-studio/Shutterstock.com; p.240: (t) © lunamarina/stock.adobe.com, (c) © oksix/stock.adobe.com, (b) © dule964/stock.adobe.com; p.246: © Sunflowerr/Shutterstock.com; p.247: © Kavalenkava/stock.adobe.com; p.250: © polya_olya/stock.adobe.com; p.269: © kovaleva_ka/stock.adobe.com; p.271: © Autumn Sky/stock.adobe.com; p.274: © Arob/Shutterstock.com; p.281: © stokkete/stock.adobe.com; p.282: (l) © kite_rin/stock.adobe.com, (c) © michael spring/stock.adobe.com, (r) © VAKSMANV/stock.adobe.com; p.284: © Syda Productions/stock.adobe.com; p.287: (c) © Zamrznuti tonovi/stock.adobe.com, (b) © WavebreakMediaMicro/stock.adobe.com; p.292: (l) © ravennka/stock.adobe.com, (cl) © sommersby/stock.adobe.com, (cr) © janaluchenko/stock.adobe.com, (r) © R. Gino Santa Maria/stock.adobe.com; p.293: © PRPicturesProduction/stock.adobe.com; p.298: © ZhouEka/stock.adobe.com; p.304: © Dzha/stock.adobe.com; p.306: © Skórzewiak/stock.adobe.com; p.309: © White bear studio/stock.adobe.com; p.310: © bilanol/stock.adobe.com; p.316: (l) © kunpisit/stock.adobe.com, (c) © Francesco Milanese/stock.adobe.com, (r) © Brooke Becker/stock.adobe.com; p.317: (c) © Green Flame/Shutterstock.com, (b) © A-R-T/Shutterstock.com; p.320: © conzorb/stock.adobe.com; p.321: © MarekPhotoDesign.com/stock.adobe.com; p.325: (cl) © Debbie Ann Powell/stock.adobe.com, (tr) © photopixel/stock.adobe.com, (bl) © TTStock/stock.adobe.com, (bc) © coachwood/stock.adobe.com, (cr) © Noey smiley/stock.adobe.com; p.328: (t) © Rido/stock.adobe.com, (c) © mimagephotography/stock.adobe.com, (b) © muratart/stock.adobe.com; p.329: © photoyh/Shutterstock.com; p.330: (t) © 22Imagesstudio/stock.adobe.com, (c) © Gary Perkin/stock.adobe.com; p.334: © renklerin kafasi/Shutterstock.com; p.335: (t) © Syda Productions/stock.adobe.com, (b) © D. Lewis/Shutterstock.com; p.336: © BillionPhotos.com/stock.adobe.com; p.338: © Solarisys/stock.adobe.com; p.344: © Vitalii Vodolazskyi/stock.adobe.com; p.345: © Ainur Mufid/Shutterstock.com; p.361: © photka/stock.adobe.com; p.362: © Ivonne Wierink/stock.adobe.com; p.364: © Brian Jackson/stock.adobe.com; p.365: © Yaroslaff/Shutterstock.com; p.367: © Nikolai Sorokin/stock.adobe.com; p.369: © Mangostar/stock.adobe.com.

To order, please visit www.hachettelearning.com or contact Customer Service at education@hachette.co.uk / +44 (0)1235 827827

ISBN: 9781036019983

First published in 2023

This edition published in 2026 by Hachette Learning (a trading division of Hodder & Stoughton Limited)

An Hachette UK Company

Carmelite House

50 Victoria Embankment

London EC4Y 0DZ

www.hachettelearning.com

Impression number 10 9 8 7 6 5 4 3 2 1

Year 2026

Cover illustration by Oxford Designers and Illustrators Ltd

Illustrations by Magriet Brink, Leo Daly, Amelia Rosato, Andy Cooke, Peter Lubach, Aptara Ltd., Oxford Designers and Illustrators Ltd. and Hyphen S.A.

Typeset by Hyphen S.A.

Printed and Bound in Great Britain by Bell & Bain Ltd, Glasgow

A catalogue record for this title is available from the British Library.

FSC™ MIX
Paper | Supporting responsible forestry
FSC™ C104740
www.fsc.org

Contents

TERM 1

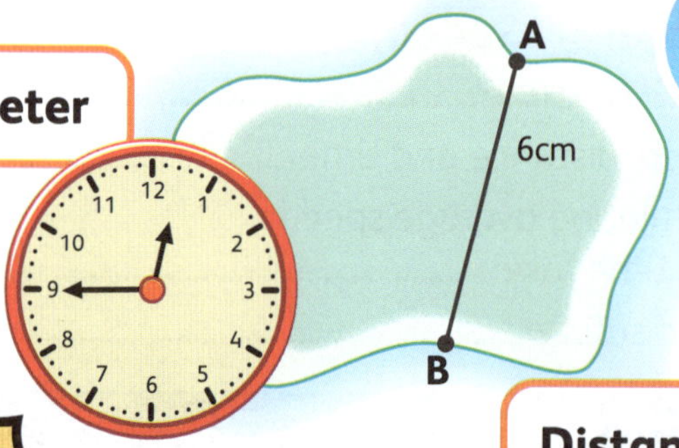

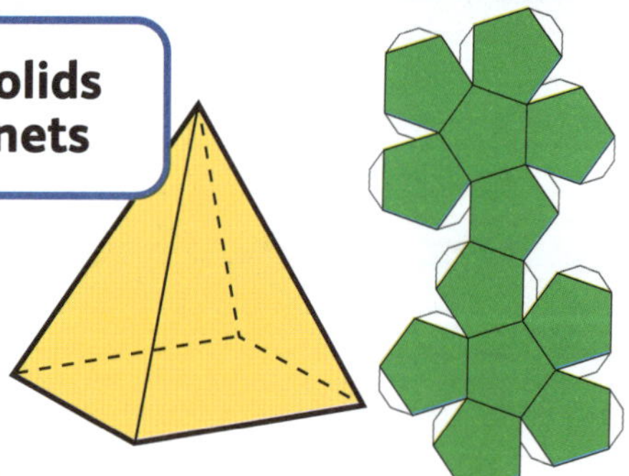

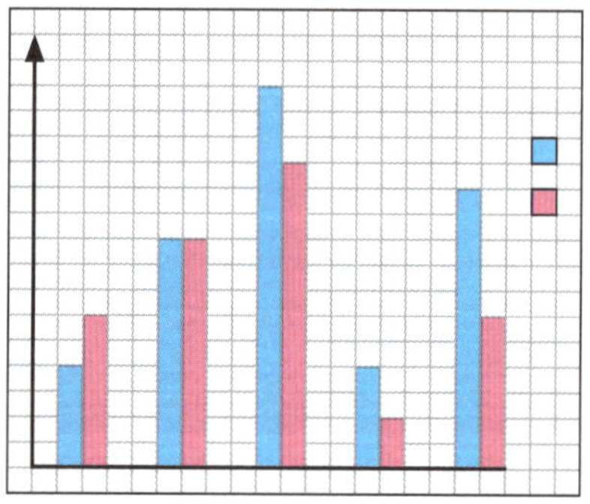

Unit 1: Number

Chapter 1 Sets

In this chapter you will explore this focus question:

> What special symbols and language do we use when working with sets?

To answer this question, you will:

- identify the members of a set
- associate the members of a set with the properties of the set
- distinguish between equal and equivalent sets
- distinguish between finite and infinite sets
- list the members in the intersection or union of two sets
- draw Venn diagrams to show the intersection of two sets
- know and use the symbols for set notation.

Starting point

1 Shawnae asked people whether they travelled by car or by bus to get to work.

She drew this diagram to show what she found out.

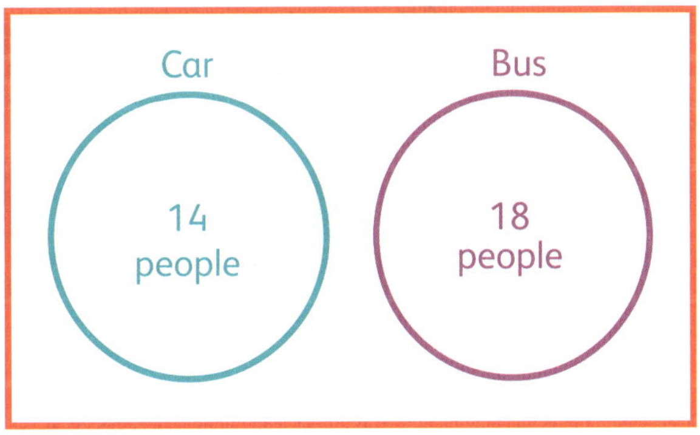

Discuss these questions with a partner.

a What information does this diagram give you?

b How is the information shown?

c Ms Byers walks to work. She doesn't use the car or the bus. How could you show this information on the diagram?

d Mr Johnston uses his car to go to work on Monday, Wednesday and Friday but he uses the bus on Tuesday and Thursday. How could you change the diagram to include 5 people who use both types of transport?

e Nobody uses the train to travel to work. How would you show this on a Venn diagram? Why?

2 A set is a collection of things that have something in common.

Look at the pictures. What do each of these sets have in common?

Set A

Set B

Set C

In maths we use special symbols and notation for sets. We can write set C like this:

C = {soccer ball, football, basketball}

C is the name of the set. It is usually a capital letter.

All the members or elements of the set are listed or described inside the brackets.

We use curly brackets { } called braces to contain the set.

3 Look at these sets:

a {1, 3, 5, 7, …}

How would you describe this set? What do you think the three dots mean?

b {0, 1, 2, 3, …, 18, 19, 20}

How would you describe this set?

This time the three dots are in a different place. What do you think this means in this set?

c {a, e, i, o, u}

How would you describe this set?

d { }

How would you describe this set? Why?

e Create your own set and write it in curly brackets. Challenge a partner to describe your set.

Members of sets

Key maths idea

A **set** is a well-defined collection or group. How would you define this set?

The set of vowels in our alphabet can be written as
Vowels = {a, e, i, o, u}

The set of vowels has five members or elements.

The five **elements** or **members** are written as a list inside **curly brackets (braces)**, each separated by a comma. This is called **set notation**.

We use the **symbol** ∈ to show that an item is a member or element of a set.

Example a ∈ {a, e, i, o, u}

→

Key words

set
element
member
curly brackets (braces)
set notation
symbol
universal set
null set (empty set)

To show that an item is not a member or element of a set we cross through the element symbol like this: $\notin$

Example $b \notin \{a, e, i, o, u\}$

The set of vowels is part of the larger set of letters of the alphabet. We can say that the alphabet is the **universal** (largest) **set** in this example. The universal set contains all the possible members you may use in a particular problem. We use the letter U to represent the universal set.

A set that has no members is called an **empty set** or **null set**. This is shown using the symbol $\varnothing$ or { }.

The set of town names in Jamaica that contain only vowels would be an example of an empty set.

Remember these important symbols:

$\in$ member of a set

$\notin$ not a member of a set

U universal set

$\varnothing$ or { } null set or empty set

1 Use set notation to list the members of these sets:

a five subjects you study

b three animals that can swim

c letters in your name.

2 How many members are in each of these sets?

a odd numbers between 1 and 10

b months of the year

c letters of the alphabet

d integers between 4 and 6

3 Daniel has this universal set of shapes.

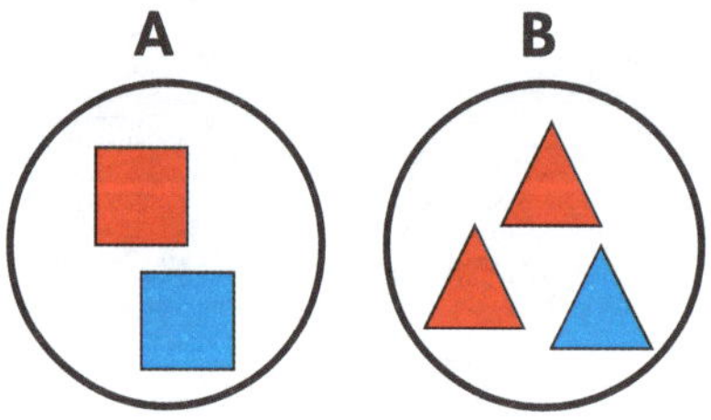

He puts them into sets A and B.

a Copy and complete these set notations:

A = { …

B = { …

b How could you reorganise the shapes to make two different sets? Use set notation to list the members of these new sets.

4 Copy and complete these set notations using the symbol $\in$ or $\notin$.

a 1 ☐ {odd integers}

b November ☐ {days of the week}

c F ☐ {consonants of the alphabet}

d 

Equal and equivalent sets

Key maths idea

Equal sets have the exact same members or elements. The order of the members does not matter as long as the members are exactly the same.

Example

{◆, ★, ●} = {★, ●, ◆}

{◆, ★, ●} ≠ {★, ●, ■}

{2, 4, 6, 8} = {8, 6, 4, 2}

Equivalent sets have the same number of members.

The members of each set do not need to be the same.

This symbol shows that sets are equivalent: $\longleftrightarrow$

This symbol shows that sets are not equivalent: $\not\longleftrightarrow$

Example

{◆, ★, ●} $\longleftrightarrow$ {A, B, C}

{★, ▲, ■} $\not\longleftrightarrow$ {Q, R, S, T}

Key words

equal sets

equivalent sets

1 Choose the pairs of sets that are equal to set A.

A = {★, ◆, ■, X, ●}

B = {■, X, ●, ★, ◆}

C = {★, ●, ■, X, ■}

D = {■, ★, X, ●, X}

E = {■, X, ★, ●}

F = {X, ★, ◆, ●, ■}

G = {■, ●, ★}

2 Copy and fill in the symbol $\longleftrightarrow$ or $\not\longleftrightarrow$ between the pairs of sets.

a {6, 3, 2, 1} ☐ {△, ○, □, ⬠, ☆}

b {◆, ◆, ◆} ☐ {C, +, ÷}

c {=, >, √, π} ☐ $\{\frac{1}{2}, \frac{1}{3}, \frac{3}{4}, \frac{3}{7}\}$

d {A, B, C} ☐ {?, !, ;}

e 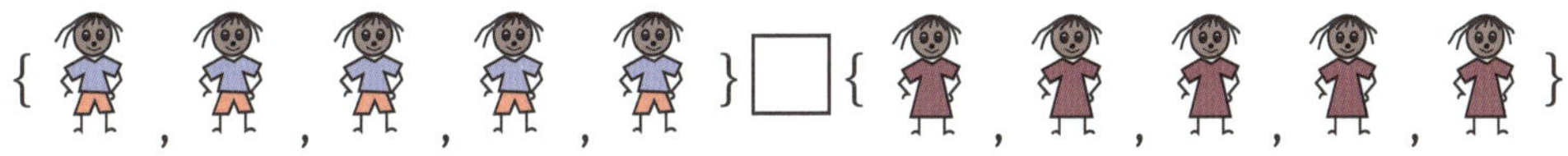

f {factors of 9} ☐ {factors of 8}

Remember: equivalent sets have the same number of members. The members do not need to be the same.

3 Draw four equivalent sets with five members.

Finite and infinite sets

Key maths idea

A set with limited members is called a **finite set**.

Example

Set B = {Nathaniel, Jada, Joshua}

Set B has three members. These are the members limited to this set.

In some sets, you may notice three little dots. These are called ellipsis and this means 'continue on'. The next set contains the first few prime numbers, but they go on infinitely. We draw three dots at the end of the set to show this.

This is an example of an **infinite set**:

{2, 3, 5, 7, 11, 13, 17, ...}

Now look at this set:

{7, 14, 21, ..., 84}

This set shows multiples of 7, up to 12 × 7. The ellipsis in this set shows that there are more multiples of 7 between 21 and 84, but these are not listed. This is a finite set.

Key words

finite set

infinite set

1 Describe each set and say if it is finite or infinite.

a {5, 10, 15, 20, 25, 30}

b {0, 1, 2, 3, ...}

c {5, 6, 7, 8, ..., 44, 45, 46}

d {..., −2, −1, 0, 1, 2, ...}

e {3, 6, 9, ..., 33, 36}

2 Write finite or infinite set for each of the following sets:

a days of the week

b hours in a day

c multiples of 6

d the set of all integers.

3 Write each of these sets in set notation. If the set is infinite, list at least five members before using the ellipsis.

a the set of factors of 24

b the set of multiples of 9

c the set of prime numbers to 31

d the set of even numbers greater than 40

e the set of positive integers with 2 in the ones place

Subsets

Key maths idea

Sets can be further organised into smaller sets.
When you put the members of a set into smaller sets, these smaller sets are called **subsets**.

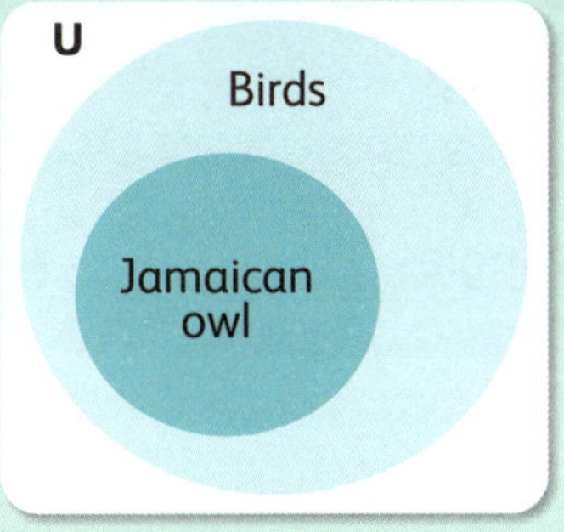

Owls are a particular type of bird, so every owl is also a bird. We can express this in the language of sets by saying that the set of owls is a subset of the set of birds.
{owls } ⊂ {birds} because every owl is a bird.

The symbol ⊂ means 'is a subset of'.

{Michael, Angela, Tony} ⊂ {Heather, Victor, Freddie, Beth, Alison, Michael, Angela, Tony}
This means that the set of the children is a subset of Angela's family.

If Set A = {1, 2, 3, 4, 5}, then sets {1, 2, 3}, {3, 4} and {1} are all subsets of Set A.

We can show subsets in set notation: {1, 2, 3} ⊂ {1, 2, 3, 4, 5}

We say this aloud as 'the set 1, 2, 3 is a subset of the set 1, 2, 3, 4, 5'.

{6} is not a subset of {1, 2, 3, 4, 5} because 6 is not in the parent (main) set.

In set notation, we write: {6} ⊄ {1, 2, 3, 4, 5}

We say this aloud as 'the set 6 is not a subset of the set 1, 2, 3, 4, 5'.

Key word
subset

1 Write down whether each statement is true or false:

a {Wednesday, Thursday} ⊂ {days of the week}

b {multiples of 4} ⊂ {even numbers}

c {  } ⊂ {things to wear}

d {vowels} ⊂ {consonants}

2 Rewrite in set notation.

a All squares are rectangles.

b Not all rectangles are rhombuses.

c A = {2, 3, 4, 5, 6, 7}
B = {2, 4, 7, 8}
C = {2, 4}

3 Look at the sets in question 2c. Copy and complete. Use the symbol ⊂ or ⊄ to make the statements true.

a B ________ A

b C ________ A

c B ________ C

d C ________ B

4 Make four possible subsets of the set of polygons.

5 **a** List five members of the set of sports in set notation.

b Give three subsets of the set of sports in set notation.

Venn diagrams

Key maths idea

A **Venn diagram** uses circles in a frame to show sets and the relationships between them. For example, if U = {the 26 letters of the alphabet}, V = {vowels} and C = {consonants}, the Venn diagram could look like this:

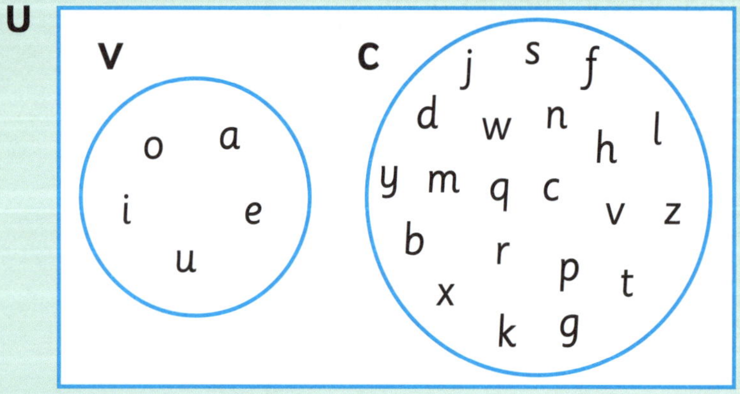

Sometimes Venn diagrams show the number of members in each set without listing them all. This Venn diagram shows the same sets as the previous one, but it uses the number of members of each set rather than the letters.

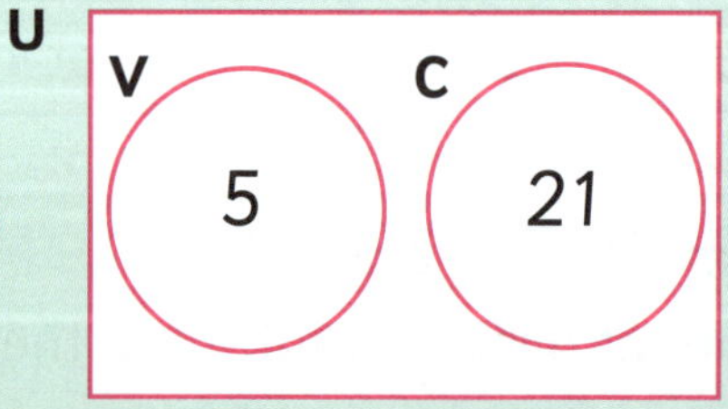

Key word
Venn diagram

If the members of the sets are shared, then the circles overlap.
The shared members are written in the overlapping section.
This Venn diagram shows:
U = {26 letters of the alphabet}
A = {letters in the word CHICKEN}
B = {letters in the word JERK}

E and K are members of set A and set B.

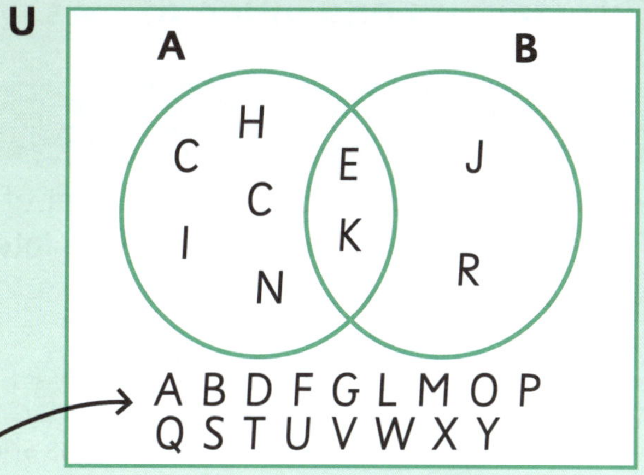

These letters are not found in set A or set B, but they are in the universal set U.

1 If U = {whole numbers from 1 to 50}, draw Venn diagrams to show these sets:

	Set A	**Set B**
a	Numbers from 1 to 10	Prime numbers
b	Odd numbers from 2 to 20	Composite numbers
c	Numbers from 1 to 20	Factors of 25
d	Prime numbers from 5 to 50	Multiples of 5
e	Multiples of 10	Multiples of 20

2 Leroy asked people whether they liked cats or dogs. He drew this Venn diagram.

Discuss these questions with a partner.

a How many people liked both cats and dogs?

b How many people didn't like cats or dogs?

c How many people liked cats?

d How many people liked dogs?

e What is the universal set in this example? How many members are in the universal set?

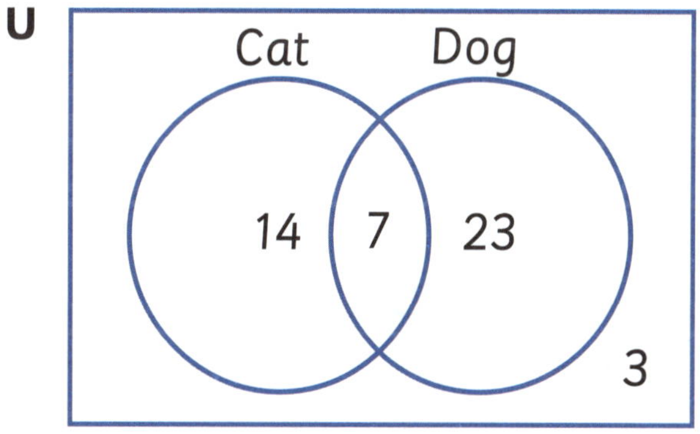

3 Look at this Venn diagram about lessons after school.

a How many members does the universal set contain?

b Describe the set of students who go to maths lessons after school.

c How many students have both maths and Spanish lessons after school?

d Use set notation to list the members of the set 'Does not do maths or Spanish lessons after school.'

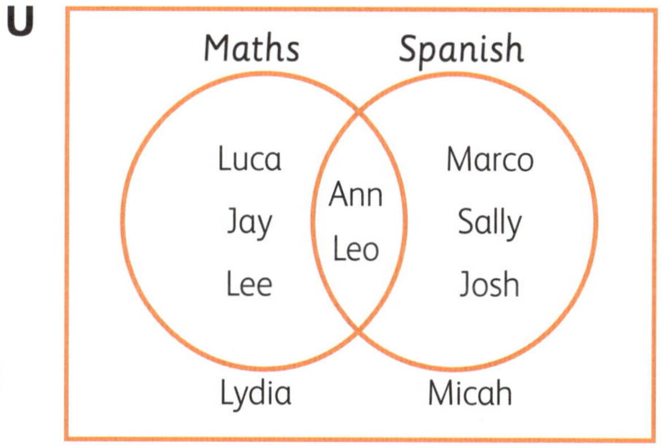

Intersection and union of sets

Key maths idea

The **intersection** of two sets is the set of members that both sets have in common. The symbol for intersection is ∩.

Example

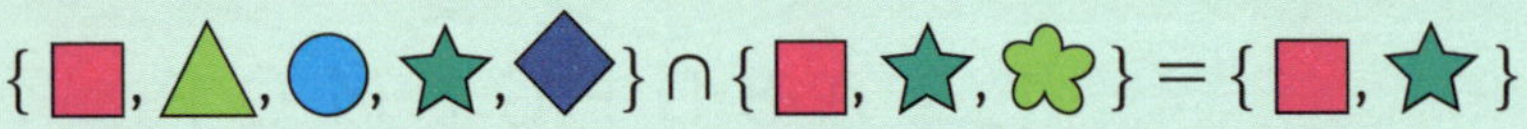

The intersection of two or more sets is shown on a Venn diagram by overlapping the circles.

U

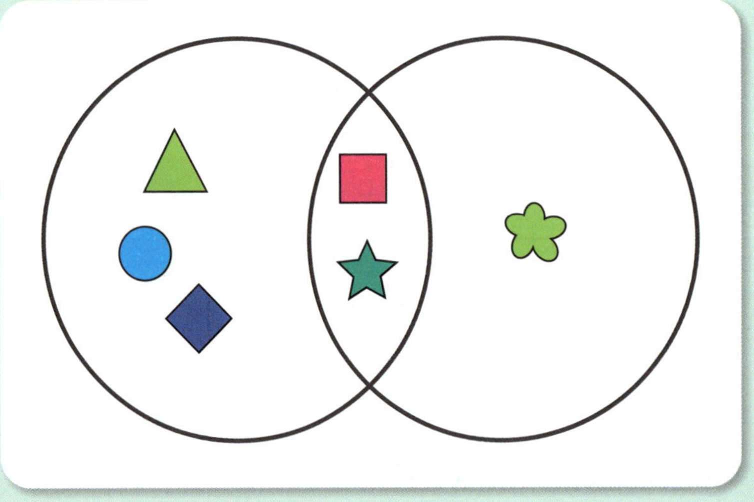

The **union** of two sets is the combined members of both sets. Each member is listed once only.
The symbol for union is ∪.

Example

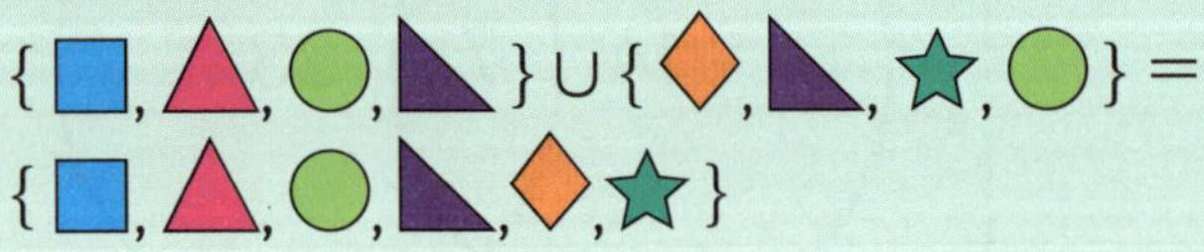

Key words

intersection
union

1 Look at the Venn diagram.

- **a** Use set notation to write the intersection of sets A and B.
- **b** Name some more members that could be in set B.
- **c** What could you name set A?
- **d** What could you name set B?
- **e** Where would you place the word 'soursop' in this Venn diagram?

U

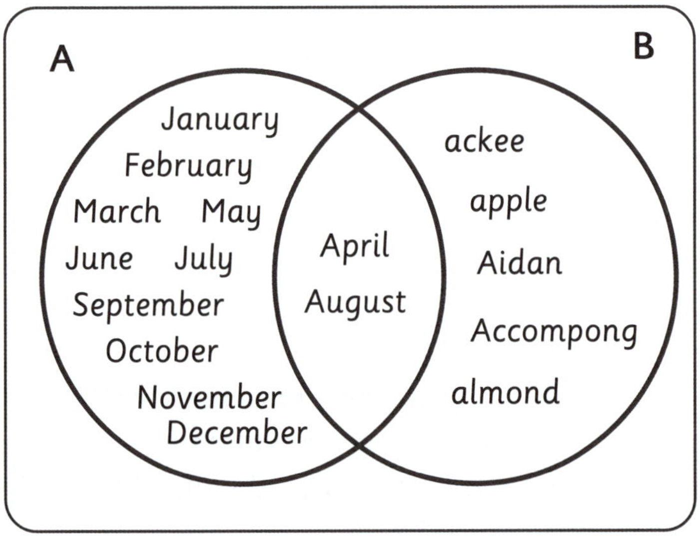

2 Write the union and intersection of the sets shown in the Venn diagrams below.

a

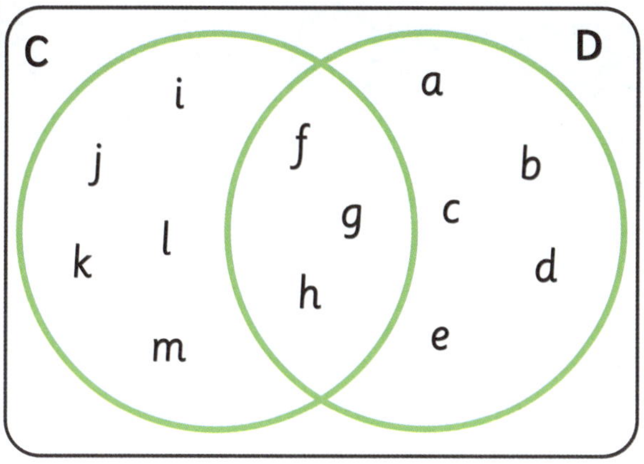

b

3 Copy and complete these statements. Represent them in a Venn diagram.

a {1, 2, 3, 4, 5, 6} ∩ {2, 4, 6, 8, 10} =

b {1, 2, 4 ,7, 11, 16, 21} ∩ {whole numbers from 0 to 15} =

c {a, b, c, d, e, f, g} ∩ {b, e, l, v, w} =

d {even numbers between 3 and 11} ∩ {prime numbers between 1 and 10} =

4 Copy and complete.

a {1, 4, 9, 16, 25} ∪ {4, 8, 10, 12, 14, 16} =

b {Liam, Kyle, Amelia, Ariana} ∪ {Liam, Amelia, Ashley} =

c {80, 90, 100} ∪ {88, 89, 90, 91, 92} =

5 Copy and complete the statements (use the symbol ∩ or ∪).

a {days of the week} ☐ {weekend days} = {Monday, Tuesday, Wednesday, Thursday, Friday, Saturday, Sunday}

b {1, 2, 3, 4, 5, 6} ☐ {2, 4, 6, 8, 10} = {1, 2, 3, 4, 5, 6, 8, 10}

c {2, 3, 4} ☐ {1, 2, 3, 4, 5, 6} = {2, 3, 4}

6 You have these sets:

U = {whole numbers less than 20}
A = {square numbers less than 20}
B = {even numbers less than 20}
C = {odd numbers less than 20}

a Draw a Venn diagram to show sets A and C.

b Draw a different Venn diagram to show sets B and C.

c What is A ∩ B?

d List the members of A ∪ B.

e Draw a Venn diagram to show the relationship between sets A and B.

Disjoint sets

Key maths idea

Set A and set B are equivalent sets because they have the same number of members. But they have no common members.

A = {2, 3, 4}

B = { 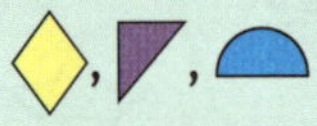 }

Two sets that have no common members are called **disjoint sets**.

We can show this in a Venn diagram:

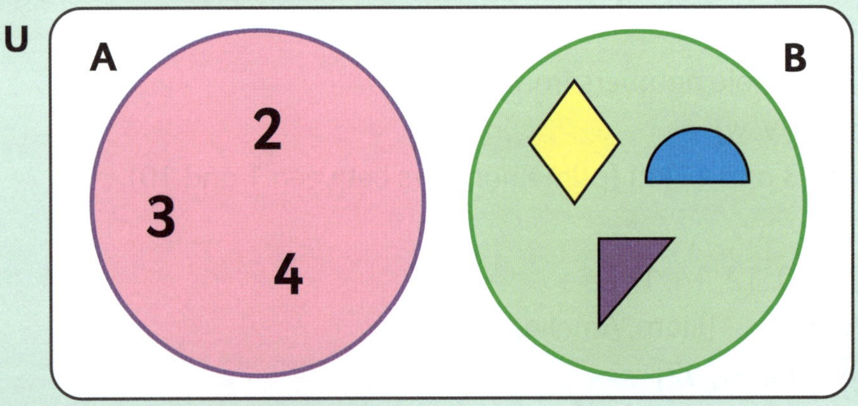

The sets S = {4, 6, 8, 10} and T = {3, 5, 7, 9} are disjoint. They have no common members.

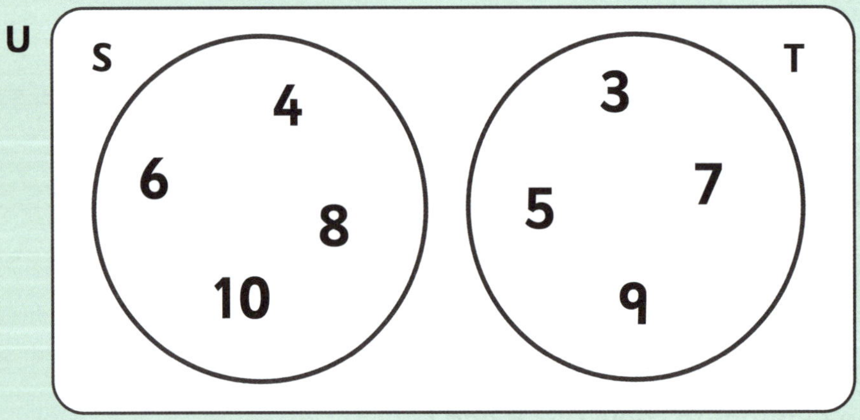

Key word
disjoint sets

Compare the sets in each pair and say whether they are disjoint or not.

1 A = {things we wear on our feet}; B = {running shoes}
2 C = {things we eat}; D = {running shoes}
3 E = {mango, pineapple, soursop}; F = {apple, banana, orange}
4 G = {Mara, Lisa, Jayden}; H = {Yohan, Elaine, Stacey}
5 I = {3, 6, 9, 12}; J = {2, 4, 8, 16}

Real-world maths: A cat food survey

VetCare Clinic in Mandeville did a survey to help them decide which cat food to buy to feed the animals in their hospital and boarding kennels.

1 The clinic tested two types of food (Kittymunch and Catzdelight) by feeding them to a set of 24 cats in their care. They drew this Venn diagram to show their results.

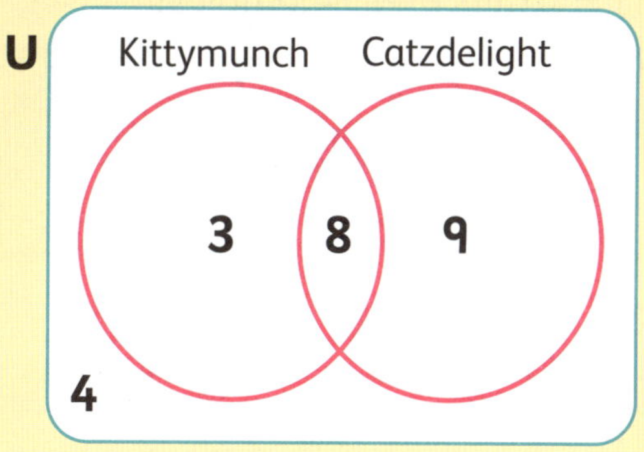

- **a** How many cats ate both types of food?
- **b** How many cats didn't eat either kind of food?
- **c** Which of these two cat foods would you recommend they buy? Why?
- **d** One of the vets says this is helpful, but that they need to try more options. Why do you think she said that?

2 The clinic director decided to do another test. She fed 25 cats in the boarding kennels Catzdelight and HappyCat. One cat didn't like either food. A quarter of the other cats liked both types of food, and two-thirds liked Catzdelight.

- **a** Use this information to draw a Venn diagram of these results.
- **b** How many cats only liked Catzdelight?

Maths detective: Things that belong together

Use the clues to help you solve each case.

Case 1

Look at the picture of items displayed at a grocery store.

Grocery stores use sets to organise and display items.

1. Make and name 5 different sets of items you can buy at the grocery store.
2. Describe each of the 5 sets.
3. How many members are in each of your 5 sets?
4. Name and list two sets that are equivalent and write down why they are equivalent.
5. Write down the union of your two equivalent sets. Remember to use set notation.

Case 2

In this part you have to complete a Venn diagram.

Look at the pictures. Then complete the Venn diagram, where:

U = {fruits and vegetables}

C = {foods we eat raw}

D = {foods we eat cooked}

ackee

banana

breadfruit

peas

cheese

mango

apple

peppers

lettuce

kallalloo

eggs

pineapple

coconut

yam

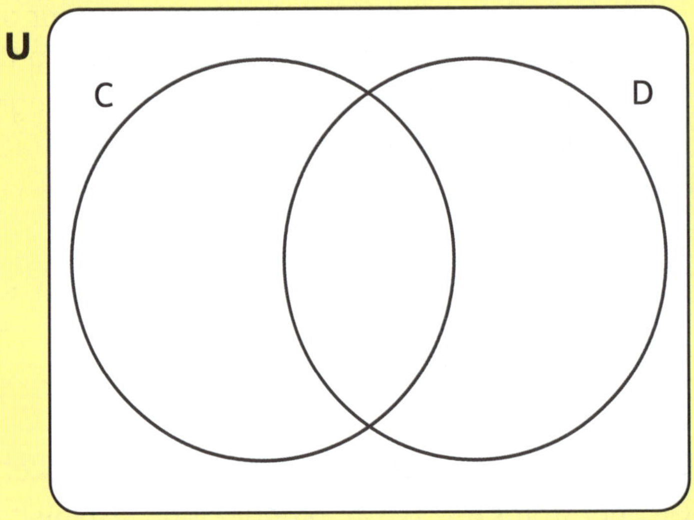

Case 3

Now use what you know about union and intersection to solve this case.

List the letters of the word CARNIVORE as set A.

List the letters of the word HERBIVORE as set B.

1 Represent set A and set B on a Venn diagram.

2 What is the intersection of A and B?

3 Use set notation to list A ∩ B.

4 Describe the members of the set of letter(s) common to both sets.

5 The members of set A and set B are combined. Write the set notation and list the members of the combined set.

Case 4

Some animals eat plants, some eat other animals (meat) and some eat both plants and other animals.

Here are two food webs showing what different types of animals eat. The arrows mean 'is eaten by'.

Food web A – Grassland

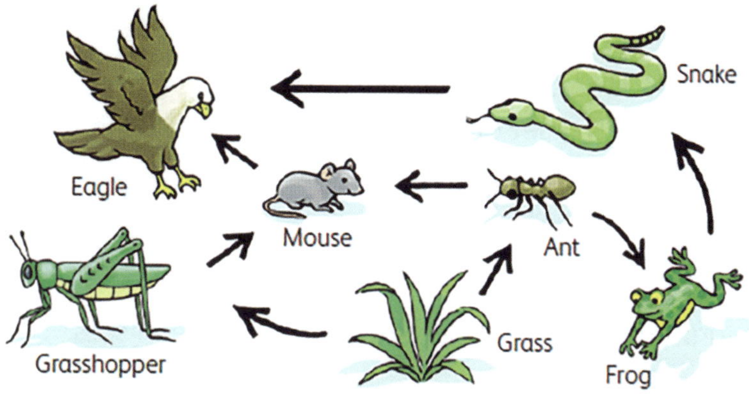

Food web B – Cold ocean

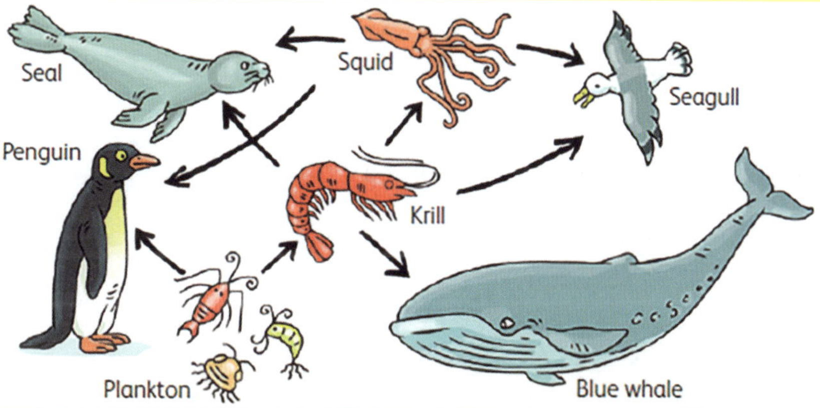

1 Answer these questions about the food webs.
   a How many of these animals only eat plants?
   b How many of these animals only eat other animals?
   c How many of these animals eat both plants and other animals?

2 Draw a Venn diagram to show this information. Use the number of members in each set.

Case 5

Ronnie was doing his Science homework and he found this Venn diagram on the internet.

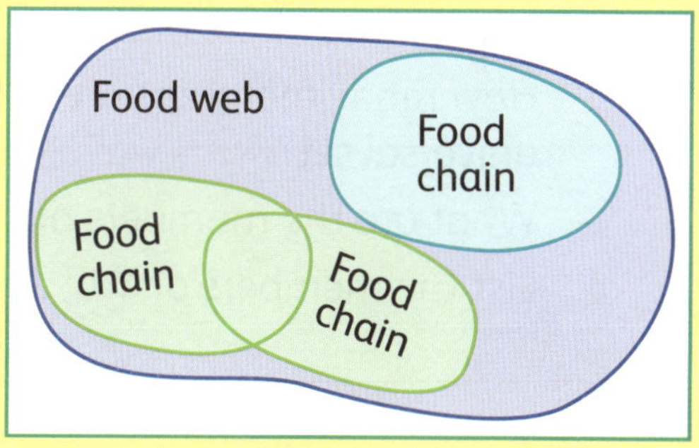

1 What does this Venn diagram show?

2 Redraw the Venn diagram and fill in plants and animals that could fit into the different sets. Use a local environment that you have learned about.

What I have learned

Read the sentences. Copy them and fill in the missing words to summarise what you learned in this topic.

1 Two sets are _____ if they have the same number of members.

2 Two sets are _____ if they have no shared members.

3 If all the members can be counted or listed, the set is _____ .

4 The _____ set is another name for the null set.

5 The _____ of two sets is the set of members common to both sets.

6 The _____ of two sets is the combined members of both sets with no members repeated.

7 A set is _____ if it has no last member.

8 A _____ is a diagram that shows the relationship between sets.

Practice questions

1 Look at the Venn diagram.

a How many members are in set R?

b What is R ∩ C? Use set notation to write the answer.

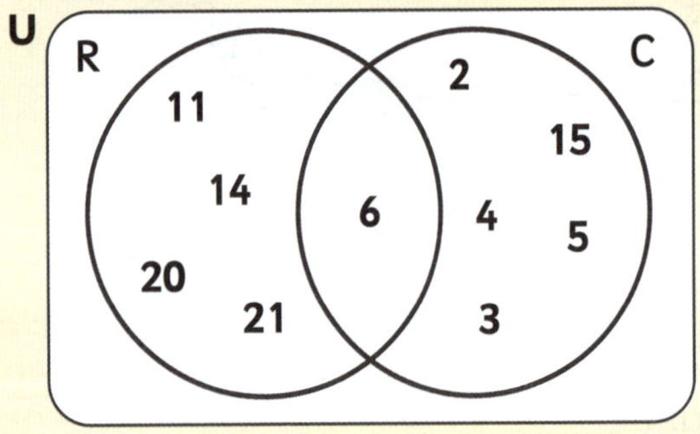

2 Use the given information to draw and label a Venn diagram.

U = {1, 2, 3, 4, …, 18, 19, 20}

A = {even numbers less than 20}

B = {the first five prime numbers}

3 Look at this Venn diagram.

a List the members of set Y.

b How many members are in the universal set?

c What are the members of Y ∩ Z?

d List the members of Y ∪ Z.

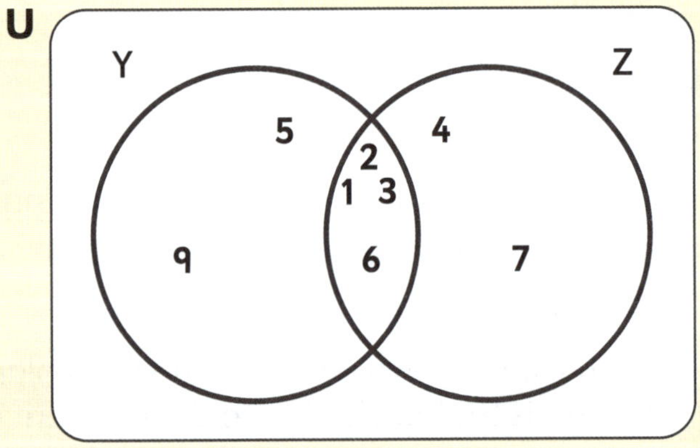

1 Self check

Check if you can do this: 🙂	If you have difficulty, try this: 😐
1 Identify the members of a set, and associate the members of a set with the properties of the set.	Go to pages 9–10 and remind yourself what we mean by 'set', and 'elements' or 'members'. Remember that the properties of a set are the characteristics or criteria that determine whether something belongs in the set or not.
2 Distinguish between equal and equivalent sets.	Go to page 11 and remind yourself what we mean by these terms.
3 Distinguish between finite and infinite sets.	Go to page 12 and remind yourself what we mean by these terms.
4 List members of the intersection of two sets, and list members of the union of two sets.	Revise page 16. Remember that the intersection is where two sets overlap, and the union is where two sets are joined.
5 Draw Venn diagrams to show the intersection of two sets.	Revise pages 14–15 and practise drawing sets that have elements in common.
6 Use the symbols for set notation.	Revise all the symbols you used on pages 13–18.

Chapter 2 Place value and exponents

In this chapter you will explore this focus question:

How do I write numbers in different number systems?

To answer this question, you will:

- use the principle of place value to read, write and use numbers
- write numbers in exponential form.

Starting point

The number of people on Earth is increasing all the time.

- There were more than 7.9 billion people on Earth in 2021. This is likely to increase to 9.7 billion by 2050. Did you know that there are more than 250 babies born on Earth every minute?
- In 1975, 2 million people lived in Jamaica. By 2021 the Jamaican population was nearly 3 million.
- Today, around half of the people in the world live in cities. 37 million people live in and around the city of Tokyo in Japan alone!

1 Use the information above to help you **estimate** how many people live in cities.

2 Now write 2 million in numerals.

3 Then write 37 million in numerals.

Key words
estimate
Hindu-Arabic number system

We use a set of 10 symbols to record numbers: 0, 1, 2, 3, 4, 5, 6, 7, 8 and 9. This is the **Hindu-Arabic number system**. The system originated in India in the 6th or 7th century, and is the most common system for representing numbers in the world.

Revising multiplication and division concepts

Key maths idea

You know that multiplication and division are **inverse operations**. That means the one operation can undo the other.

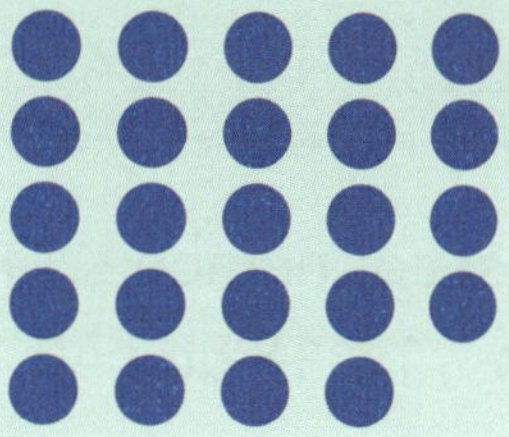

Example

$4 \times 6 = 24$ $\quad$ $6 \times 4 = 24$

$24 \div 6 = 4$ $\quad$ $24 \div 4 = 6$

A **factor** divides an exact number of times into another number without leaving a **remainder**. 4 and 6 are both factors of 24.

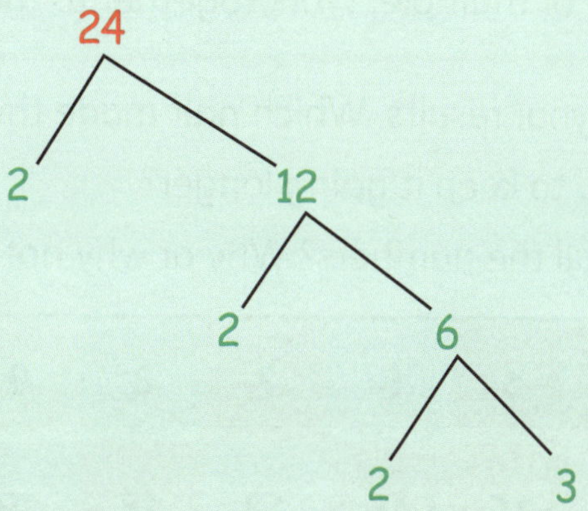

24 is a **composite number**. That means it has more factors than 1 and itself.

A **prime number** can only be divided by one and by itself. You can use division to help you break down composites into their **prime factors**.

Key words

inverse operation
factor
remainder
composite number
prime number
prime factor

For this activity you will need a beanbag.

Stand in a circle with your classmates. One student throws the beanbag to another student and calls out a division question, for example, '48 divided by 8'. The student who catches the beanbag must answer correctly, and then throw it to another student with a new question. If someone answers incorrectly, they sit down. The last one standing is the winner.

Reviewing HCF and LCM

Key maths idea

24 is a **multiple** of 6. It is also a multiple of 4.
These are the first ten multiples of 6: 6, 12, 18, 24, 30, 36, 42, 48, 54, 60
These are the first ten multiples of 4: 4, 8, 12, 16, 20, 24, 28, 32, 36, 40
24 is a **common multiple** of 6 and 4, but the **lowest common multiple (LCM)** is 12.

Key words

multiple
common multiple
lowest common multiple (LCM)
highest common factor (HCF)

Look at these two numbers: 24 and 32. What factors do they have in common? The common factors are circled:

Factors of 24: (1), (2), 3, (4), 6, 12, 24

Factors of 32: (1), (2), (4), 8, 16, 32

The **highest common factor (HCF)** is 4.

1 List the factors of each set of numbers. Circle the common factors and identify the HCF.

a 10, 25 and 30 **b** 12, 18 and 40 **c** 24, 36 and 16

2 Use this 100 x 100 number grid and some counters, beads or coins. Work in pairs. One player chooses a number. Cover it with a counter and write it on a piece of paper. The next player then chooses another number that either has the first number as a factor or multiple. Work together to make the longest sequence of numbers you can.

- As a class, compare your results. Which pair made the longest sequence?
- What strategies help to keep it going longer?
- Is it possible to use all the numbers? Why or why not?

1	2	3	4	5	6	7	8	9	10
11	12	13	14	15	16	17	18	19	20
21	22	23	24	25	26	27	28	29	30
31	32	33	34	35	36	37	38	39	40
41	42	43	44	45	46	47	48	49	50
51	52	53	54	55	56	57	58	59	60
61	62	63	64	65	66	67	68	69	70
71	72	73	74	75	76	77	78	79	80
81	82	83	84	85	86	87	88	89	90
91	92	93	94	95	96	97	98	99	100

If you have access to the internet, you can search for 'interactive factors and multiples game' to find an online version of this game.

Exploring place value

Key maths idea

Even though we only have ten number symbols, we can represent far more than ten numbers. We do this by using **place value**. Look at these examples of some different numbers.

1 This represents the number one.

10 In this number, the 1 represents 1 ten, and the 0 is a placeholder, so that we can see the place value of the 1.

150 In this number, the 5 represents 5 tens, and the 1 represents 1 hundred.

2 050 This number has a 2 in the thousands place and a 5 in the tens place.

Key words
- place value
- digit
- expanded form

You can use a place value chart to remind you what each **digit** represents.

Hundred millions	Ten millions	Millions	Hundred thousands	Ten thousands	Thousands	Hundreds	Tens	Ones
100 000 000	10 000 000	1 000 000	100 000	10 000	1 000	100	10	1

×10 ×10 ×10 ×10 ×10 ×10 ×10 ×10 ×10

On the place value chart, each digit to the left is 10 times the digit to its right.

For this activity, you will need a pile of index cards. Each card should have a number from 0 to 9. You can use cards from a number game like the one in the picture, if you have those, or make your own.

1 **a** Work in groups. Each player draws a large place value chart with seven columns.

b Shuffle the cards and place the top card facing up on the table. Write the digit in any column of your choice in your place value chart. Repeat this for another six cards, until you have filled your table. You may not move the place of a digit once you have written it down.

c Once you have all made a 7-digit number, the player with the highest number wins.

2 As a class, take turns writing each 7-digit number into a place value chart on the board. Read out your number. Use two blank index cards. On one card, write the number in numerals, for example:

1 234 567

one million,
two hundred thirty-four thousand,
five hundred sixty-seven

> Remember: to write a number in **expanded form**, you write it as a sum of the value of the digit in each place.
> For example:
> 1 234 567 = 1 000 000 + 200 000 + 30 000 + 4 000 + 500 + 60 + 7.

3 In groups, play this game. Draw a grid with 6 columns and 4 rows, big enough for an index card to fit into each block of the grid. Your teacher will give you 12 pairs of index cards (numbers and matching words). Shuffle them and place them face down on the grid. Take turns to flip two cards at a time. If the cards match, the player keeps them. The winner is the player with the most cards at the end.

4 Revise expanded notation (expanded form). Use the numbers that you created in the games above. Write out three of the numbers in expanded notation.

5 As a class, look at the numbers you created in these games. Talk about how you would order the numbers from least to greatest. Which place value would you look at first?

6 If possible, watch a video tutorial on place value up to millions. If you have access to a computer, use a spreadsheet to make place value charts that show what you have learned.

7 Write the value of the underlined digits. The first one has been completed for you as an example.

a	4 1$\underline{9}$6 9 tens = 90	**b**	19$\underline{3}$ 408	**c**	1 900 00$\underline{4}$	**d**	43$\underline{9}$ 054
e	$\underline{9}$ 245	**f**	4$\underline{5}$ 275	**g**	$\underline{8}$46 781	**h**	8 6$\underline{5}$4 023

8 Write two different numbers for each of the following:

- **a** 4-digit numbers with a 9 in the thousands position
- **b** 5-digit numbers with a 3 in the tens position
- **c** 6-digit numbers with a 2 in the hundred thousands position
- **d** 7-digit numbers with a 4 in the millions position
- **e** 8-digit numbers with an 8 in the ten millions position.

9 Write these numbers in digits.

- **a** twelve thousand, three hundred eighty-five
- **b** ninety-five million
- **c** two million, nine hundred fifty-two thousand
- **d** three hundred seven million, four hundred ninety-three thousand, two hundred twenty-nine
- **e** five hundred eighty thousand, two hundred three
- **f** six hundred seventy-three million, one hundred eleven thousand, three hundred seven
- **g** seven hundred twelve thousand, nine hundred sixty-six
- **h** thirty-seven million, nine hundred twenty-six thousand, sixty-one
- **i** nine hundred nine thousand, one hundred five
- **j** seventeen thousand six

Large numbers in real life

Key maths idea

The areas of three countries in South America are given in the table.

Country	Area in square kilometres
Colombia	1 138 910
Bolivia	1 098 580
Peru	1 285 220

How would you compare the areas to work out which country is the biggest and which is the smallest?

Which of the two numbers is larger: 1 250 004 or 1 150 382?
Start by comparing the digits with the greatest value:
1 250 004 > 1 150 382
Both numbers have 1 million, but the hundred thousand digit is different, so the value is different. 200 000 > 100 000, so we know that the number with 2 in the hundred thousands place is greater.

Explain to a partner how to use place value to compare the numbers 999 456 and 999 501.

1 Compare the monthly salaries of the people in the picture.

a Who earns the highest salary?
b Who earns the lowest salary?
c Write the salaries in order, starting with the lowest.

2 This table shows the population of some countries in 2021.

Country	Population in 2021
Costa Rica	5 094 118
India	1 380 004 385
Guatemala	17 915 568
United States	331 002 651
Jamaica	2 961 167
Bermuda	62 278
United Kingdom	67 886 011

(Source: https://www.worldometers.info/world-population/population-by-country/, as of January 2021)

a Which country has the largest population?
b Which country has the smallest population?
c Which countries have a population of less than ten million?
d Which countries have a population of more than twenty million?
e Write the population of India in words.
f Write the population of Bermuda in words.
g Write the population of the United Kingdom in words.
h Write the population of the United States in expanded form.
i Write the population of Guatemala in expanded form.
j Write the population of Costa Rica in expanded form.

Exponential form

Key maths idea

$10 \times 10 = 100$

10×10 can also be written as 10^2
We say: 'ten squared' or 'ten to the power 2'

$10 \times 10 \times 10 = 1\,000$ or 10^3
We say: 'ten cubed' or 'ten to the power 3'

$10 \times 10 \times 10 \times 10 = 10\,000$ or 10^4
We say: 'ten to the power 4'

Think about the patterns in the place value chart.
Each place is 10 times greater than the one to the right of it.

Ten thousands	Thousands	Hundreds	Tens	Ones
10 000	1 000	100	10	1
$10 \times 10 \times 10 \times 10 = 10^4$	$10 \times 10 \times 10 = 10^3$	$10 \times 10 = 10^2$		

Key words

exponential form
base
exponent
power
index

If you follow the pattern you can see that the powers go down in value as you move to the right.
10 is not multiplied by itself. It is one ten. The pattern shows that this is 10^1.
1 is not a power of ten. We say that $1 = 10^0$.

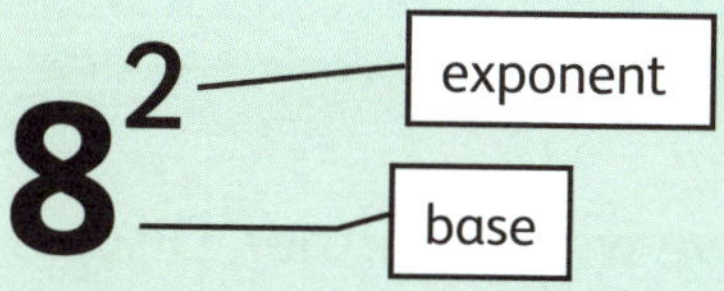

A number written like this in **exponential form** has a **base** and an **exponent**. The exponent is also called a **power** or an **index**.
In 8^2 the number 8 is the base and the number 2 is the exponent (or power or index).
An exponent tells us how many times the base is multiplied by itself.
8^2 means $8 \times 8 = 64$

Remember:
- Any base with an exponent of 1 is the base itself, for example, $10^1 = 10$ and $5^1 = 5$.
- Any base with an exponent of 0 is 1, for example, $10^0 = 1$ and $5^0 = 1$.

1 What is the value of:

a 10^5	**b** 2^3	**c** 4^2	**d** 3^3
e 5^3	**f** 3^4	**g** 4^3	**h** 2^5
i 5^4	**j** 6^3	**k** 2^0	**l** 7^1

2 What is the value of:

a 10 to the power 4	**b** 2 to the power 4
c 3 to the power 2	**d** 4 to the power 4
e 5 to the power 2	**f** 6 to the power 2
g 12 to the power 2	**h** 7 to the power 3
i 8 to the power 4	**j** 1 to the power 5

3 Complete the sentences.

a Ten thousand is 10 to the power _____.
b One million is 10 to the power _____.
c One is 10 to the power _____.
d One hundred thousand is 10 to the power _____.
e One thousand is 10 to the power _____.
f Ten is 10 to the power _____.
g Ten million is 10 to the power _____.
h One hundred is 10 to the power _____.

Scientific notation and standard form

Key maths idea

Standard notation or **standard form** is the way we usually write numbers when we use place value. **Scientific notation** is a different way of expressing larger numbers as the **product** of a number with a value between 1 and 10, and a power of ten.

Look at these examples:

$45 = 4.5 \times 10^1$

$3\,280 = 3.28 \times 10^3$

$389\,499 = 3.89499 \times 10^5$

Look at the examples above with a partner. Discuss how you convert a longer number from standard form to scientific notation.

Key words

standard notation
standard form
scientific notation
product

You will need a calculator for this activity.

1 Refresh your calculator skills with this challenge.

- **a** Pretend that some of the keys on your calculator do not work. Try to make all the numbers from 1 to 20 using the following keys: 2, 3, +, −, = and the clear button.
- **b** The 0 key is out of bounds. Is it possible to make 260, 206, 2 062 and 20 602 using the other keys?
- **c** Use only the 3, 7, +, − and = keys. Try to make 20, 57, 93 and 109.
- **d** Is it possible to make the answers 1 to 20 using only 2, 4 and the +, −, × and ÷ keys? Explain your findings.

2 Remind yourself of the powers of ten by expressing these numbers as powers of ten.

a 100 **b** 1 000 **c** 10 000 **d** 100 000 **e** 1 000 000

3 Use a calculator to work out the value of these numbers.

a 4 to the power of 5 **b** 8 to the power of 4

c 4.5 to the power of 6 **d** 1.24 to the power of 3

4 Look at these numbers expressed in scientific notation. Express them in standard notation.

a 2.13×10^3 **b** 3.499×10^5 **c** 9.1897×10^6 **d** 5.0001×10^8

5 Work with a partner. Express each of these numbers in scientific notation. Use these steps to help you.

- Look at how far you need to move the decimal point to get a number with only one digit before the decimal point. Use this to help you work out which power of ten to use.
- Express the number as a decimal multiplied by that power of ten.

a 15 **b** 450 **c** 9 828 **d** 84 735 **e** 400 289

Maths detective: Number skills

Use your number skills to solve these cases.

Case 1

Write the correct calculations to replace a to i.

Number	Expanded form	Expanded form using powers of ten
5 027	5 × 1 000 + 0 × 100 + 2 × 10 + 7 × 1	$5 \times 10^3 + 0 \times 10^2 + 2 \times 10^1 + 7 \times 10^0$
27 345	2 × 10 000 + 7 × 1 000 + 3 × 100 + 4 × 10 + 5 × 1	**a**
316 980	**b**	$3 \times 10^5 + 1 \times 10^4 + 6 \times 10^3 + 9 \times 10^2 + 8 \times 10^1 + 0 \times 10^0$
48 914 768	4 × 10 000 000 + 8 × 1 000 000 + 9 × 100 000 + 1 × 10 000 + 4 × 1 000 + 7 × 100 + 6 × 10 + 8 × 1	**c**
d	**e**	$7 \times 10^5 + 2 \times 10^4 + 0 \times 10^3 + 4 \times 10^2 + 1 \times 10^1 + 6 \times 10^0$
2 655 402	**f**	**g**
406 841	**h**	**i**

Case 2

Jayvon and Samira play a game with a die to see who can make the greatest possible number.

They take turns to roll the die and enter the score as a digit in any place in the place value chart. After 7 rolls, the tables look like this:

Jayvon:

Ten million	Million	Hundred thousand	Ten thousand	Thousand	Hundred	Ten	One
4		3	1	1	2	3	1

Samira:

Ten million	Million	Hundred thousand	Ten thousand	Thousand	Hundred	Ten	One
	4	3	2	1	1	1	3

1 Who would win if the next roll of the die was:

a 4 **b** 5 **c** 2?

2 Play a game like this with a partner, but try to make the smallest possible number.

3 What strategy did you use to try and make the smallest number? How does the strategy change if you want to make the greatest possible number?

Case 3

After a lesson on exponential form, Chrissie went home and said to her mom, 'I learned that 10^6 is the same as 600 000'.

Is Chrissie correct in her thinking?

Use words, numbers or a place value chart to explain your answer.

Case 4

Sonia and Yohan invented their own number game using exponential form.

It works like this:

- They each roll a die.
- They use the number they roll themselves as the base and their partner's number as the exponent.
- They use a calculator to work out the value of their number.
- The person who gets the biggest value scores a point.
- If the values are equal, they both score a point.

These are first five rolls of the dice.

Sonia	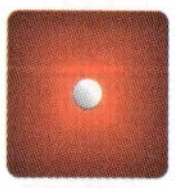	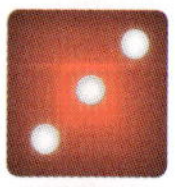	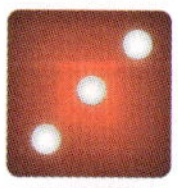	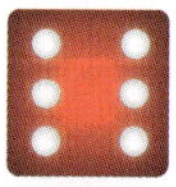	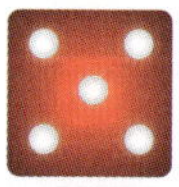
Yohan	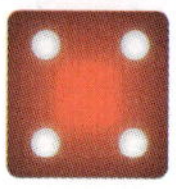	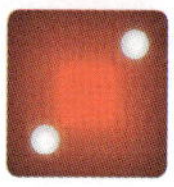		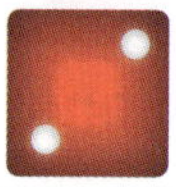	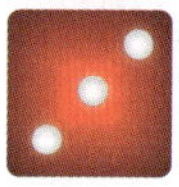

Work out how many points Sonia and Yohan would each have at the end of the five rounds.

Real-world maths: Working with large numbers

People who work with very large numbers (or very small numbers) often write the numbers in a shortened form called scientific notation.

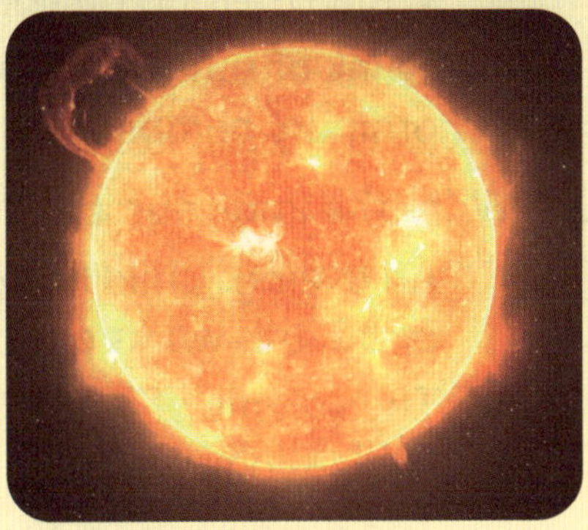

Scientific notation involves writing numbers as a multiplication using powers of ten. Look at this example to see how this method works.

Example

Scientists use a unit called Kelvin to measure temperatures. The temperature at the centre of the Sun is 15 600 000 Kelvin
In scientific notation this is written as

$$1.56 \times 10^7$$

1.56: A number between 1 and 10 (can include decimal part)

10^7: 10 to the power of a whole number.

When you are given a number in scientific notation you can convert it back to the original large number.

The distance from the Earth to the Moon is 4.0×10^5 kilometres.

$10^5 = 10 \times 10 \times 10 \times 10 \times 10$, which is 100 000 (the power of 5 tells us there will be 5 zeros in the product)

$4 \times 100\,000 = 400\,000$

1 Shelley-Anne says you can also work this out using what you know about place value. She writes these two examples on the board to show what she means.

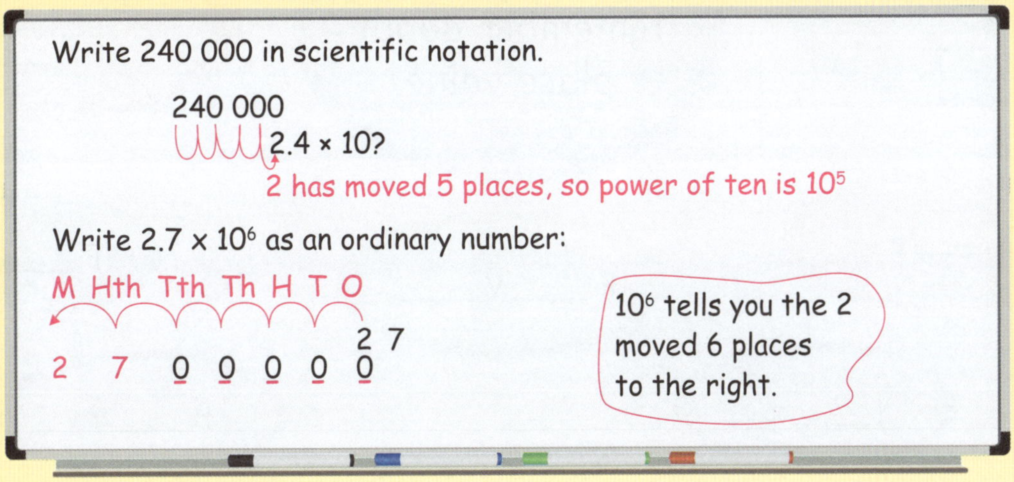

a Discuss with your partner how Shelley-Anne has worked this out.
b Which method do you find easier? Why?
c Think of some other ways of converting between scientific notation and ordinary numbers.

2 Write the large number in each example using scientific notation.
a An average child has approximately 270 000 germs on their hands. In a class of 30, this works out to approximately 8 100 000 germs.
b The volume of water in the Earth's oceans is about 135 000 000 cubic kilometres.
c The distance from the Earth to the Sun is about 150 000 000 kilometres.

3 Write these numbers out in full. You don't have to be able to say the numbers.
a The mass of an elephant is 5.5×10^5 grams.
b In 2021 the population of China was approximately 1.36×10^9.
c The Earth has a mass of 6×10^{24} kilograms.

4 Why do you think scientists, astronomers, geographers and engineers might find it easier to work with numbers in scientific notation? Share your ideas with the class.

What I have learned

Copy the diagram and fill in the missing letters in the key terms.
Under each one, give an example to show what it means.

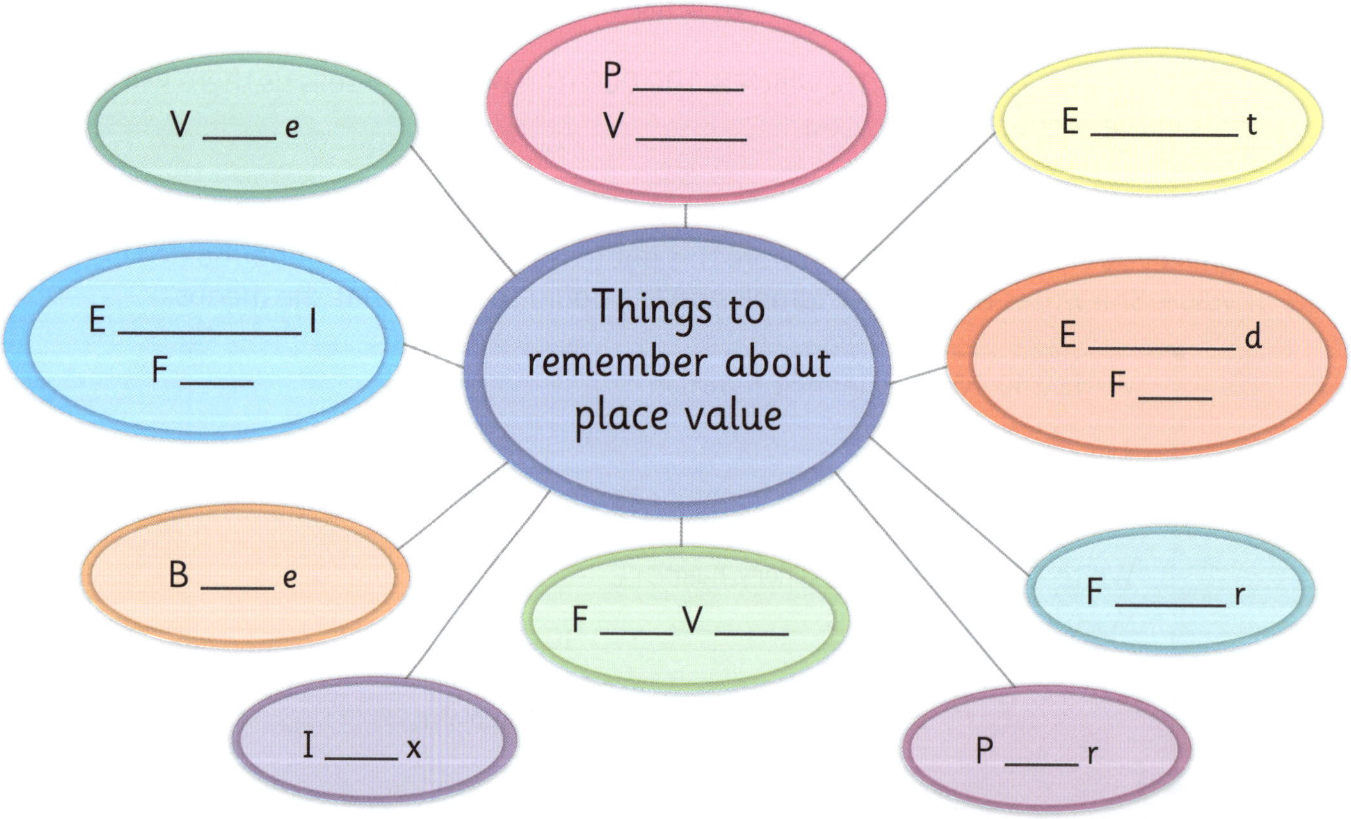

Practice questions

1. a Write the first ten multiples of 5, 7 and 10.
   b Use your list to identify the lowest common multiple of 5 and 7.
   c Write the highest common factor of 54 and 20.
2. Write three examples of:
   a composite numbers
   b prime numbers
   c multiples of ten
   d numbers in scientific notation.
3. True or false? Give reasons for your answers.
   a Odd numbers cannot be composite numbers.
   b Even numbers greater than 2 cannot be prime numbers.
4. Write each number in expanded notation.
   a 14 569 588
   b 33 098
   c seventeen million, eight hundred fifteen thousand, eighty-nine
   d forty-five thousand, nine hundred sixteen
5. Write two examples of each type of number.
   a a 6-digit number with a 5 in the hundreds place
   b a 7-digit number with a 9 in the ten thousands place
   c an 8-digit number with a 5 in the ten millions place
6. Write the number that is:
   a 1 less than 10 000
   b 1 more than 99 999
   c 1 less than 100 000 000
   d 1 more than 999
   e 1 less than 1 000
   f 1 more than 9 999
   g 1 less than 100 000
   h 1 more than 999 999
   i 1 less than 1 000 000
   j 1 more than 9 999 999
7. Give three real-life examples of situations where we need to use numbers that are greater than one million.
8. Write the value of:
   a ten to the power of 5
   b ten to the power of 6
9. Use your answers to Exercise 8 to explain the meanings of these terms:
   a exponent
   b base
10. Explain in your own words what we mean by scientific notation, and give an example.

2 Self check

	Check you can do this:	If you have difficulty, try this:
1	Remember your multiplication and division tables.	Go back to page 25 and remind yourself about × and ÷ facts.
2	Find the HCF or LCM of given sets of numbers.	Go back to pages 25–26. Make sure you recall what factors and multiples are, and how to compare lists of factors to find the highest common factor or lowest common multiple of a given pair or set of numbers.
3	Use place value to write larger numbers.	Revise page 27 and remind yourself of the place value of each digit in larger numbers that have six, seven or eight digits.
4	Write numbers in exponential form.	Go back to pages 30–31 and remind yourself how to write numbers using a base and index.

Unit 2: Measurement

Chapter 3 Distance and scale

In this chapter you will explore this focus question:

How do I calculate and use distance and scale measurements?

To answer this question, you will:

- use measurements in the environment
- estimate and measure distances to include scale drawing
- interpret a simple scale drawing and calculate the actual distances using the scale on a road map or floor plan.

Starting point

1 a What message does this sign convey?
  b Explain what you understand by the term distance.
  c Read these words. What kinds of distances would you use these terms for?

width | height | length | depth | thickness

  d What does '>1 m' mean?
  e How long is 1 metre?
  f What could you use to make sure you were standing 1 metre apart from someone?
  g Why is it sometimes necessary to measure the distance between people like this?

2 Choose the unit of measure that would be the most appropriate to measure these items.

The mass of a hippopotamus	litres
The volume of water in a bathtub	grams
The length of an ant	metres
The length of a football pitch	millimetres
The mass of a banana	tonnes

3 Choose the most suitable measurement for a glass of juice from the four options below.

250 ml   2 litres   0.5 litres $\frac{1}{2}$ kg

4 Estimate the height of a classroom door.

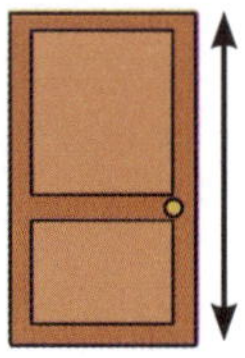

20 mm   20 cm   20 m   2 m   0.2 km

5 Work with a partner. Use a ruler or a tape measure. Find three things in your classroom that are:

a shorter than 10 cm
b about half a metre long
c longer than 1 metre.

Revising fractions and ratios

Key maths idea

Read the questions and answers about the towers of blocks.
How many blocks out of the total are blue?
In tower A, 1 out of 3 blocks are blue. In tower B, 2 out of 6 blocks are blue.

What **fraction** of the blocks are blue?
In tower A, $\frac{1}{3}$ of the blocks are blue.

In tower B, $\frac{2}{6}$ of the blocks are blue.
We can simplify this fraction to $\frac{1}{3}$.

What is the **ratio** of blue blocks to pink blocks?
A ratio describes a relationship between quantities. There is 1 blue block for every 2 pink blocks. The ratio of blue blocks to pink blocks is 1 to 2. We write this as 1 : 2.

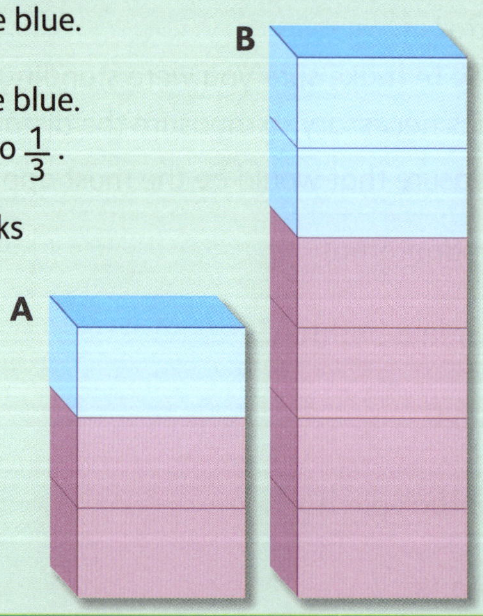

Key words
fraction
ratio

1 Look at this set of shapes. Copy and complete the statements and answer the questions.

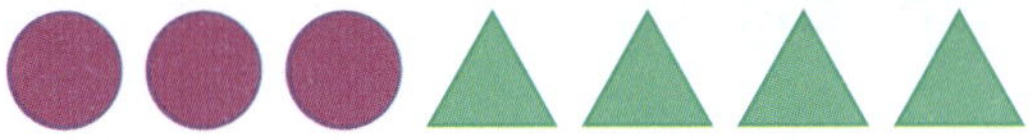

a The total number of shapes in the set is ____ .

b $\frac{3}{7}$ of the shapes in the set are ____ .

c What fraction of the shapes are triangles?

d What does the ratio 4 : 3 describe?

2 A map has this diagram to show the scale of the map.

a Measure the bar that represents 1 000 m on the diagram in centimetres. Use it to complete this statement:

____ cm on the map = 1 km in real life

b Work out what distance 1 cm on the map represents.

3 Discuss with a partner and explain how you work this out. For the map scale given in Exercise 2, which of these ratios describes the scale of the map to the distances in real life?

a 1 : 1 000 **b** 1 : 4 000 **c** 1 : 25 000 **d** 1 : 250 000

Revising units of measurement

Key maths idea

When we measure how long, wide, tall or far away something is we are measuring length. Height, width and **distance** are all types of length. The standard **unit** of length is a metre. We also use smaller and larger units that are based on the metre.

1 metre (1 m) = 100 centimetres (100 cm)

$\frac{1}{2}$ m = 50 cm = 500 mm

1 cm = 10 millimetres (10 mm)

1 kilometre (1 km) = 1 000 m

1 decimetre (1 dm) = 10 cm or $\frac{1}{10}$ of a metre

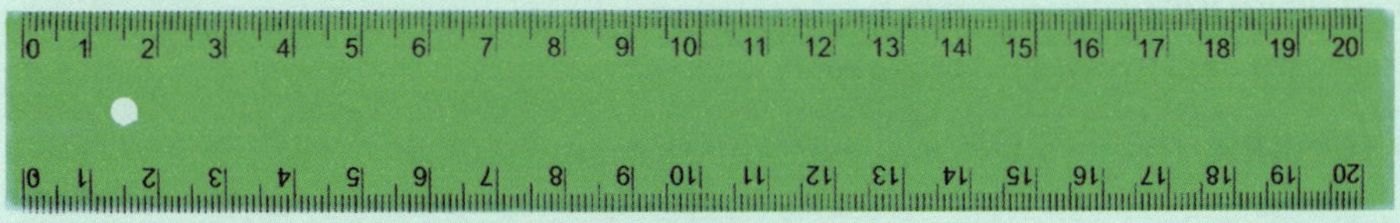

This ruler is marked in centimetres.

Key words

distance

unit

convert

Sometimes you may need to **convert** between units of length.
You can use this diagram to help you:

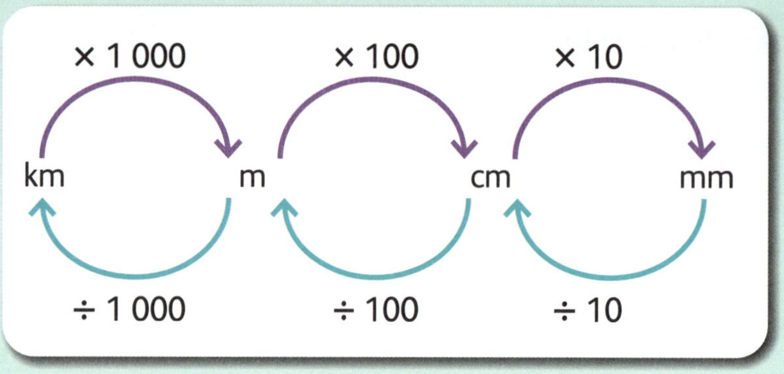

1 Work with a partner.

- **a** Discuss how the diagram works. Express the units as ratios:
  - the ratio of metres to kilometres
  - the ratio of centimetres to metres
  - the ratio of millimetres to metres.
- **b** Explain how to convert 2 km to metres.
- **c** How many metres are in 150 cm?
- **d** What operations do you need to do to change 8 kilometres to centimetres?

2 Measure the length of each object, first to the nearest centimetre and then in millimetres. Write your answers in your notebook.

a 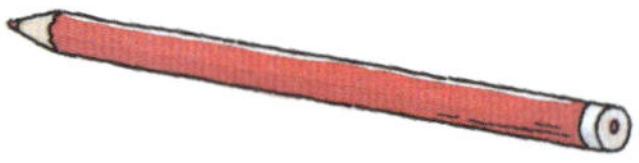 **b**  **c**  **d** 

3 Measure the length of each line in centimetres. Write your answers using decimals.

a **b**

c **d**

4 Estimate each of these lengths. Compare your estimate with a partner's.
Are they similar? How did you arrive at your answers?
Discuss your methods.

- **a** The height of the tallest tree near your school.
- **b** The length of your little finger.
- **c** The distance from school to the nearest airport.
- **d** The height of your teacher.

5 Work again with a partner. Estimate each other's heights in centimetres. Then use a tape measure or metre stick to measure. Write your heights to the nearest centimetre. Keep these measurements. You will use them again later in this chapter.

Reading a map

Key maths idea

Have you ever zoomed into an online map or GPS? When you zoom in or out, you are changing the **scale** of the map you are looking at. You cannot do this in an atlas or map book, because the image is printed at a particular scale.

A **map scale** is given as a ratio. It shows the relationship of the distance between two points on the map and the same two points in real life. You can always find the scale in the map key. The key may express the ratio as a diagram, as a number relationship or as a ratio, for example:

1 cm = 100 m
1 : 10 000

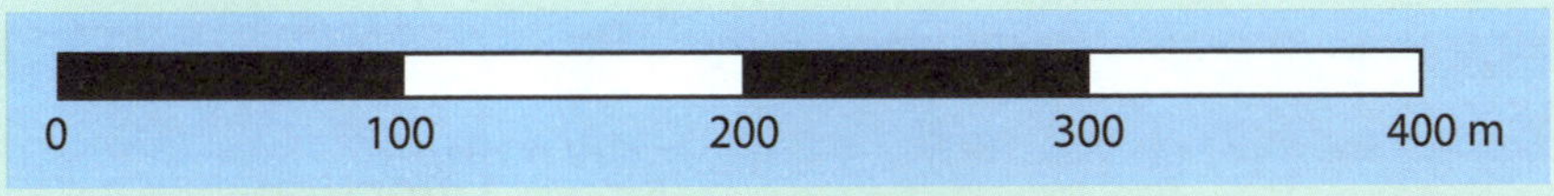

Key words

scale
map scale

Real-world maths: Working out map distances

1 What does this map show? Discuss with a partner how you would work out the real distance between two places on the map.

2 Use the scale on the map to estimate:
- the real distance between Belize and Jamaica
- the real straight line distance from Puerto Rico to Nicaragua
- the length of the north coast of Jamaica.

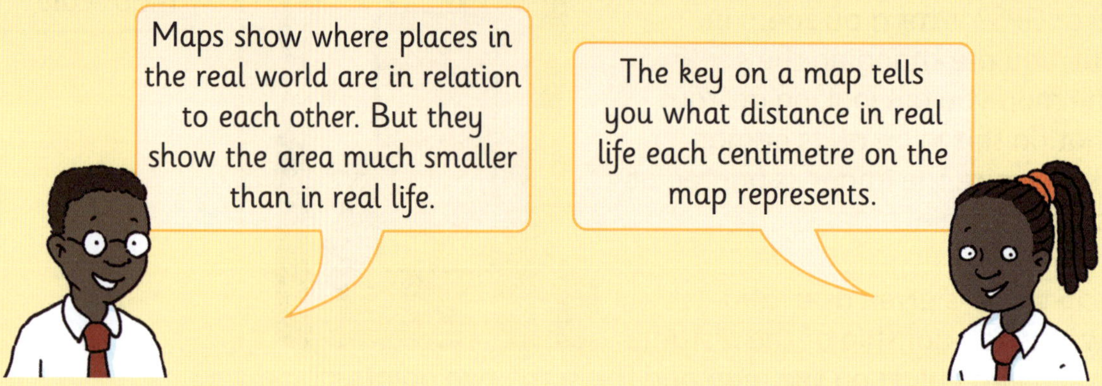

3 Discuss with a partner how you can use a ruler to measure the real-life distances shown on the map.

Scale measurements

Key maths idea

The distances on a map are a **fraction** of the real distances. The scale on a **map** or **plan** compares the size of the map with the size of the real area it represents.

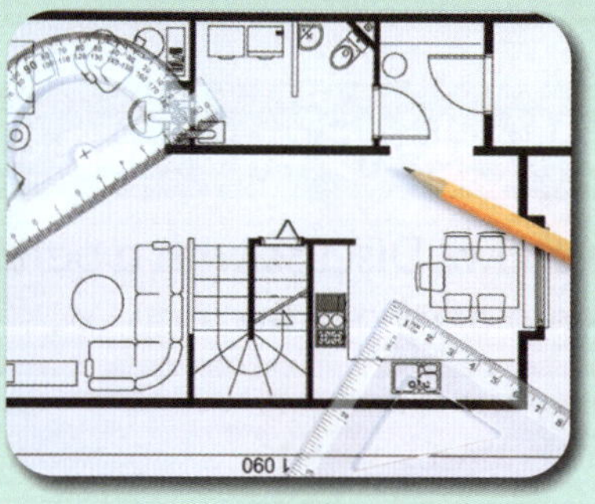

You have seen that scale can be shown in different ways:

- in words, for example, 1 centimetre on the map represents 100 kilometres in the real world.
- as a bar or line, for example, this line shows that 1 cm on the map is equivalent to 200 km in the real world.

  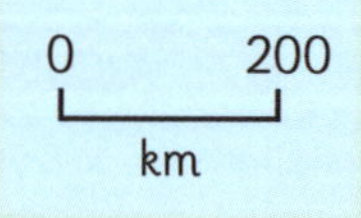

  This bar shows that 1 cm on the map is equivalent to 10 km in the real world.

  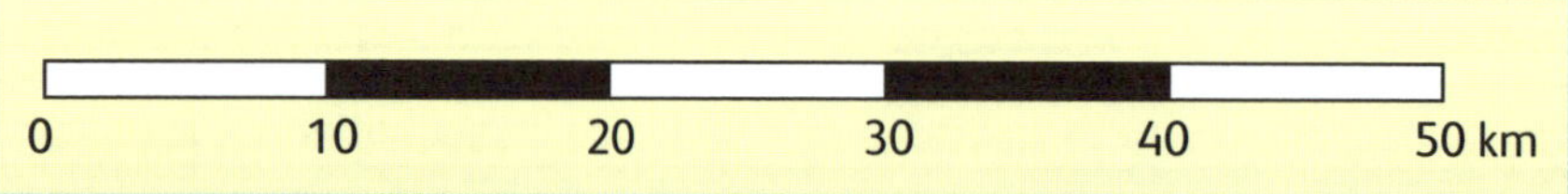

- As a ratio. For example, 1 : 20 000, this tells you that 1 unit on the map is equivalent to 20 000 of the same units in the real world.

Key words
- fraction
- map
- plan
- scale model

We often need to use the scale to convert the measurements of a drawing, plan or model to the real-life distances or dimensions they represent. Pay attention to the units you are working with, for example cm, m or km.

Look at this plan of a pool in a garden. On this plan, 1 cm on the plan represents 5 m in the real world.
The pool is 2 cm long on the plan.
$2 \times 5\text{ m} = 10\text{ m}$
So, the real pool is 10 metres long.

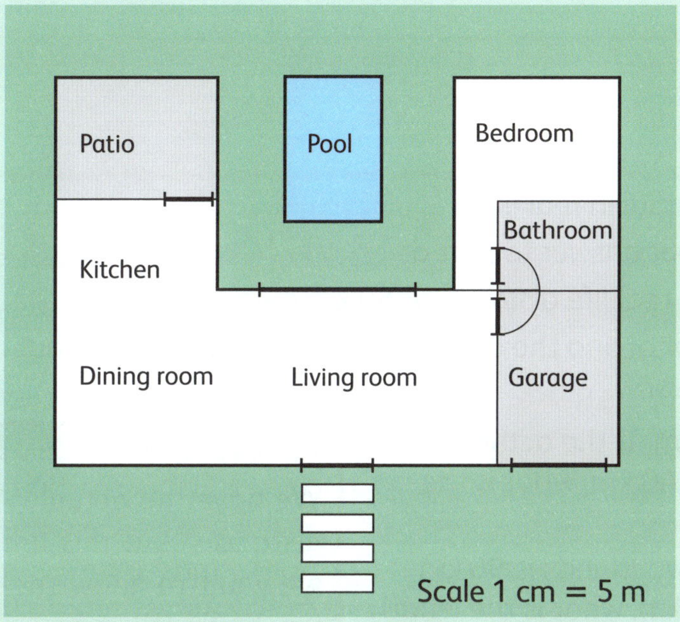

This child is playing with a 1 : 72 **scale model** of a real digger. The ratio scale tells you that 1 cm on the model represents 72 cm on the real digger.
If the digging arm on the model is 12 cm long, the length on the real digger is 864 cm, because $(12 \times 72)\text{ cm} = 864\text{ cm}$.
We can convert this to 8.64 m.

1 Work with a partner.

a Measure each other's height to the nearest centimetre.

b Draw two stick figures using a scale of 1 : 20 to compare your heights.

2 On a map, 1 cm represents 10 km.

a The distance between Town A and Town B on the map is 7 cm. What is the actual distance in kilometres?

b The distance between Town C and Town D in real life is 28.5 km. If you measure the length between the two towns on the map, what length will you measure?

3 A different map has a scale of 1 : 25 000. Julia measures the lengths of some roads on the map. These are the measurements on the map. Write the real-life distances in km.

a	Main Road	17 cm
b	St Johns Road	35 cm
c	Nanny Road	8.2 cm

4 Look at this bar scale.

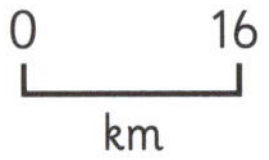

a What does 1 cm on this map equal in real life?

b Explain how you worked out your answer to part a.

c On this map, how long would a real-life distance of 50 km measure in cm?

d The distance between the hospital and the airport on the map is 4.5 cm. What is the actual distance in kilometres?

5 **a** A map has a scale of 1 cm : 5 km. If the actual distance between two towns is 40 km, what is the length on the map?

b The actual distance between two islands is 96 km. If 1 cm on a map represents 6 km, what is the length on the map?

c The distance between two football clubs is 55 km. If 1 cm on a map represents 10 km, what is the length on the map?

> Be careful!
> We have the actual distances, but this time we want to know the length on the map. Use your knowledge of inverse operations to help you answer the question.

Measurement

Real-world maths: Finding distances on maps

The names of some countries are shown on this map of South America.

The capital cities of these countries are shown by a dot.

Scale: 1 cm = 600 km

To find the actual distance between Caracas and Brasilia:

- Read the scale of the map. The scale of this particular map tells us that 1 cm on the map represents an actual distance of 600 km on the ground.
- To enlarge from 1 to 600 we multiply by 600, so to convert the map distance to actual distance we multiply by 600.
- Use a ruler to measure the distance between the dots on the map.
- Multiply your measurement by 600: this is the actual distance in kilometres!

1 **a** What is the distance on the map between Santiago and Brasilia?

b What is the actual distance between Santiago and Brasilia?

2 **a** What is the distance on the map between La Paz and Lima?

b What is the actual distance between La Paz and Lima?

3 Approximately how far is it in kilometres from Georgetown to the southern tip of South America?

4 Approximately how wide is South America at its widest point?

Scale drawings

Key maths idea

Real places are too big to fit onto paper so we use **scale drawings** for plans and maps. A plan of a house or a neighbourhood is sometimes called a **blueprint**. This is because architects and town planners used to print these plans on blue paper. But today a blueprint is a word meaning an accurate representation or plan showing a particular place. Look at this floorplan of a house.

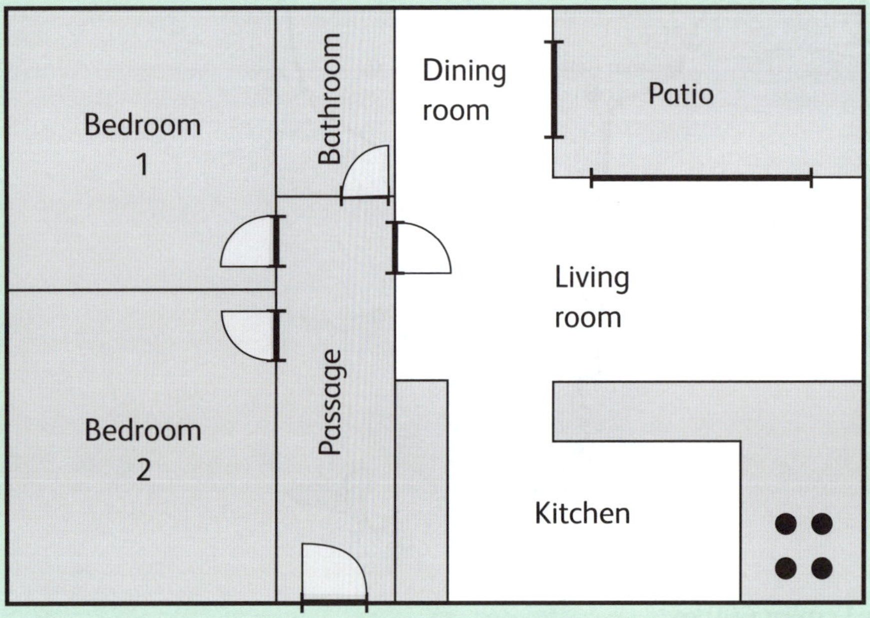

Scale: 1 : 80

The scale of this floorplan is shown as 1 : 80.
This means that 1 cm on the sketch represents 80 cm in real life. To convert a line on the drawing to its real-life measurement, multiply by 80. In other words, each dimension on the house is 80 times bigger than the corresponding dimension on the drawing!

Key words

scale drawing
blueprint

1 Complete the table.

Measurement	Length on floorplan	Length in real life (cm)
Length of passage		
Width of passage		
Width of Bedroom 1		
Width of bathroom		
Width of house (at widest point)		

2 Convert each of the lengths from Exercise 1 into metres.

Maths detective: Scale drawings and models at work

Use your new skills to solve these cases.

Case 1

How do different people use scale drawing or models in their work?

1 These pictures show how an interior designer, a product engineer and a jewellery designer use scale drawing.

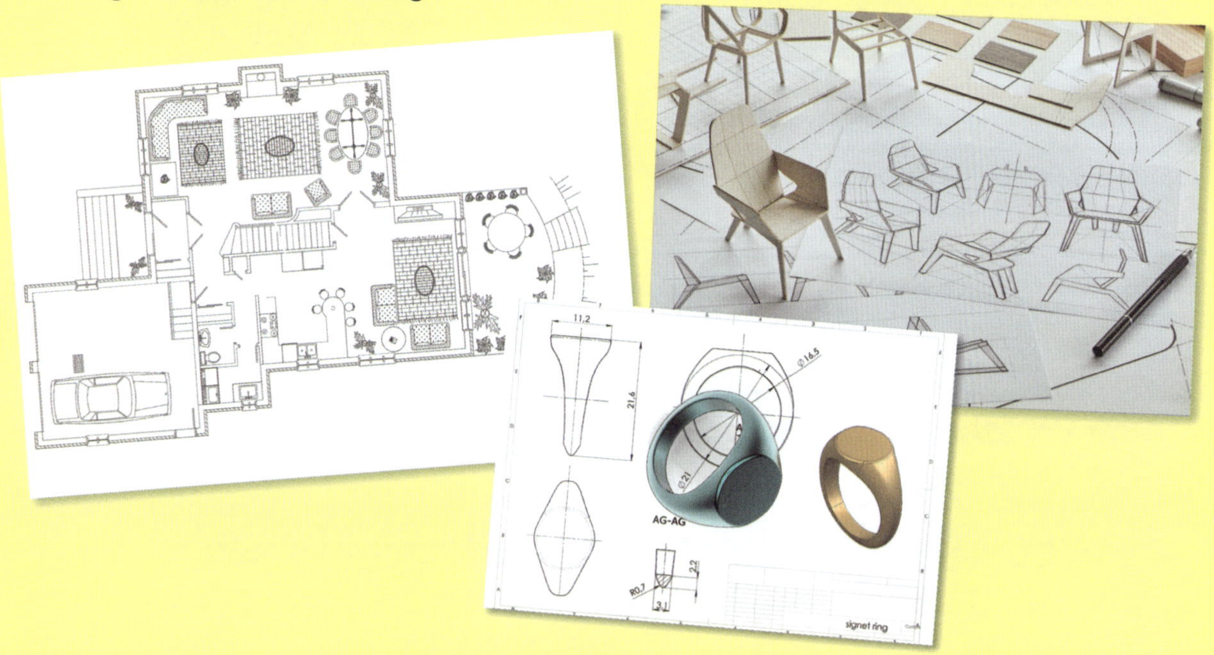

a What does each drawing show you?

b Why are these kinds of scale drawings important in each profession?

2 Jerome made this list of some professions that make use of scale drawings, blueprints or scale models.

architect
fashion designer
landscaper
engineer
construction worker
jewellery designer
CAD operator
event manager
multimedia animator
electrical draftsperson
cartographer
model maker
industrial designer
technical illustrator
surgeon
car designer

a Choose three of these professions and find out how and why scale drawings or models are useful.

b Can any of the professions on the list be done without scale drawing?

c In which profession is scale drawing the most important? Give reasons for your choice.

Case 2

How can knowledge of scale and how it works help people plan and make decisions?

The islands of the Greater Antilles are Cuba, Jamaica, Hispaniola (Haiti and the Dominican Republic) and Puerto Rico. You are going to suggest which cities in this region a delegation of health officials should visit on a research trip. Start by making a list of each country in the Greater Antilles and its capital city.

1 What is the scale of the map?

2 Explain in your own words how you would use the scale on this map to work out the distance between each pair of cities in the table.

3 Copy and complete the table.

From (city)	To (city)	Distance on map	Actual distance
Kingston	Havana		
Kingston	San Juan		
Kingston	Port-Au-Prince		
Kingston	Santo Domingo		
Havana	Port-Au-Prince		
Santo Domingo	San Juan		
San Juan	Port-Au-Prince		
Port-Au-Prince	Santo Domingo		

4 Write a paragraph to suggest the most sensible travel itinerary for the researchers. Give at least two reasons why you chose the route that you suggested.

Real-world maths: Design a space

Work in a group. Choose one activity.

Activity A

Your school wants to build a computer laboratory for students and has announced a competition for the design. You have to submit your design in the form of a poster that includes:

- a heading
- a scale drawing of the proposed building
- information about how many students the room can accommodate and how much space they will have.

Your scale drawing should show:

- appropriate dimensions of the building (length and breadth)
- how and where the chairs, desks and power points will be placed
- the number of desktop computers and laptops that will be available
- the teacher's workstation
- location of a storage area or storage cupboard
- a scale that is clear and easy to read.

Use a separate project page to:

1. explain how and why you chose the scale you used
2. show your calculations for the elements in the scale drawing
3. list the instruments you used to complete your design
4. give a reason why you used these instruments
5. write a short paragraph to convince the school board why your design should be chosen.

Activity B

You are a landscape architect at ABC Landscaping Ltd.
Your client has asked you to design a garden with different sections.
The garden area has a length of 15 m and a width of 10 m.

- Section A is square.
- Section B is a rectangle, but not a square.
- Section C is triangular.

1. Draw your design.
2. Clearly label the three sections on grid paper using a suitable scale.
3. Be sure to include the dimensions on your design.
4. Explain to your client why you chose the scale you used in your design.

5 Calculate the actual distance around each section of the garden to work out the length of the borders you will need to pave.

6 Paving slabs cost $1 500 per running metre. How much would your garden design cost to pave? Show your calculations clearly. (Note: Some perimeter lengths will need to be rounded up to the next whole metre and then multiplied by the unit price to find the total price.)

What I have learned

1 In your notebook, answer these questions to summarise what you have learned in this chapter.

2 Why is it important to include a scale on a map or a floor plan?

3 How do you use a scale on a map or floor plan to work out real distances?

4 What is the difference between the scale on a measuring instrument and the scale on a map?

5 Give three real-life examples of scale diagrams and how they are used.

Practice questions

1 a There are _____ mm in one centimetre.

b There are _____ cm in one metre.

c There are _____ m in one kilometre.

2 Use the facts in Exercise 1 to calculate the values in the table.

mm	cm	m	km
66 000	a	b	c
d	2 650	e	f
g	h	25.5	i
j	k	l	1.5

3 On the map, Town A and Town B are 6 cm apart. The map is not shown to scale. What is the actual distance between the towns?

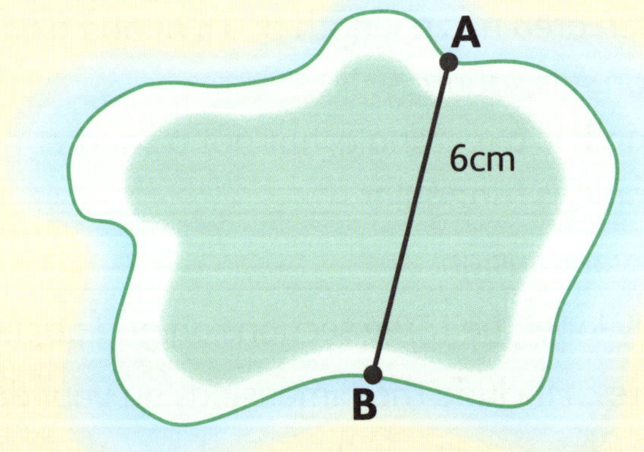

Scale: 1 cm = 10 km

4 A map has a scale of 1 cm = 100 m.
Calculate the missing distances in the table to show the distance on the scale drawing and the actual distance.

Distance on scale drawing	Actual distance
4 cm	**a**
7.5 cm	**b**
c	900 m

5 Bailey made a scale drawing of a neighbourhood park. The scale he used was 1 centimetre : 4 metres. If the actual length of the volleyball court is 16 metres, how long is the volleyball court in his drawing?

6 Chloe measured a garden and made a scale drawing. The scale she used was 1 centimetre : 6 metres. In the drawing, the front patio is 4 centimetres long. What is the length of the actual patio?

3 Self check

	Check you can do this:	If you have difficulty, try this:
1	Know the relationship between units of measurement for length or distance.	Go back over the activities on pages 39–40, and remind yourself how to use millimetres, centimetres, metres and kilometres. You can also go over pages 41–42.
2	Understand the concept of fractions and ratios.	Go back to page 40 and remind yourself of the difference between a fraction and a ratio.
3	Interpret a simple scale drawing and calculate actual distances using the scale on a road map or floor plan.	Revise the work on map scales and reading maps and plans on pages 43–45.

Chapter 4 Time

In this chapter you will explore this focus question:

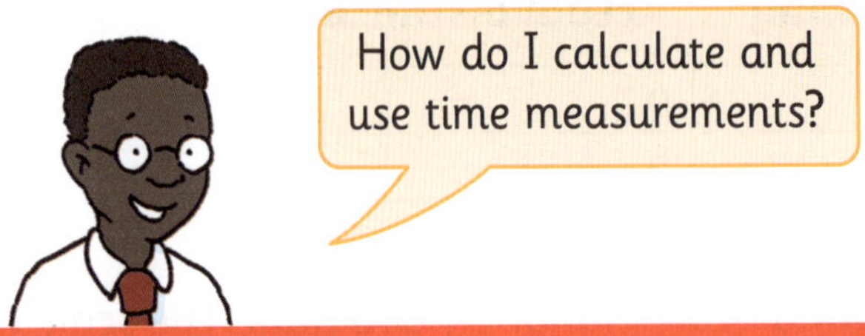

To answer this question, you will:

- read and write times using 12-hour and 24-hour notation
- convert times written in one format to another
- solve problems involving time written in 24-hour notation.

Starting point

1 Taniyah was looking for an app that she could use to plan her day. She found a clock-face planner that she can download and use on her phone.

a How does this diagram show what the person is doing at different times of the day?

b How is time written on the app?

c The actual time of day is shown in the centre of the diagram. What time is this? What is the person doing at this time?

d How much time did this person spend on meals?

2 Taniyah used the app to plan her day. This is what her plan looked like:

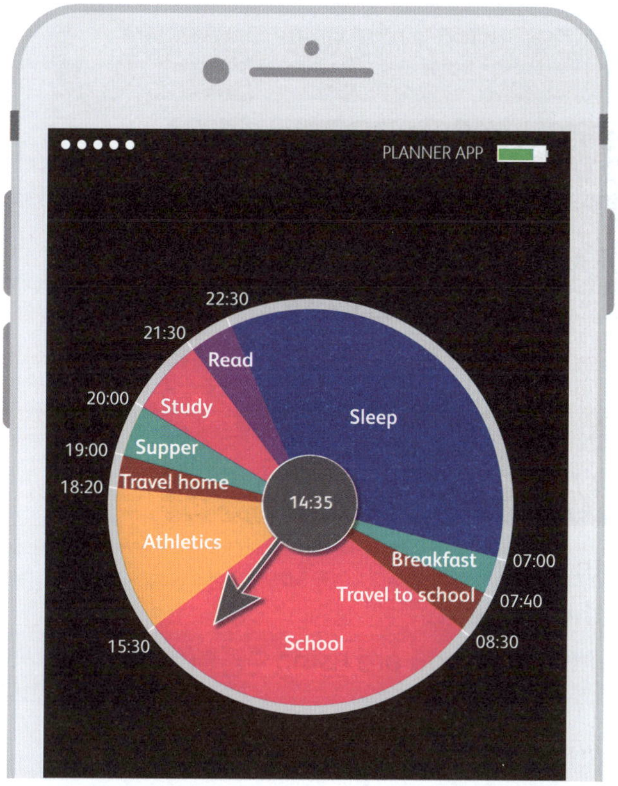

a What time is shown in the centre? How else could you write this?

b What is Taniyah currently doing? When will this activity finish?

c Compare Taniyah's day with your day. How is it similar? How is it different?

3 Five friends took part in a race. Their times are shown in the table.

Name	Time
Gabrielle	239 seconds
Jada	199 seconds
Jordan	100 seconds
Leon	202 seconds
Femi	119 seconds

a Who finished the race closest to two minutes?

b Which two runners crossed the finish line closest together?

c What was the difference, in seconds, between the fastest time and the slowest time?

d Convert each time in seconds to minutes and seconds.

The 24-hour time system

Key maths idea

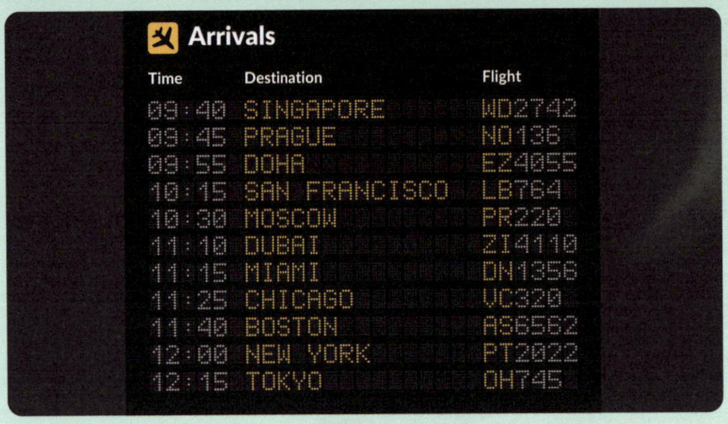

9 a.m. means 9 o'clock in the morning.

9 p.m. means 9 o'clock in the evening.

When you use **a.m.** and **p.m**. times, you are using **12-hour notation**.

Notation is the way you write sometimes – the symbols or abbreviations you use.

Another type of time notation is **24-hour notation**. When you use this notation, you write the time using four digits. The first two digits tell the hour, and the second two digits tell the number of minutes past the hour. There is usually a colon or full stop between the hour digits and the minute digits. In this system, the morning hours run from 00:00 (midnight) to 11:59 (1 minute before noon), and the afternoon hours run from 12:00 (noon) to 23:59 (1 minute before midnight).

Example

01:00 means 1 o'clock in the morning.

13:00 means 1 o'clock in the afternoon.

14:15 means quarter past 2 in the afternoon.

22:45 means quarter to 11 in the evening.

Key words

a.m.

p.m.

12-hour notation

notation

24-hour notation

How would you write the time 7 hours after midnight in 12-hour notation? And 11 hours after noon in 24-hour notation?

Talk about

How could a clock face like this one help you to write times using the 24-hour format?

1 Use the clock face in the Talk about box to help you work out how each of these times is written and read using the 24-hour format.

a 11:15 p.m. **b** 1:00 p.m. **c** 8:30 p.m.

2 Copy and complete the table.

12-hour time	24-hour time
5:00 a.m.	05:00
8:00 a.m.	**a**
b	07:30
c	08:15
11:35 a.m.	**d**
12:00 noon	**e**

3 Match the equivalent times shown on the analogue and digital clocks.

4 Write each of these times in the 24-hour format.

5 Write these 24-hour times as a.m. or p.m. times.

a 10:14 **b** 13:30 **c** 18:15 **d** 21:17 **e** 19:08

6 Write the time each activity starts in the 24-hour format.

Activity	12-hour time	24-hour time
Netball	 p.m	**a**
Football	 a.m	**b**
Cricket	 p.m	**c**
Swimming	 a.m	**d**

7 Write down the missing times in the table.

12-hour time	24-hour time
6:00 p.m.	18:00
8:00 p.m.	**a**
11:00 p.m.	**b**
c	21:45
11:55 p.m.	**d**
12:00 midnight	**e**

8 Lamar converts the time on the analogue clock to 24-hour format.
This is his answer.

Explain whether Lamar is correct or incorrect.

Solve problems involving time

Key maths idea

When we talk about different times, how much time has passed between events, or how long something takes, we talk about **time intervals**. A time interval is the time that passes between two events.

Example

A journey begins at 07:57 and ends at 10:04.

Calculate how long it took.

You can use a number line to help you.

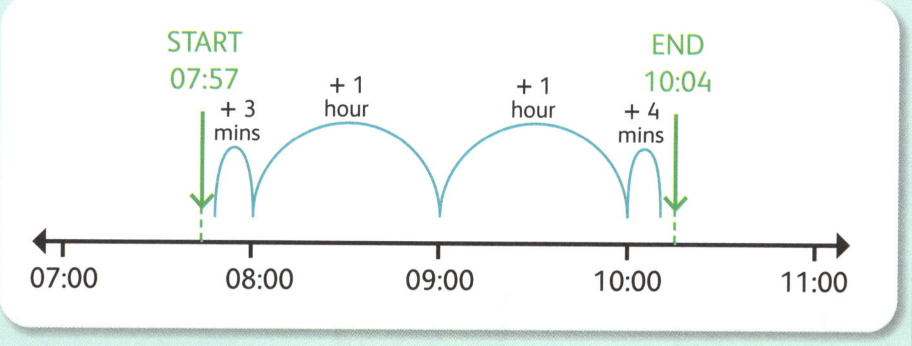

Jump to the next hour:

Think: How many minutes after 7:57 is the next hour?

Then jump forward in whole hours:

Think: How many hours after 8:00 will be 10:00?

The final jump goes to the end time:

How many minutes after 10:00 is 10:04?

Next, add the hour jumps and the minute jumps.

3 minutes + 1 hour + 1 hour + 4 minutes = 2 hours and 7 minutes.

So: 07:57 to 10:04 is 2 hours and 7 minutes.

Key word

time interval

1 Use the timetable to answer the questions.

DEPARTURES					
Kingston–Norman Manley Airport	06:05	06:35	07:05	07:43	08:15
Harbour View	06:15	06:45	--	07:53	--
Bull Bay	06:35	07:05	07:23	08:13	08:50
Grants Pen	06:50	07:20	07:38	08:28	09:05
Albion	07:10	07:40	07:58	--	09:25
South Haven	07:25	07:55	08:13	09:03	09:30

a On the 06:35 bus, how long does it take to get from Harbour View to South Haven?

b Can you travel to Albion on the 07:43 bus from the airport? How do you know?

c Anthony travels on the 08:15 bus from the airport to Grants Pen. How long should his journey take?

d The 06:05 service is running 12 minutes late. What time will it arrive at South Haven?

e Joshua arrives at the Bull Bay bus stop at 07:08. What time will he get to Albion?

> **!** Why do you think some slots in the timetable are empty?

2 Calculate the time intervals between these times.

a Start:

a.m.

End:

b Start:

p.m.

End:

c Start:

a.m.

End:

d Start:

p.m.

End:

e Start:

p.m.

End:

f Start:

a.m.

End:

3 Sophia leaves school at the time shown on the clock. She arrives home one hour later. What would the time be on a 24-hour digital clock?

4 Look at these television listings. Calculate the length of each programme.

a Rising Starz

b Quiz Show

c Weather Report

d Sports Time

e Why can't you calculate the duration of Nightly News?

Start time	Programme
20:30	RISING STARZ
22:05	QUIZ SHOW
22:55	Weather Report
23:00	SPORTS TIME
23:45	Nightly News

Talk about

To change any time after noon from 12-hour to 24-hour, just add 12 to the hours. Will this always be true? Are there any examples when this rule will not work?

Maths detective: Time

Use your knowledge of time and read the clues to work out these cases.

Case 1

Read the clues carefully! Then use what you know about time to work out which watch belongs to each person. Write the name of the person and then the letter.

1 Glenroy is on his way home from a football match that ended at quarter past seven in the evening.

2 Amelia is reading in bed. She has to turn the light off at 9:30 p.m.

3 Oniel's alarm has just gone off. He set it for quarter past six in the morning.

4 Jada looked at her watch and saw that she had 15 minutes to get to her 9 a.m. appointment.

5 Yendi is at basketball practice. She practices from 2:30 p.m. to 3:30 p.m. every day.

6 Junior is half an hour late for a lesson that started at 1 in the afternoon.

Case 2

Peta-Gay looked at the time.
She noticed that the four digits in the time were consecutive numbers in ascending order.

1 What is the next time that the digital clock will show the time as four consecutive digits in ascending order?

2 Junior says the time can also have consecutive 2-digit numbers in descending order, for example 12:11 or 15:14. How many times like this can you find?

Case 3

24-hour times are used in many different occupations. The 24-hour system is especially important where an error in the time given can result in problems.

1 Choose three of the occupations from the list.

2 Find out more about each occupation and how and why they use the 24-hour system.

3 Write a short article for a maths journal sharing what you found out.

pilot   computer programmer   military planner
transport worker   meteorologist   hospital staff

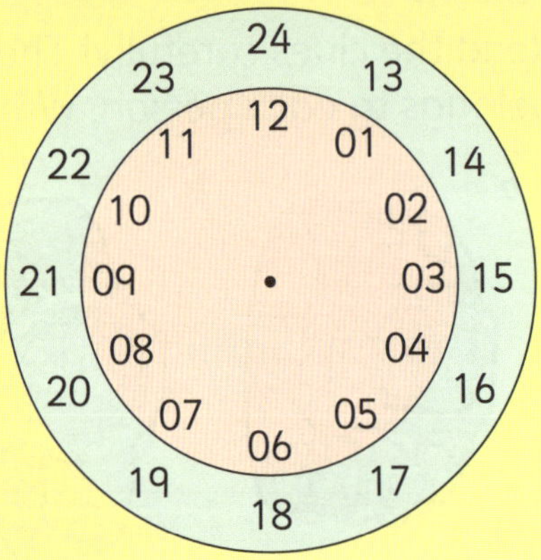

Real-world maths: Time to travel

1 Class 6A were at the airport waiting for a flight from Montego Bay to Kingston. They heard an announcement that the flight would take off at fourteen forty-eight.

  a Stephen's younger brother Rodney asked him what 'fourteen forty-eight' means. How should Stephen answer this question?

  b What do you need to remember when you tell time using the 12-hour and 24-hour time system?

  c How are these two systems of telling time the same? How are they different?

2 The flight arrived in Kingston at fifteen fifty-six. It was 12 minutes later than scheduled.

  a At what time was it scheduled to arrive?

  b Is it correct to say the flight arrived at 56 past 3 in the afternoon? Give a reason for your answer.

3 At the airport in Kingston you are not allowed to park in front of the building from 2:15 p.m. to 6:05 p.m.

  a Design and draw a 'No Parking' sign showing this information. Use 24-hour times on the sign.

  b For how long are vehicles not allowed to park outside the building?

What I have learned

1 Draw an analogue clock and a digital clock and write notes to explain how each shows the time.

2 Draw a flow diagram with at least three steps to teach someone how to convert times from the 12-hour system to the 24-hour system.

Practice questions

1 Alex starts cooking at quarter past 6 in the evening. When he finishes, his clock display shows: **18:49** How long did Alex cook for?

2 Danielle looks at her alarm clock. **23:59**

a What time does it show?

b What will the time be in 1 minute?

3 A journey starts at 10:23. It ends at 13:15.
Is the journey shorter or longer than 1.5 hours? By how much?

4 Here are some activity sessions for a sports club.

Activity	Start	End
Swimming	11:30	13:00
Badminton	13:15	13:45
Football	13:50	14:45
Tennis	14:55	15:25
Gym	15:45	17:15

a Which two sessions last 0.5 hours?

b Which two sessions last 1.5 hours?

c Which session lasts more than 0.5 hours but less than 1.5 hours?

5 How many hours and minutes are there from 7:47 p.m. to 11:21 p.m.?
Choose the correct answer.

a 4 h 16 min b 3 h 34 min c 4 h 34 min d 3 h 16 min

6 Match the 24-hour times to their equivalent in words.

20:55	Ten minutes to two
13:50	Five minutes to ten
09:55	Ten minutes to three
15:05	Ten minutes past four
02:50	Five minutes past three
16:10	Five minutes to nine

7 The cooking times for barbecue meats are shown as follows:

Cooking times	
Pork	20 min per 500 g
Chicken	19 min per 500 g
Beef	28 min per 500 g

a How long would a 1.25 kg piece of beef take to cook?

b What is the difference in cooking time between 2.5 kg of chicken and 1.5 kg of pork?

c A 2 kg chicken was placed on the barbecue at 4:20 p.m. What time should the chicken be ready? Give your answer using 24-hour time.

4 Self check

Check you can do this:	If you have difficulty, try this:
1 Read and write time.	Go back to look at some of the analogue and digital clocks in this chapter. Remind yourself how to find the hour and the minutes. Remember, analogue clocks have a face, with a short hand pointing to the hour and long hand pointing to the minutes. Digital clocks use numbers.
2 Explain the relationships among units of time, using 24-hour time notation.	Go back to page 56, and remind yourself how to write time in this way.
3 Use 24-hour time in problem situations.	Go back to page 59. Make sure you can work out how long an event took, given the starting and end time. You should also be able to read timetables, and answer questions involving times and durations.

Chapter 5 Perimeter

In this chapter you will explore this focus question:

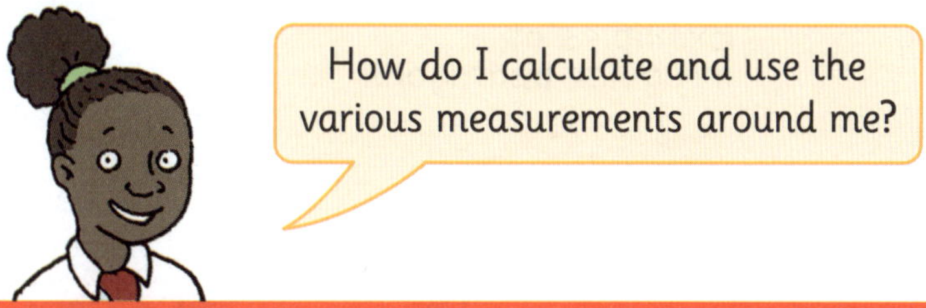

To answer this question, you will:

- measure and calculate the perimeter of regular and irregular polygons
- use the appropriate formula to calculate the perimeter of regular polygons
- calculate the measurement of one side of a polygon when given the perimeter and the lengths of the other side.

Starting point

1 These three vegetable patches need a fence around all four sides.

- **a** For each vegetable patch, how could you calculate the total length of the fence?
- **b** Why do you only need the measurements of two sides for vegetable patch B?
- **c** Why do you only need one side measurement for vegetable patch C?
- **d** What do you call any shape that has all sides the same length and all angles the same size?

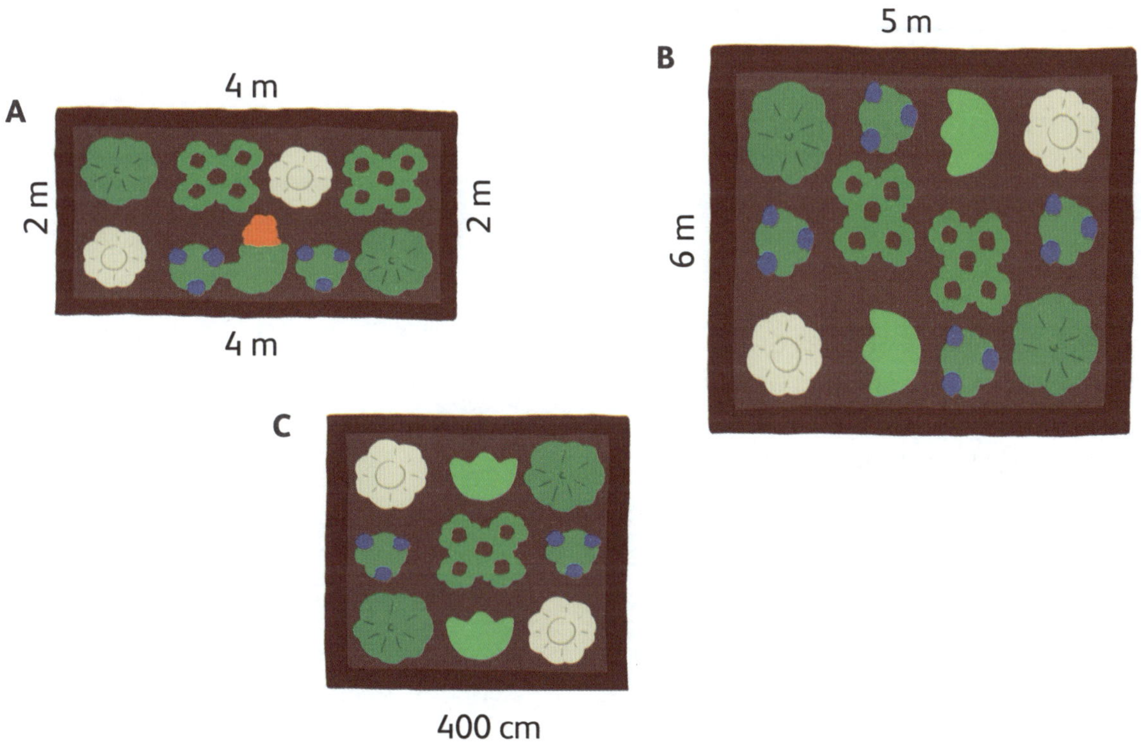

2 Stacy made fact cards for different kinds of polygons. Here is the card she made for a 5-sided shape:

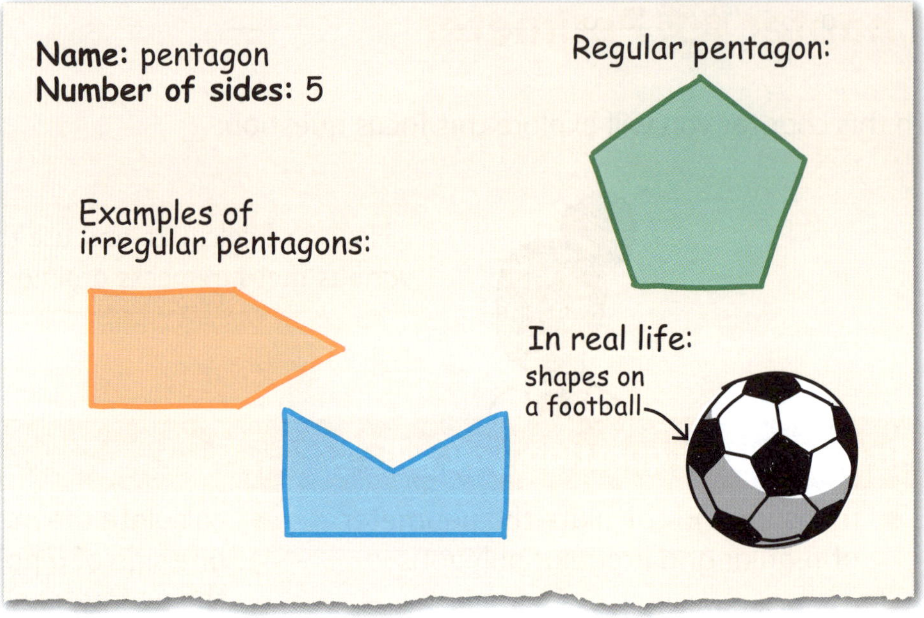

3 A polygon is a many-sided shape. Each of the sides is straight and the shape is 'closed.' Regular polygons have all their sides equal in length. For each shape, say whether it is a polygon, and if so, if it is regular or irregular.

> Remember: 'poly' means 'many'.

a parallelogram

b semi-circle

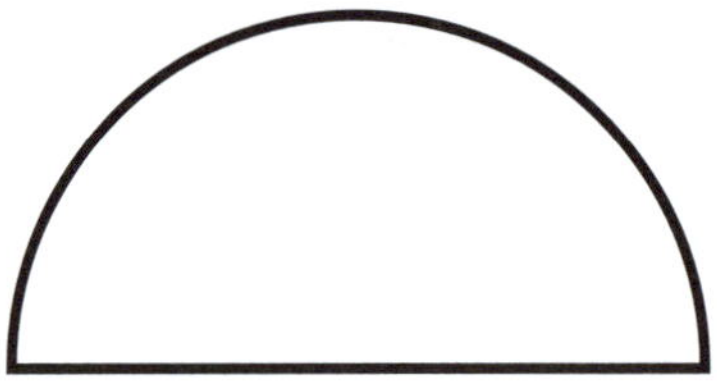

c open path

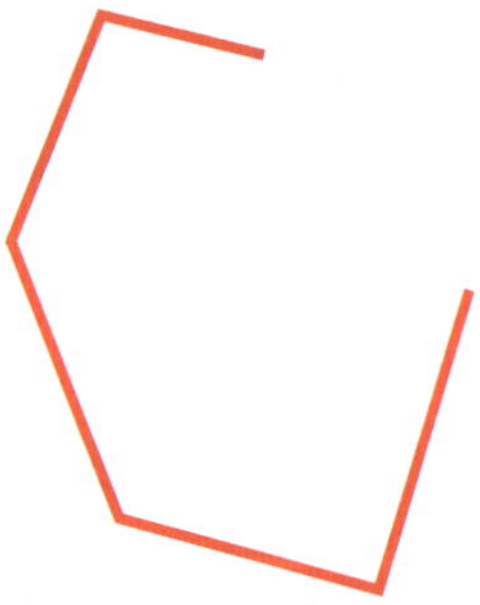

Measuring and calculating perimeter

Key maths idea

Perimeter is the distance around a **plane shape**. Plane shapes are two-dimensional or **2-D shapes** because they have only two dimensions we can measure: length and width.

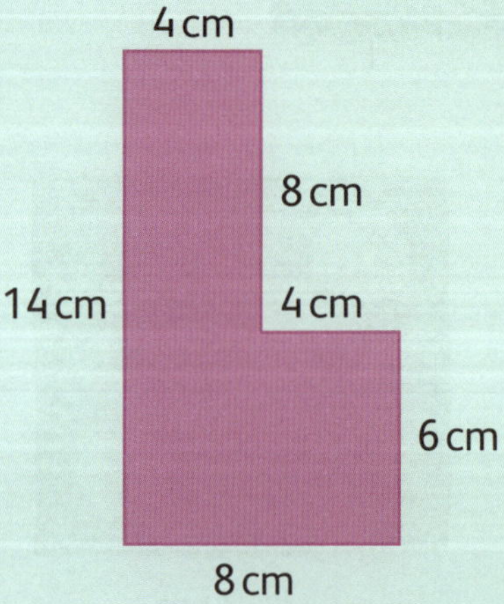

We can think of the perimeter as the sum of the lengths of all the sides.

The perimeter of this shape is:

4 cm + 8 cm + 4 cm + 6 cm + 8 cm + 14 cm = 44 cm

Key words

- perimeter
- plane shape
- 2-D shape

1 Measure the length of the sides of each shape. Work out its perimeter.

a

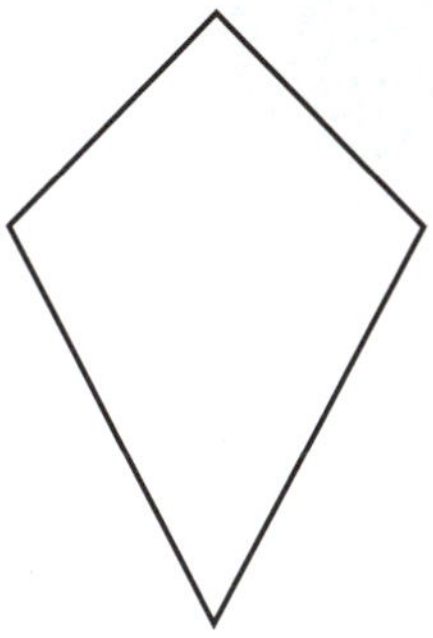

b

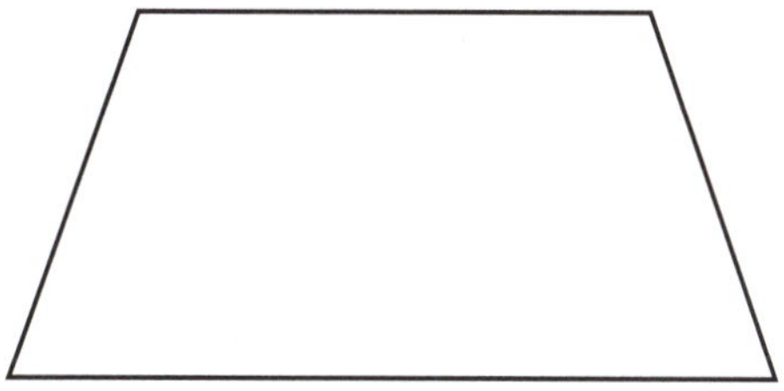

c

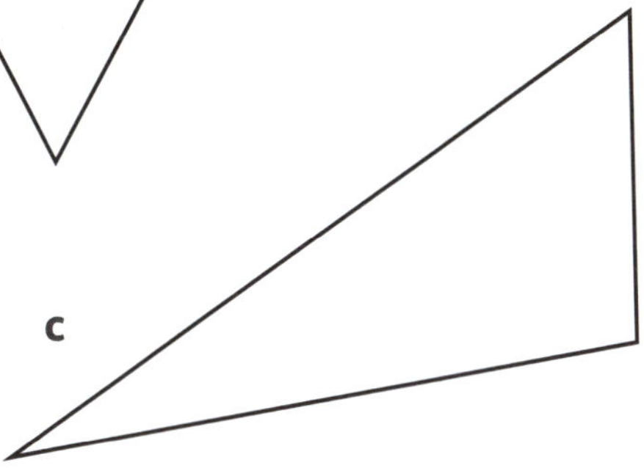

2 Measure these shapes in millimetres to calculate the perimeter.

a

b

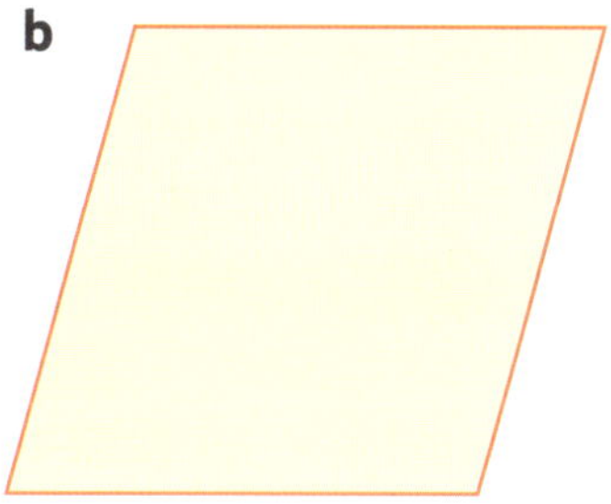

c

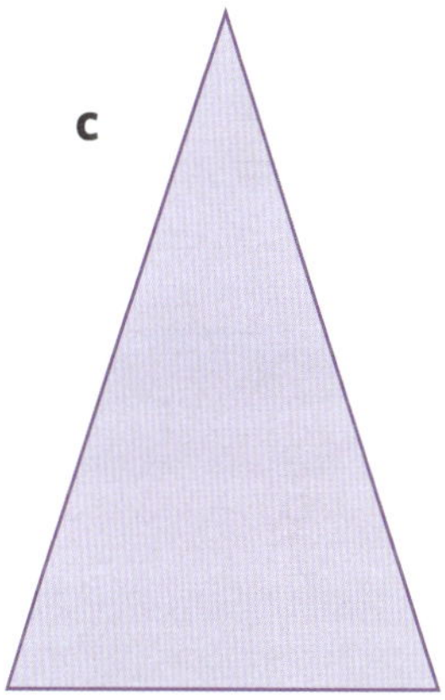

Using properties of shapes to work out perimeter

Key maths idea

These diagrams show that the sides with matching marks are equal in length. Discuss with a partner how you can use this information to identify each polygon and to work out the perimeter of each shape.

A

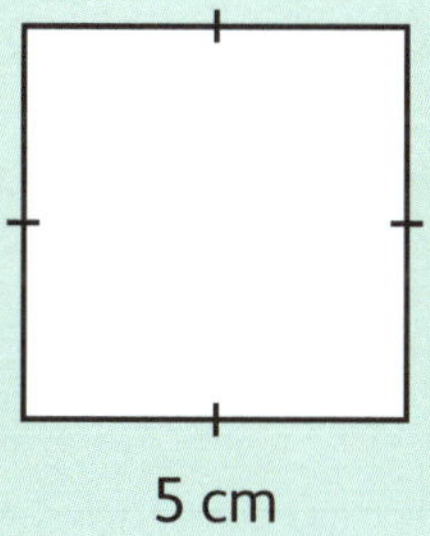

B

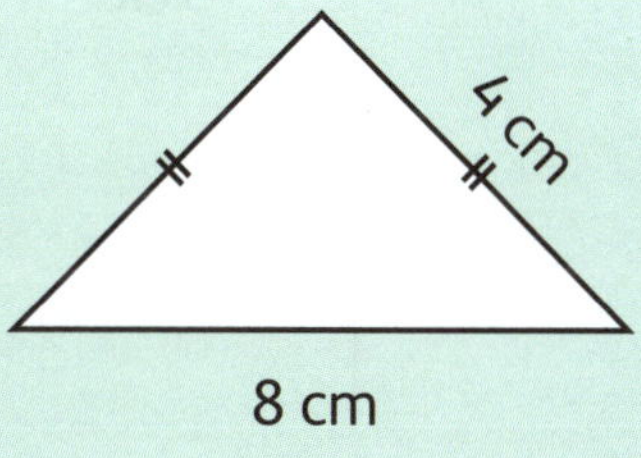

C

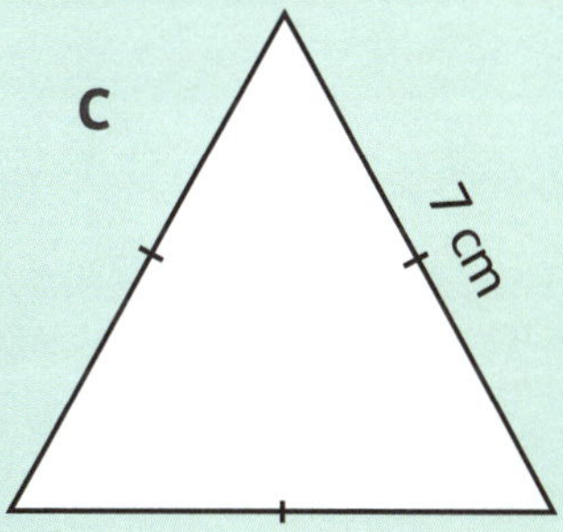

Key words

isosceles
equilateral
scalene

A triangle with two of its sides equal in length is called an **isosceles triangle**. A triangle with all three sides equal in length is **equilateral**. A **scalene** triangle has all three sides different in length.

Work out the perimeters of these shapes. Discuss with a partner how you worked it out. Remember: the diagrams are not drawn to scale, so you should use the information given in each sketch.

a

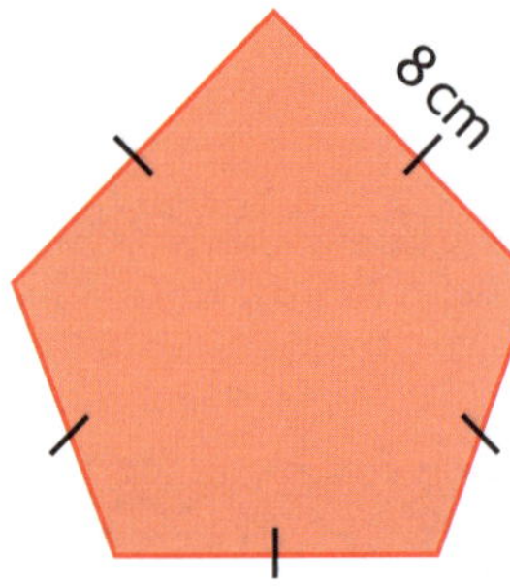

b

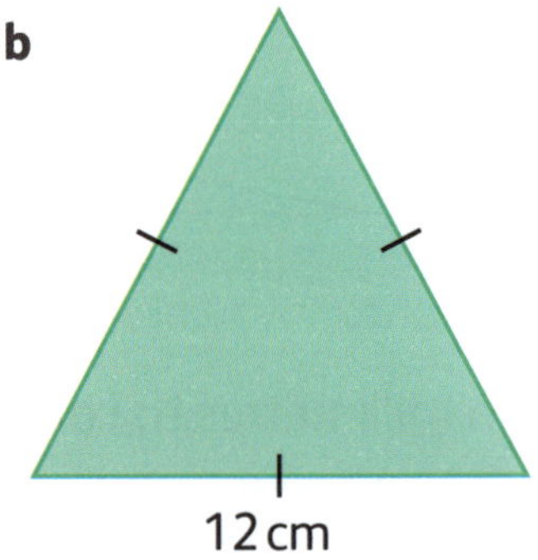

c

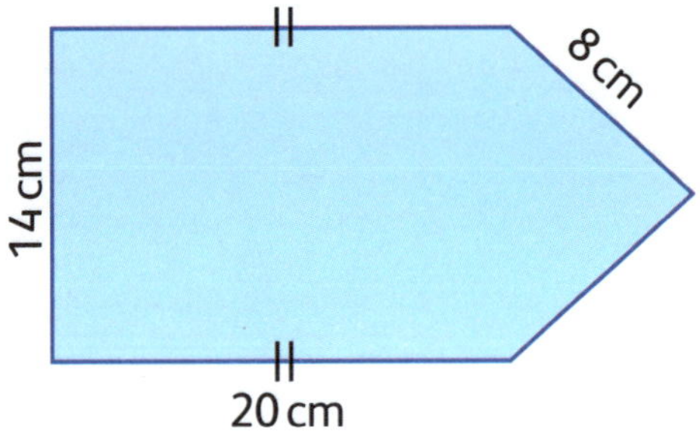

d

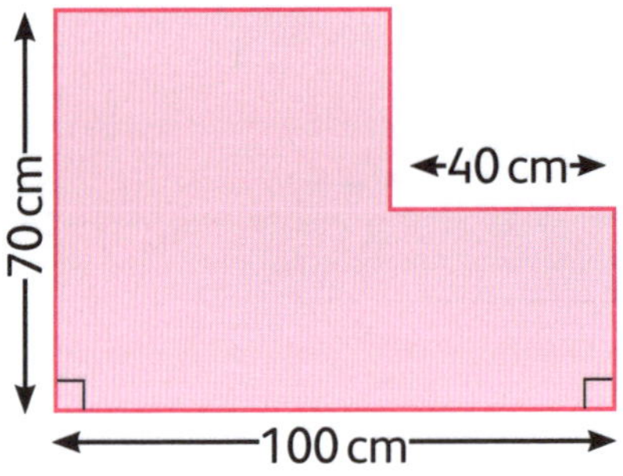

e

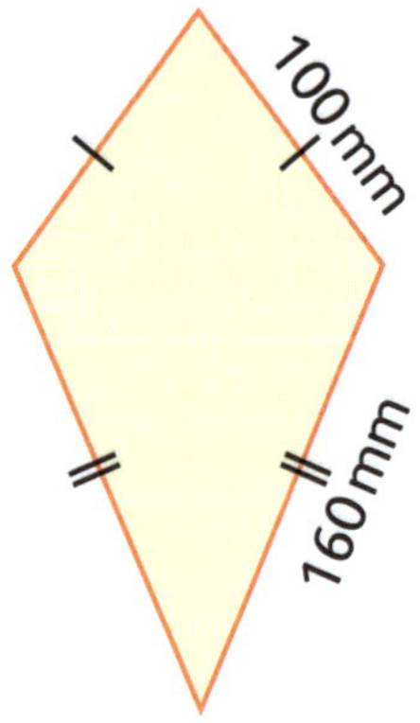

f

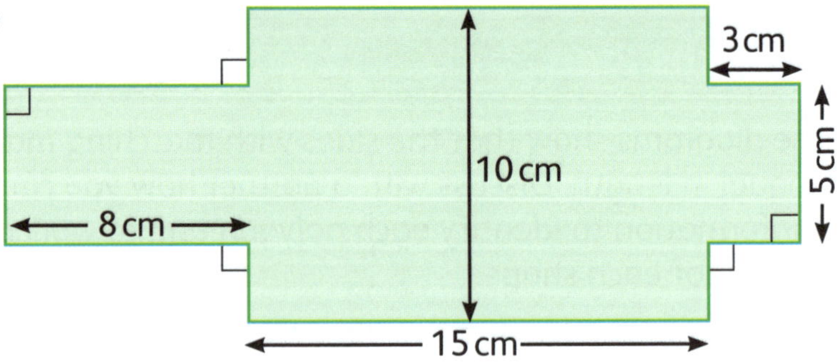

Finding the perimeter of regular polygons

Key maths idea

A **polygon** is **regular** when all sides are equal in length and all angles are the same size. So if you know the length of one side of a regular polygon, you can work out the perimeter. You can use this understanding to write a **formula** for each type of regular polygon.

For example, this square has one side 6 cm long.
We know that a square has four sides of equal length.
Therefore, the perimeter of this square is:
$6\text{ cm} + 6\text{ cm} + 6\text{ cm} + 6\text{ cm} = 24\text{ cm}$
Or:
$4 \times 6\text{ cm} = 24\text{ cm}$
We can use this to develop a formula for the perimeter of a square.
$P = 4 \times s$
where P = perimeter and s = side length

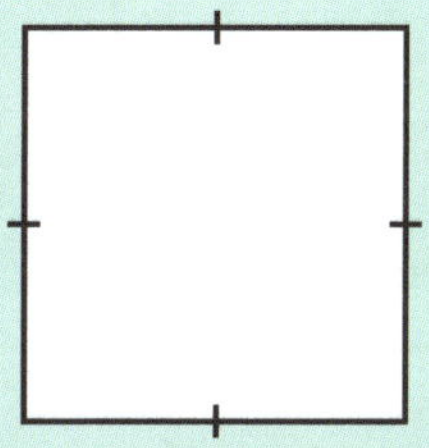

6 cm

Key words

polygon
regular
formula
pentagon
hexagon
heptagon
octagon
irregular

1 Work out the formula for each regular shape. Let P = perimeter and s = side length.

a regular **pentagon**

b regular **hexagon**

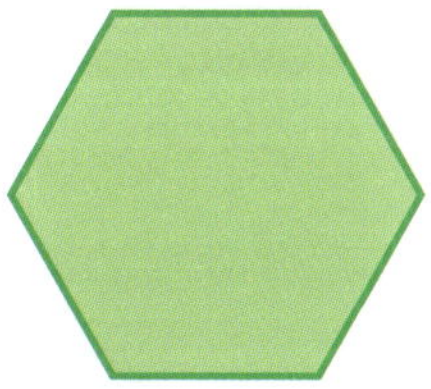

c regular **heptagon**

d regular **octagon**

2 A parallelogram is an **irregular** polygon, because not all its sides and angles are equal. However, it has two pairs of opposite sides equal in length. Use this information to work out the formula for a parallelogram's perimeter.

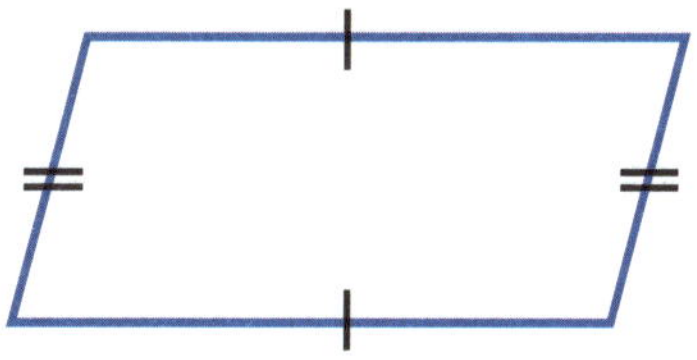

3 Work out:

a the perimeter of a regular hexagon with sides each 4 cm in length
b the perimeter of a regular pentagon with each side 10 cm in length
c the side length of a square with a perimeter of 100 cm
d the side length of a regular heptagon with a perimeter of 35 cm.

4 Draw the following shapes:

a two different triangles, each with a perimeter of 12 cm
b two different rectangles, each with a perimeter of 24 cm.

Solving problems involving perimeter

Key maths idea

When you are solving problems that involve perimeter, it can be helpful to draw a sketch of the shape you are working with. The sketch does not have to be accurate or even to scale.

For example, the perimeter of a rectangle is 200 mm. The longer side measures 80 mm. How long is the shorter side?

Here is a sketch of the information:

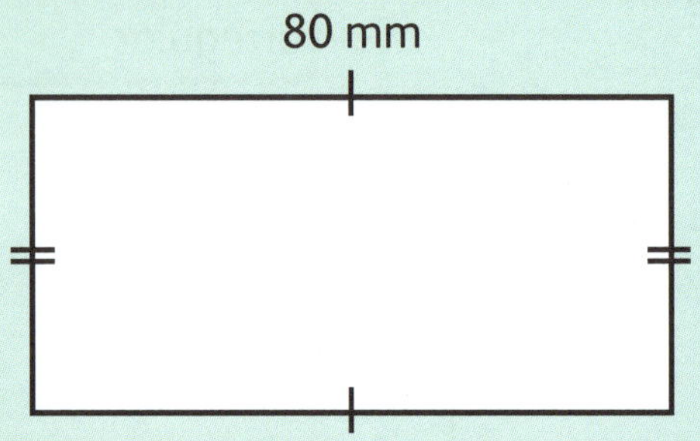

What do the marks on the sides tell you?

How can you work out the remaining side lengths?

1 Anthony has made a drawing of a square photo frame. The outside edge of each side will be 7.5 cm. What is the perimeter of the outside edge of the photo frame?

2 The width of a rectangular field is 8 metres. Its length is twice the width. What is the perimeter of the field?

3 A sketchbook is 20 cm long and 14 cm wide. Iris uses some ribbon to make an edge around the front cover. She has 12 cm left afterwards. How long was her piece of ribbon?

4 The perimeter of the outside edge of a rectangular swimming pool is 44 metres. The length of the pool is 12 m.

a What is the width of the pool?

b If Jayden wanted to put a fence around two short sides and one long side of the pool, how many metres of fencing would he need?

5 A trapezium-shaped desk has a longer length of 2.1 m, a shorter length of 1.8 m and sides of 1.3 metres.

a What is the perimeter of the desk?

b Brianna puts two desks together with the longest sides of the two desks touching, as shown on the far right.

What is the combined shape called? Work out the combined perimeter of the desks.

6 Malcolm makes wooden picture frames. One is a rectangle which measures 60 cm by 30 cm.

a What is the perimeter of the rectangular picture frame?

b If he makes a square frame that has the same perimeter as the rectangular one, what is the length of one side?

7 James and Carol are making gift cards. The length of James's rectangular card is 7.4 cm and the width is 4.8 cm. The total perimeter of Carol's rectangular card is 14.9 cm. How much longer is the perimeter of James's rectangle than Carol's?

8 Michael and Tadeen each make a shape using five 1 cm squares. Michael said his shape had a perimeter of 12 cm while Tadeen said her shape had a perimeter of 10 cm. Draw what their shapes may look like and calculate the perimeter to work out if:

a both Michael and Tadeen are wrong

b both Michael and Tadeen are correct

c only Michael is correct

d only Tadeen is correct.

Maths detective: A pool enclosure

Case 1

The Rhonda family has a rectangular swimming pool in their back yard, which is in the shape of a pentagon. For safety reasons, the area must be surrounded by a fence. The length of the fence will be 62.68 metres.

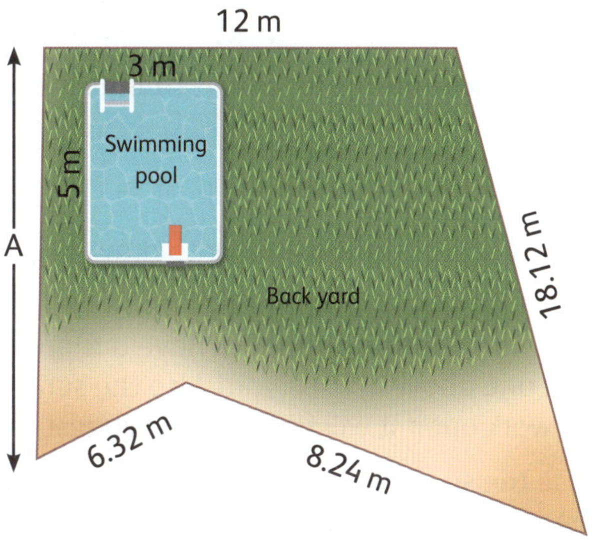

1 Calculate the length of fence for the section labelled A.

2 Rodney Rhonda walked along the outside edge of the pool twice.

a How many metres did he walk?

b How many more metres would he need to walk until he has walked the same measurement as the length of the wire fence?

3 The cost for 1 metre of the fence is $4 462.00. What is the total cost of the fencing?

4 The Rhondas pay for the fencing using sixty $5 000-dollar notes. How much change will they get?

5 Write your answer to Exercise 4 in words.

Case 2

The perimeter of this rectangular area is 2 x (5 + 7) metres = 24 metres.

1 Draw two diagrams of rectangles that have the same perimeter as the one shown here. One of these rectangles must be a square.

2 Explain how you worked out the side lengths for the diagram of the square.

3 If the Rhonda family wanted to install a shaded and paved area in the top right corner of the yard, so that they can sit in the shade, which of the two designs would work better? Or could both work? Give a reason for your answer.

Case 3

1 This diagram shows a right-angled triangle.

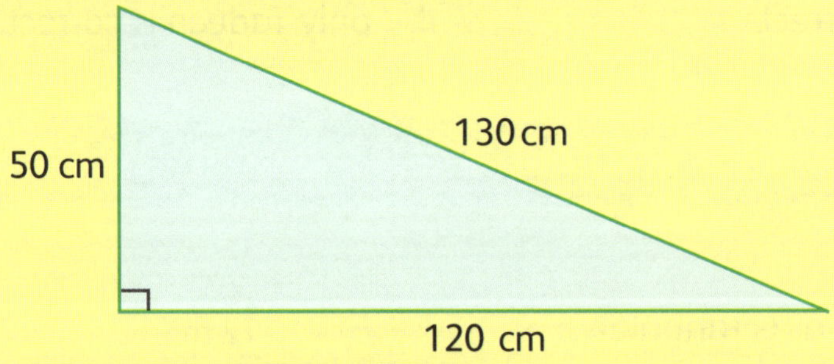

2 What is the perimeter of this triangle?

3 Draw a regular hexagon with the same perimeter as the right-angled triangle.

4 Label the length of each side of the regular hexagon.

5 Use graph or squared paper to draw a scaled plan of the Rhonda's yard. Colour in the area covered by the pool.

6 The Rhonda family want to have a triangular flower bed the same shape and size of the one above in their yard. Use the same scale to make a model triangle and use it to work out if it will fit. If it fits, suggest the best place for it.

7 Use the same scale to make a model of the hexagon. How many hexagonal flower beds would the family be able to fit in so that there would still be space to move around them? Show this on your plan.

Real-world maths: A fence around a playground

1 Your school wants to fence off a playground area that has the shape of an irregular octagon. The cost of fencing is $680.00 per metre. If this diagram represents the playground, calculate the total cost of fencing the playground area.

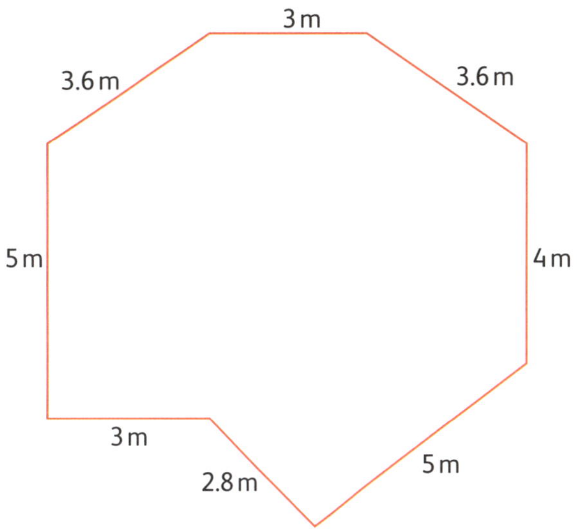

2 The diagram represents a section of the walkway that leads to the playground. The principal would like to have a border of paver stones installed around it.

a How many metres of stone edging will be needed to complete the border?

b The supplier sells edging at a cost of $2 438.50 per 5 metres. How many 5 metre lengths will the school need to purchase?

c The school has a budget of $60 000.00 to build the border. Do they have enough money to add the edging? Explain your answer using mathematical terms and calculations.

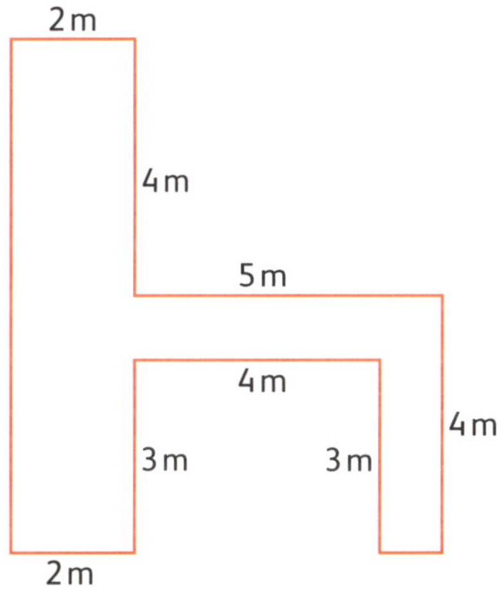

What I have learned

1 Copy the sentences into your notebook. Fill in the missing words to summarise what you learned in this chapter.

a The total distance around a 2-D shape is called its _____ .

b The perimeter of a _____ polygon is equal to the length of one side multiplied by the number of sides.

c Strings, rulers and metre sticks are instruments we can use to measure the _____ of an object.

d The perimeter of a polygon is determined by adding the lengths of the _____ of the shape.

2 Draw three different shapes that all have a perimeter of 36 centimetres. One should be a triangle, one should be a rectangle, and one should be a square.

3 Explain the rule for finding the perimeter of a regular polygon.

Practice questions

1 The total perimeter of each shape is given. Calculate the length of each unknown side.

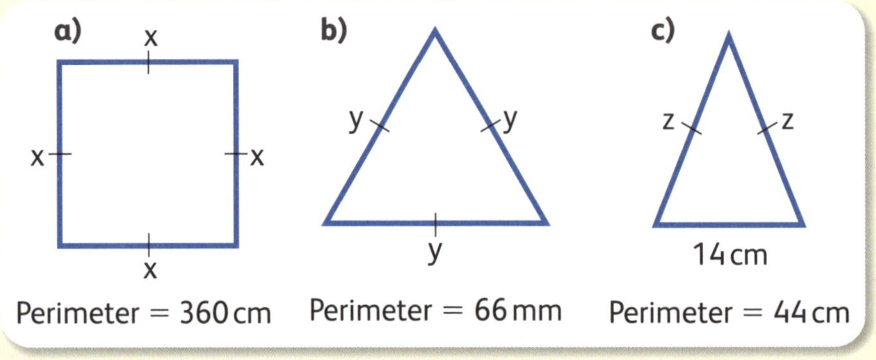

2 The length of a rectangle is 4 cm longer than its width. If the perimeter is 28 cm, how long are the sides?

3 A regular hexagon has a perimeter of 3 456 mm. What is the length of each side?

4 The figure on the right is made up of identical regular hexagons, each with sides measuring 2 cm. What is the perimeter around the outside of the figure? Choose one answer.

- a 14 cm
- b 36 cm
- c 20 cm
- d 24 cm

5 This is the plan for a triangular flower bed. Mr Campbell wants to plant rose bushes around the edges of the bed. The rose bushes will be planted 0.8 m apart.

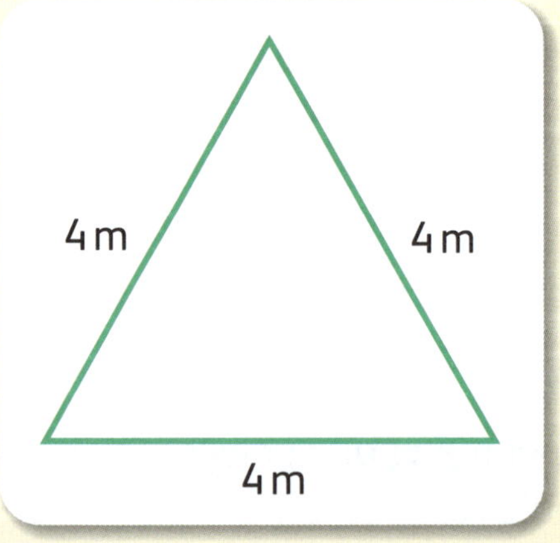

- a How many rose bushes will Mr Campbell need?
- b When he had planted the rose bushes, he planted two marigold plants between each rose bush to keep pests away. How many marigold plants did he plant?

5 Self check

Check you can do this:	If you have difficulty, try this:
1 Measure and calculate the perimeter of regular and irregular polygons.	Go back to page 66. Make sure you remember the definition of a polygon, and the definition of perimeter. Remind yourself of the units of measurement for length, and use a ruler to measure and add up the lengths of the sides of a polygon.
2 Use the appropriate formula to calculate the perimeter of regular polygons.	Go back to pages 67–69. Remind yourself what we mean by perimeter, and make sure you know the definition of a regular polygon. Also look at the diagrams on page 68 and identify sides of equal length. Remind yourself how we multiply the length of a side by the number of sides to find perimeters of regular shapes.
3 Calculate the measurement of one side of a polygon when given the perimeter and the lengths of the other sides.	Revise page 70. Remember: we calculate perimeter by adding up the lengths of the sides of a shape. If the sides are equal in length, we can multiply the length of one side by the number of sides of that length. When you know the perimeter, you can use inverse operations to help you solve questions about the side lengths.

Unit 3: Geometry

Chapter 6 3-D solids and nets

In this chapter you will explore these focus questions:

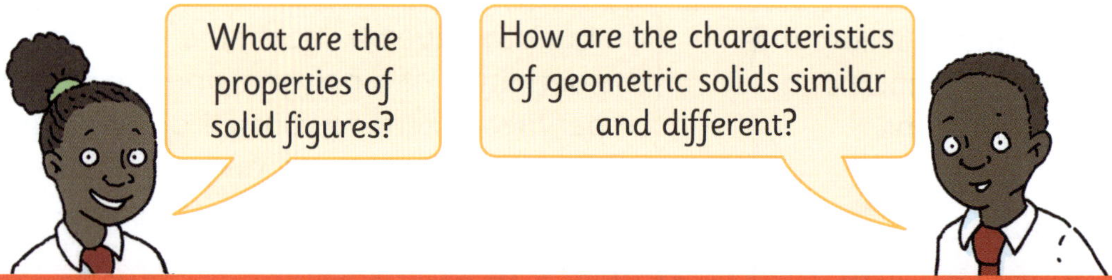

To answer these questions, you will:

- recognise faces, edges and vertices of a solid
- classify solids according to the number and shape of their faces
- represent and solve problems using geometrical models
- explore geometric forms in the world around you
- draw and describe nets of triangular-based prisms
- identify and create polyhedra (tetrahedron, hexahedron and octahedron)
- classify solid shapes (prisms, pyramids and polyhedrons) according to their properties.

Starting point

1 The photographs show a concert hall called the Harpa, which has won international architectural awards for its modern and unique design. The outside walls are made of thick Perspex and metal frames.

a How did the architects use mathematical solids and other shapes in their design?

b Which part of the building looks most interesting to you? Sketch it and say why you find it interesting.

2 Work with a partner. Find the following on the photograph of power pylons:

a parallel lines

b a pentagon

c a right-angled triangle

d a trapezium

e an obtuse-angled scalene triangle.

3 With your partner, design and draw a small electricity pylon using a combination of only regular polygons.

Revising what you know about shapes and their properties

Key maths idea

Remember: a **polygon** is a 2-D closed shape that has sides made of straight lines that join at vertices. If a shape is not closed, or if its sides are curved, it is a non-polygon.

Look at this artwork. What do you notice about the shapes? Does it look 2-D or 3-D to you? Explain your ideas to a partner.

Key words

polygon
quadrilateral
trapezium
parallelogram
rhombus
kite
parallel
adjacent
line of symmetry

1 What shapes is each pattern made from? Say whether they are polygons or not.

a

b

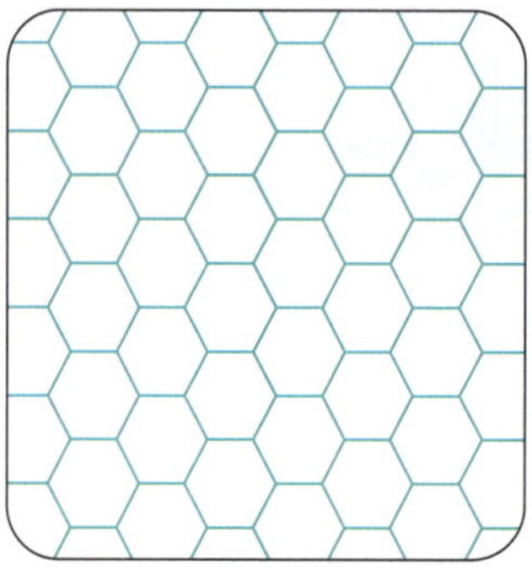

c

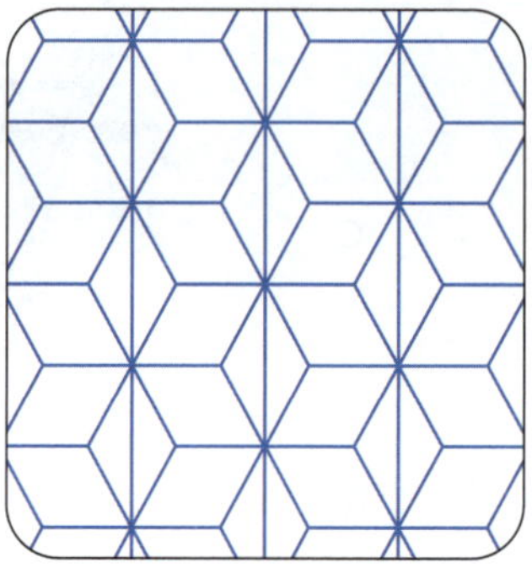

d

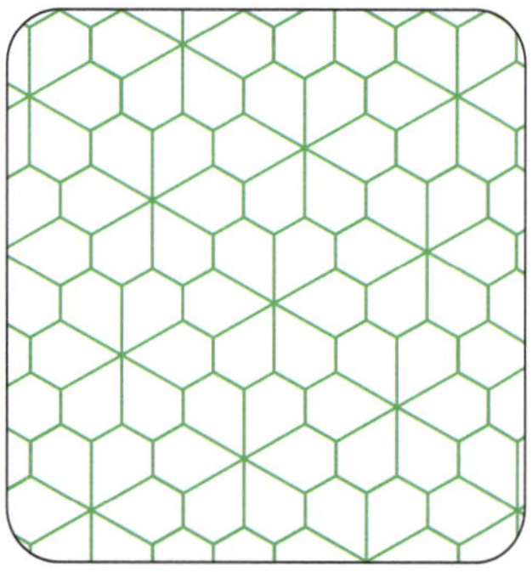

e

f

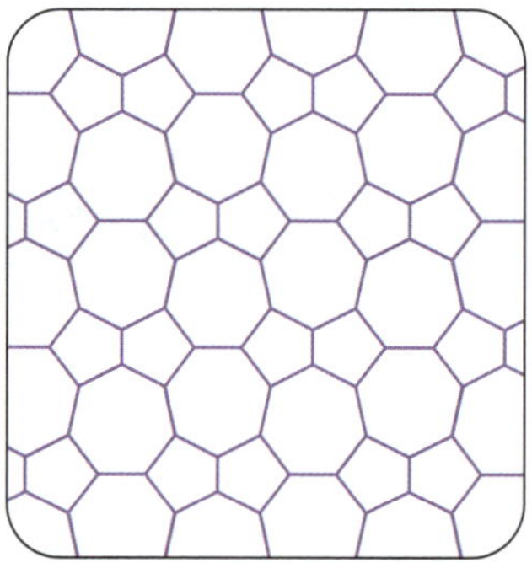

2 A **quadrilateral** is a polygon with four sides. Look at these different types of quadrilaterals, and match each shape to all the properties that apply.

Square

Rectangle

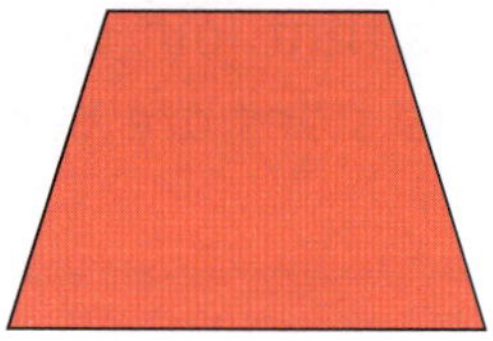

Trapezium

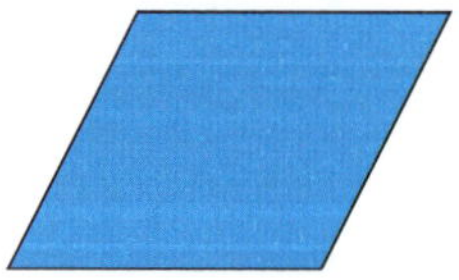

Parallelogram

Rhombus

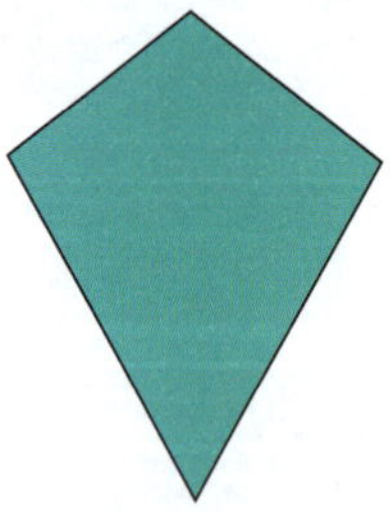

Kite

Properties:

2 pairs of opposite sides equal in length

all sides equal in length

1 pair of **parallel** sides

2 pairs of parallel sides

2 pairs of **adjacent** sides equal in length

4 right angles

1 **line of symmetry**

2 lines of symmetry

3 Draw Venn diagrams to show the intersections of these sets:

a the set of quadrilaterals and the set of regular polygons

b the set of triangles and the set of regular polygons.

Solid shapes

Key maths idea

Objects such as boxes and tins are **3-D solids** or **solid shapes**. They have three dimensions: length, width and height. Solids such as **cubes** and **cuboids** have flat **faces**. Their **edges** are straight. Solids such as cylinders and cones also have **curved surfaces** that join their faces along curved edges. A **sphere** is a ball shape with a single curved surface and no faces.

Prisms have identical parallel end faces, joined by rectangular faces. A prism has the same cross-section all the way along its length. The shape of the cross-section gives the prism its name.

Some different cubes, cuboids and cylinders.

Pyramids have a polygon base and triangular faces that meet at one of the vertices, called an **apex**. The base of the pyramid helps to give the shape its name, for example a square-based pyramid.

Triangular prism

Rectangular prism

Pentagonal prism

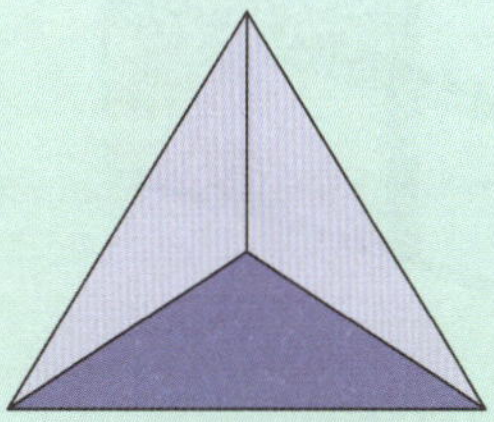

Triangular-based pyramid

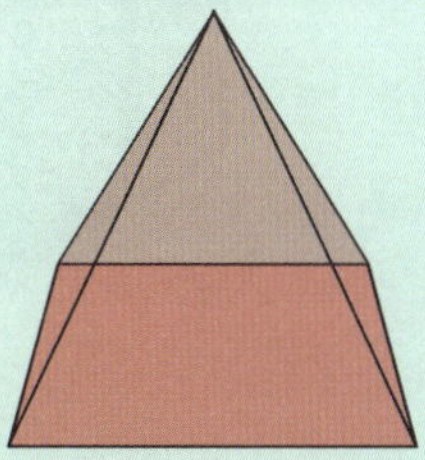

Square-based pyramid

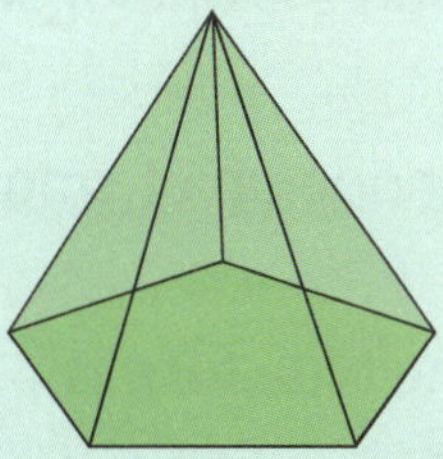

Pentagonal-based pyramid

Key words
- 3-D solid
- solid shape
- cube
- cuboid
- face
- edge
- curved surface
- sphere
- prism
- pyramid
- apex

1 Real-life objects sometimes have shapes very much like geometric solids. For each picture, say which 3-D solid it looks like, and which shapes make up the faces or curved surfaces.

a 

b 

c 

d 

e 

f 

2 Your teacher will show you (or you can collect) some different boxes and tins. Sort them into different types of solids. Talk about the shapes of the faces, and how many faces each one has. Draw up a table of your findings.

3 Discuss with a partner the difference between 2-D and 3-D shapes. What are the main similarities and differences between:

a a triangle and a pyramid

b a rectangle and a cuboid

c a cube and a square

d a prism and the 2-D shapes that make up its faces?

Maths detective: A puzzle cube

Have you solved one of these types of cubes before? When it is solved, each face has all the smaller squares the same colour. Use the picture to help you work out whether each statement is true or false. Explain how you know.

1 Each face is made up of four smaller squares.

2 Each of the smaller squares is one of six colours.

3 The shape is a cuboid made up of smaller cubes.

4 On the finished cube the blue face is opposite the white face.

5 The total number of smaller squares on the cube is 36.

Exploring polyhedra

Key maths idea

A **polyhedron** (plural **polyhedra**) is a solid shape with flat faces. Each face is a polygon. A **vertex** (plural: **vertices**) is where three or more faces meet at a point.

The word 'polyhedron' comes from 'poly-' meaning 'many' and '-hedron' meaning 'face'.

We classify polyhedra according to the number of faces, and the shape of the faces. A **tetrahedron** has four triangular faces. It is also called a triangular pyramid.

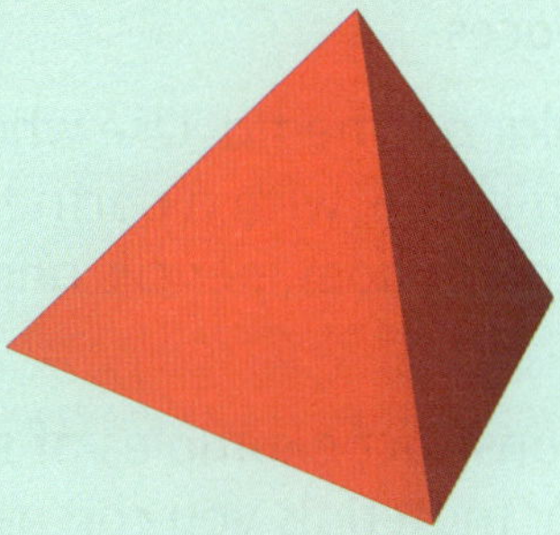

Tetrahedron

A **hexahedron** is a polyhedron with six flat sides. Cubes, cuboids and many types of prisms are all examples of hexahedra.

Cube

Hexagonal prism

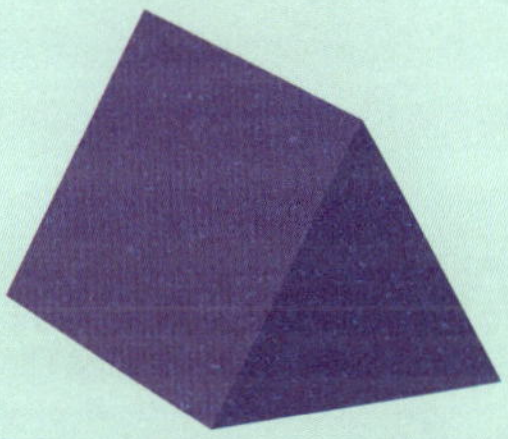

Triangular prism

Pentagonal prism

Cuboid

Look at these pictures of an **octahedron** and a **dodecahedron**. Try to work out how many faces they have?

octahedron

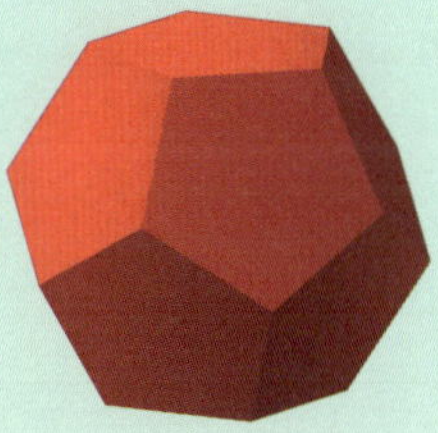

dodecahedron

Key words

polyhedron
polyhedra
vertex/vertices
tetrahedron
hexahedron
octahedron
dodecahedron

1 For each 3-D solid, describe how many faces, how many edges and how many vertices it has, as well as the shape of its face.

- a cuboid
- b square-based pyramid
- c cube
- d triangular prism
- e triangular-based pyramid

2 For each statement, say whether it is always true, sometimes true or never true.

- a A cuboid is the same as a cube.
- b Prisms have a cross-section that is always the same, so a cuboid is a prism.
- c Cubes and cuboids have the same number of vertices.
- d A pyramid has five faces.

3 There are only five regular polyhedra. Use what you know about the definitions of 'regular' and 'polyhedra'. Work with a partner to draw these five solids. You can use dotted paper or square grid paper, pencils and rulers, or you can work on a computer using software.

4 Use craft materials to construct examples of some of the polyhedra you have learned about. Here are ideas for materials you can use:

- thin coloured card, scissors, glue
- fudge sticks or cotton buds and tape
- modelling clay.

Real-world maths: A recycled robot

A robotics team used **recycled** boxes to build a small robot. They **designed** and programmed it so that its arms could move and use a paintbrush.

1 Use found 3-D objects from your local **environment** to build your own robot.

2 Write short notes explaining what you used, where you found the materials and why you chose these shapes.

3 If you could program the robot to do one task for you, what would it be?

Key words
recycle
design
environment

Real-world maths: Ayesha's nesting box

Ayesha wants to build a nesting box for the birds in her garden. She found this picture online.

1 Ayesha has a flat sheet of wood. Draw a pattern to show the pieces she will need to cut out to build the nesting box.

2 What could she use to draw the opening on the front piece before she cuts it out?

3 What could she use to make the perch?

4 Write a set of instructions to help her build the nesting box once she has cut out the parts she needs.

Nets

Key maths idea

Imagine that you could unfold the faces of a 3-D solid into a 2-D shape. This shape forms the **net** of the solid. You have worked with nets in earlier grades. Here is a net of a cube.

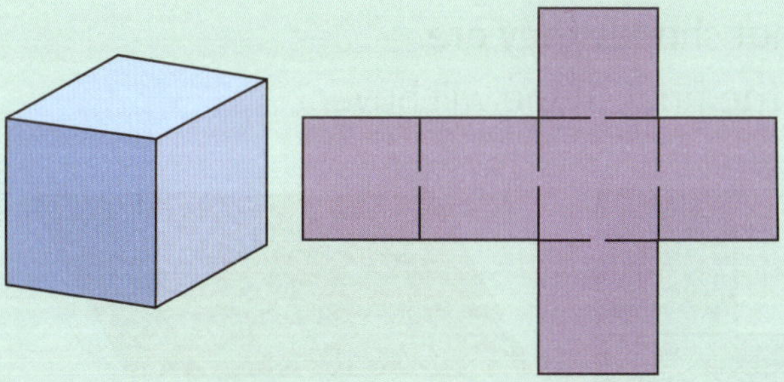

You can also use a net as a pattern that you can fold into a 3-D solid.

Key word

net

1 Draw nets for two different triangular prisms. One should have an isosceles triangle at each end.

2 Hayley drew a net for a pentagonal prism. How many polygons did she have to draw and what were their shapes?

3 Layla drew this net for a box 12 cm long, with square ends.

a What mistake did she make in her design?

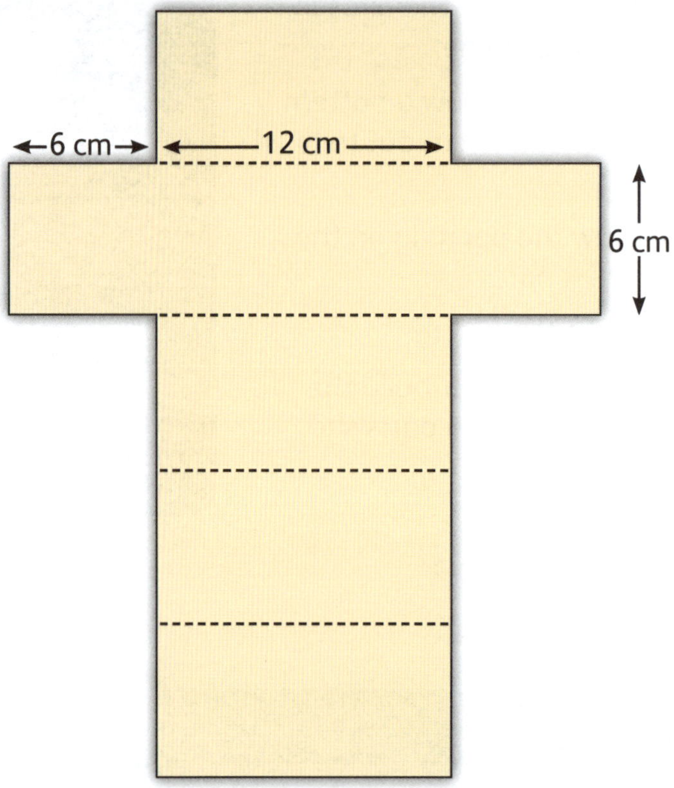

b Sketch two other possible nets that would make the correct shape.

4 Look at these nets for 5 different polyhedra. For each polyhedron, say:

a how many faces it has, and what shapes they are

b how many edges and vertices the final shape will have.

5 Look at these paper decorations. What kind of polyhedra have been used to form these shapes?

Maths detective: Investigate solids

Case 1

Use what you know about sets and solids to work out where each of these shapes will fit into the Venn diagram.

1 Look at these 3-D solids.

a
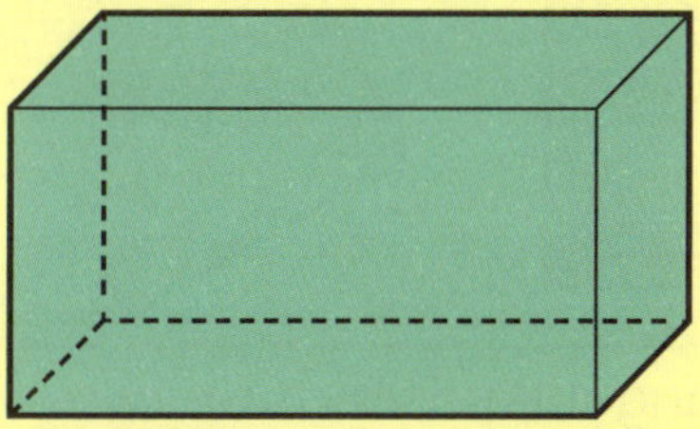

b
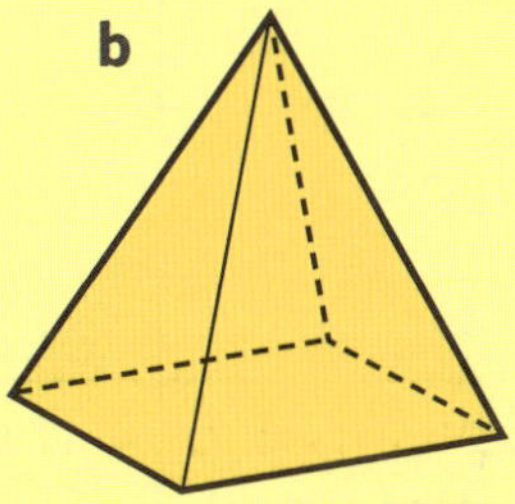

c
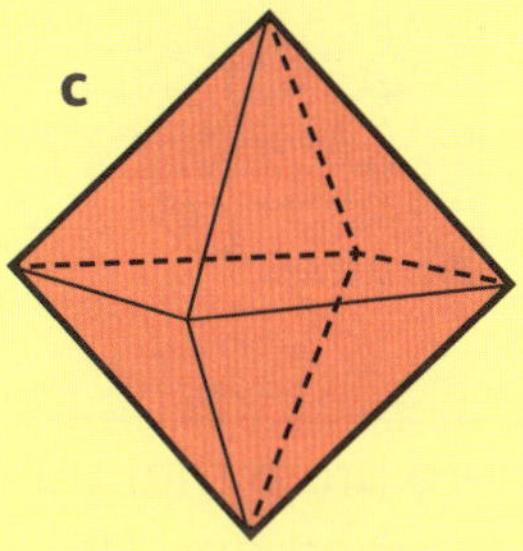

d
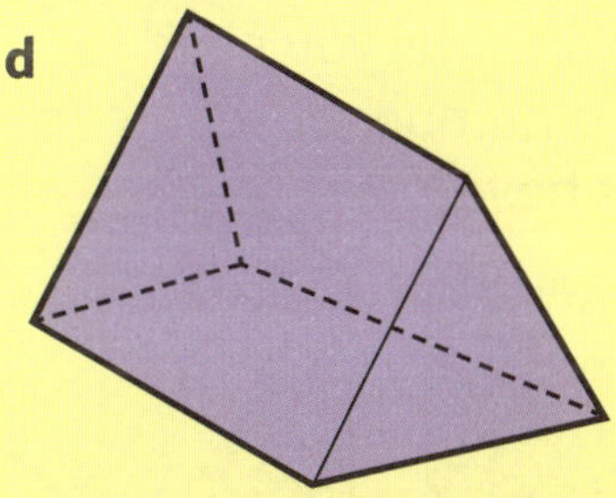

e
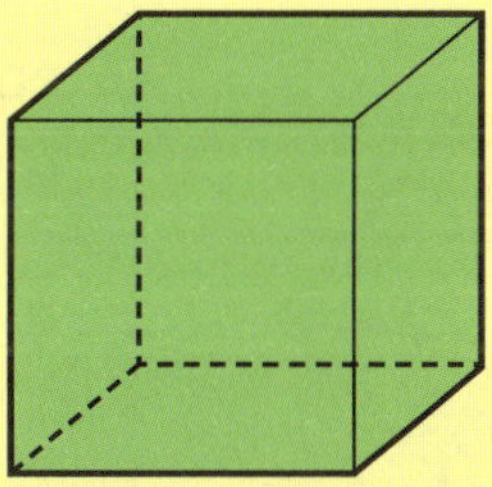

f
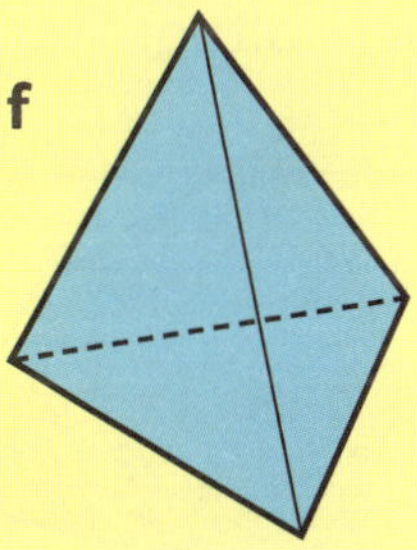

Copy the Venn diagram and write the letters of the shapes in the correct sections.

U = {polyhedrons}

Set A = {solids with rectangular faces}

Set B = {solids with triangular faces}

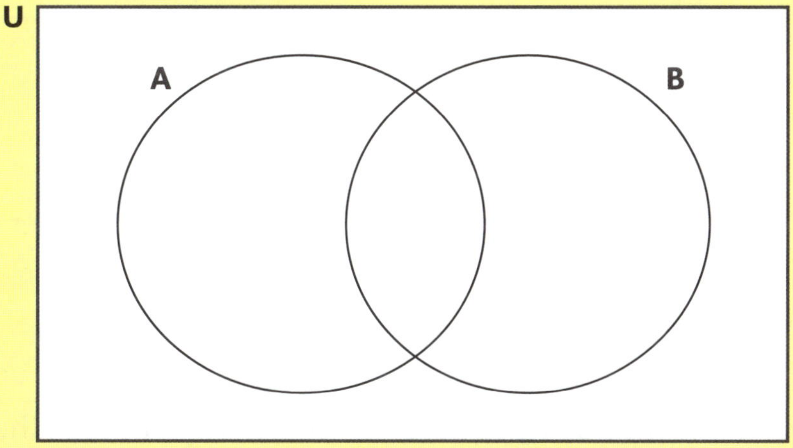

2 Describe the set A ∩ B.

Case 2

Use the information and what you know about different shapes to solve each problem in the table. Show your working.

1 There are 9 squares, 5 triangles and 7 hexagons. How many sides are there altogether? 	**2** Sage has 13 rectangles and 12 trapeziums. She gives 8 rectangles away to a friend. How many vertices are left? 
3 There are 8 cubes and 14 square based pyramids. How many faces are there in total? 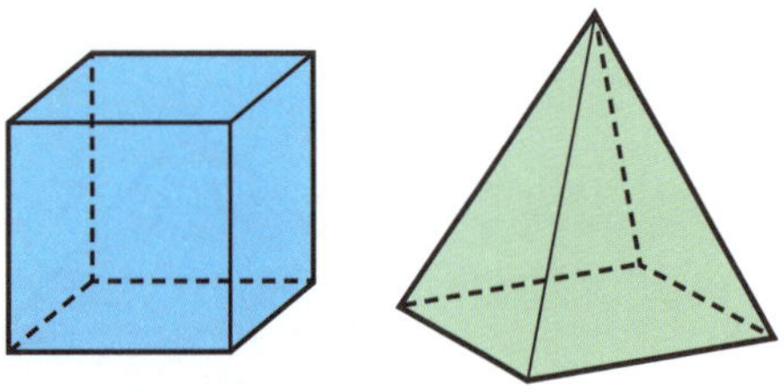	**4** Salim has 13 cuboids and 7 pyramids. He gave 4 pyramids to his teacher. How many edges are left? 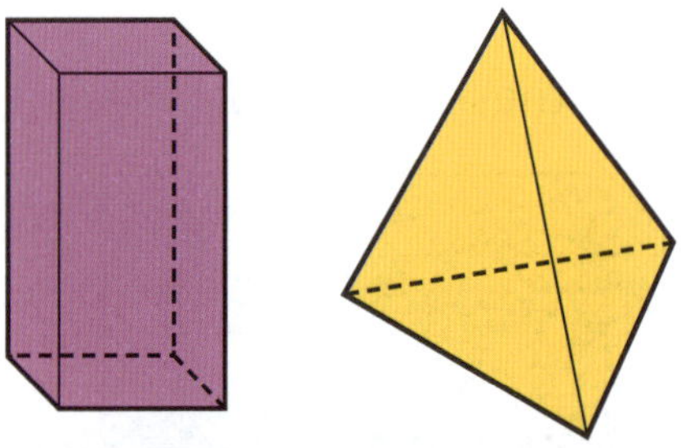
5 There are 8 triangular-based prisms and 6 octahedrons. How many vertices are there altogether? 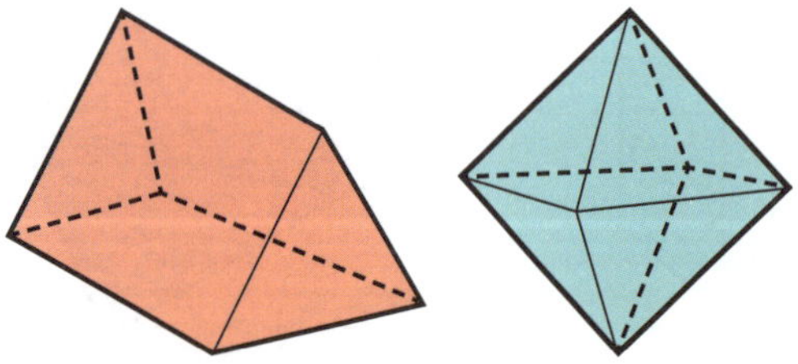	**6** The rectangular prisms have two square faces. There are 9 cubes. How many rectangular prisms will you need to equal the same number of square faces that the cubes have? 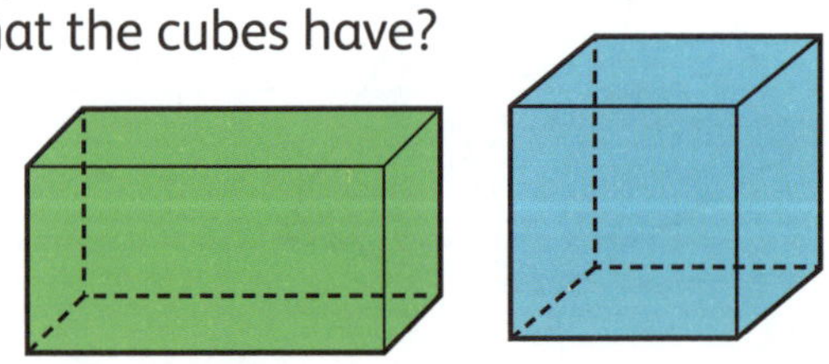

Case 3

Use a recycled 3-D container for these questions.

1 Measure the dimensions of the container. Remember to choose the most sensible measuring unit.

2 Draw the net for the container.

3 Label the faces, vertices and edges of the container on the net.

4 Write a paragraph about the relationship between the number of faces, edges and vertices of the container.

5 Give three examples of where we use this shape in the real world.

Case 4

The number of dots on opposite faces of a normal die always add up to 7.

This means that if 6 is facing up, the number on the bottom will be 1.

If 1 is facing up, the number on the bottom will be 6.

Some Grade 6 students drew these nets to build model cube-shaped dice.

1 Which ones can be folded up to make a cube?

2 Which of the cubes will give dice with the opposite faces adding up to 7?

What I have learned

1 Copy these maths words into your notebook.
Write a definition or draw a labelled diagram to show the meaning of each word.

a polygon **b** solid **c** face **d** vertex **e** edge
f prism **g** pyramid **h** polyhedron **i** net

2 Write an entry in your notebook to explain what new things you learned about shapes and solids in this unit.

3 Prepare a short slide show or make a poster to show pictures of different types of polyhedra that are used in real-life. Your presentation should include the mathematical names of the polyhedra.

Practice questions

1 Why is a circle not a polygon?

2 Why is a rectangle not a regular quadrilateral?

3 Sketch five different types of quadrilaterals.

4 If possible, write the names of one or more names of polygons that fit into each part of the table.

Hint: think of different types of polygons!

	4 equal sides	2 pairs of equal sides	1 pair of parallel sides
4 right angles	a	b	c
No right angles	d	e	f

5 Which of these is not a polyhedron? Give a reason for your answer.

cube sphere dodecahedron

6 For each net, identify and describe the faces, and then say what kind of solid the net makes.

a
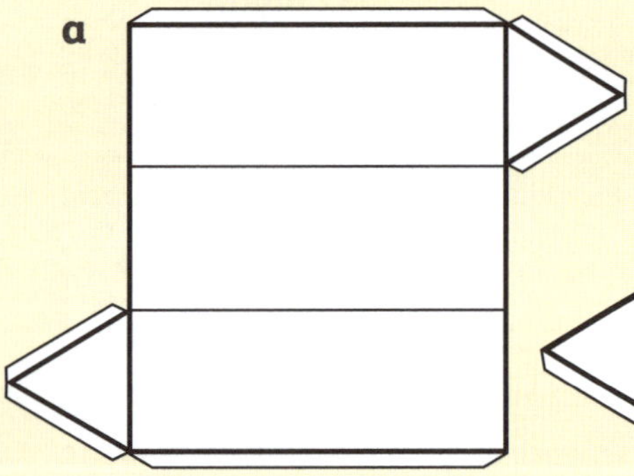

b
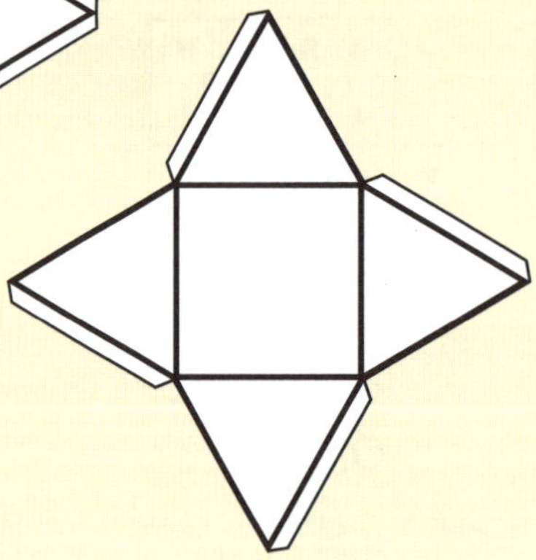

c
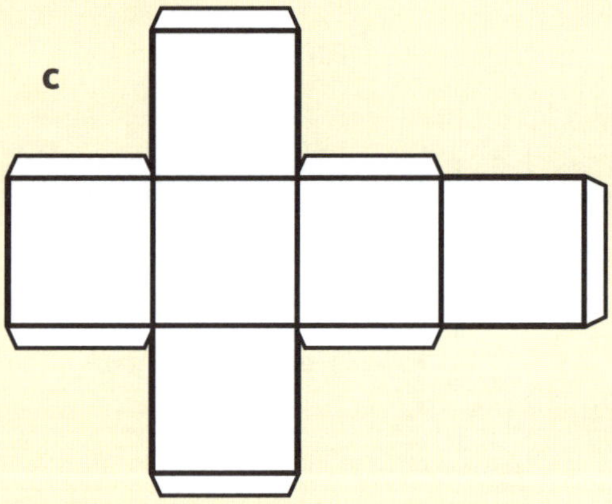

7 Draw possible nets for each of these 3-D shapes.

a

b

c

6 Self check

Check you can do this:	If you have difficulty, try this:
1 Recognise faces, edges and vertices of a solid, and classify solids according to the number and shape of faces.	Go back to page 79, and make sure you remember how do describe and classify solids.
2 Represent and solve problems using geometrical models.	Go back to the questions on the puzzle cube on page 80. Think about how you worked out the number of faces, and the number of smaller cubes that make up the whole cube.
3 Explore geometric forms in the world around you.	Look back at the pictures on page 80, and identify real objects that have shapes of 3-D solids and polyhedra.
4 Draw and describe nets of triangular-based prisms.	Revise page 83 and look back at the prisms you drew for Exercise 1 in the activity.
5 Identify and create polyhedra (tetrahedron, hexahedron and octahedron).	Look back at page 81 and make sure you can identify these three types of polyhedra, as well as make them from simple materials such as sticks or card.
6 Classify solid shapes (prisms, pyramids and polyhedrons) according to their properties.	Revise the work on pages 79–81 and make sure you can identify the face shapes and numbers of faces of each of these polyhedra.

Unit 4: Statistics and probability

Chapter 7 Data handling

In this chapter you will explore this focus question:

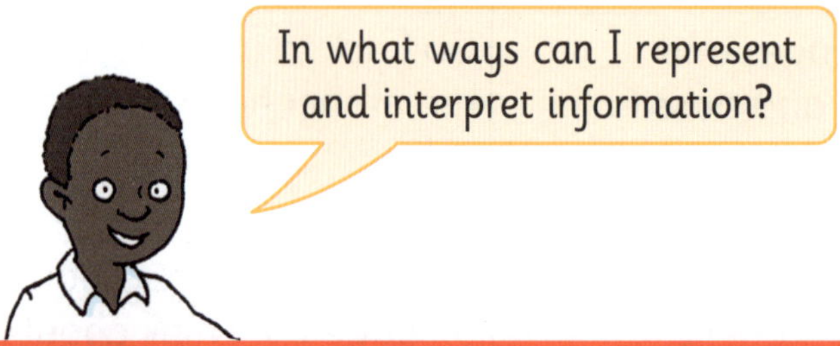

To answer this question, you will:

- identify patterns, and describe and predict outcomes from data collected
- collect data using direct observation, experiments, interviews and questionnaires
- identify patterns in data
- interpret these patterns
- represent data in a stem-and-leaf plot
- discuss the appropriate uses of various tables and graphs
- solve problems in which data is given by means of a graph or diagram.

Starting point

1 Each year the National Oceanic and Atmospheric Administration (NOAA) publishes the Hurricane Season Outlook. This graph predicts the likelihood of strong storms and hurricanes for the coming season.

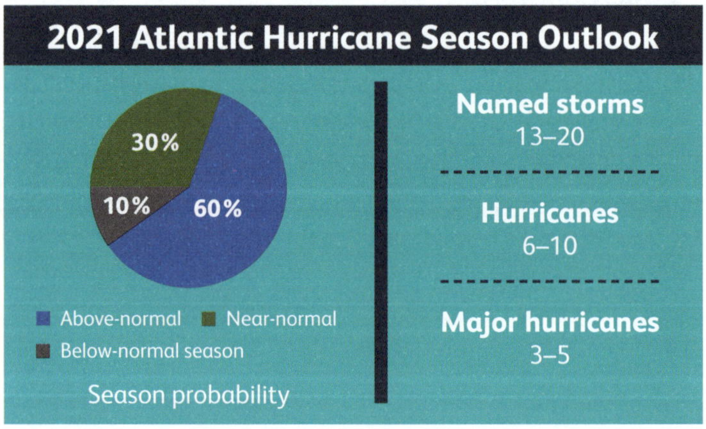

(*Source:* NOAA)

a Look at the pie chart. What were the chances that 2021 season would have fewer hurricanes than usual? Choose the best description.

1 in 10 chance | **6 out of 10 chance** | **3 out of 10 chance**

b What was the most likely outcome for the 2021 season?

c The numbers on the pie chart add up to 100. Why is that important for a pie chart?

d Why do you think charts like this are important for everyday life?

> An average hurricane season produces 14 named storms, of which 7 become hurricanes, including 3 major hurricanes.

2 Read the tweet above. If a season produces 10 named storms, of which 7 become hurricanes, including 6 major hurricanes, would you say that is above or below normal? Discuss this question with a partner and give reasons for your answer.

3 During the Covid-19 pandemic, data became available as each country reported how many new cases of coronavirus were diagnosed each day.

a This graph is a combination of a bar graph and a line graph. What do the bars show?

b The line graph plots a 7-day moving average. This means that the data scientists took an average of the last 7 days to plot the new point each day. Why do you think they did this?

c In real life, the numbers of people with coronavirus were probably higher than the people reflected in graphs like this. Why do you think that was the case?

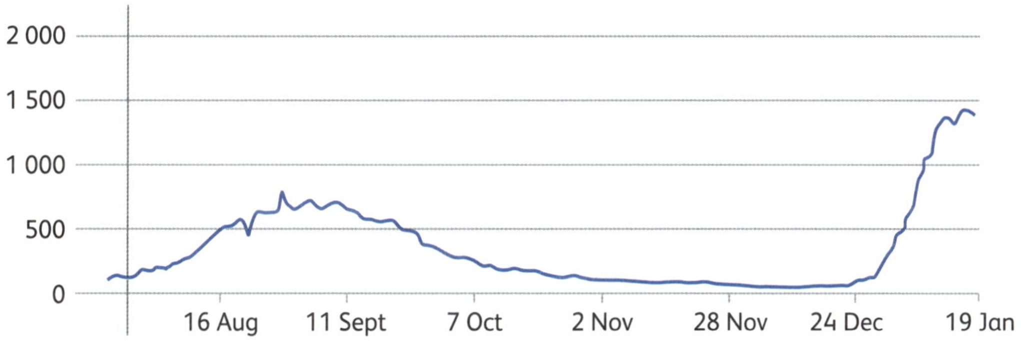

Source: https://ourworldindata.org/coronavirus/country/jamaica

4 Why do you think this graph shows the number of cases per 100 000 people? How does this help us to compare the numbers of cases in different countries? (Hint: think about population sizes.)

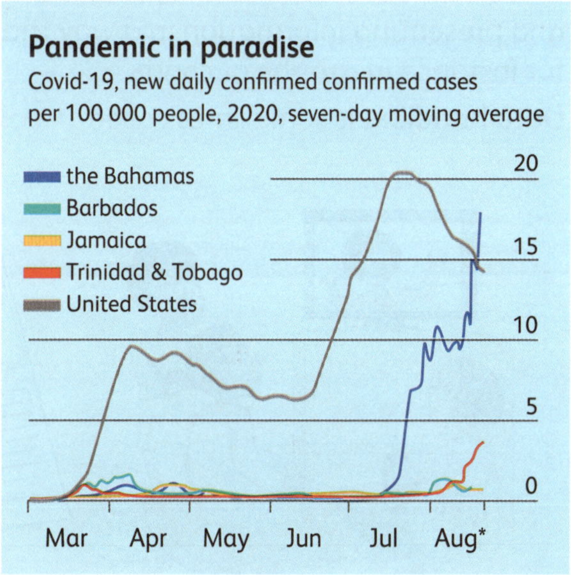

Source: Johns Hopkins University CSSE
*To August 19th

Statistics and probability

5 Look at this graph.

Why must we flatten the curve?
By following protective measures in our personal lives, our communities and our environments, we can slow down the speed at which the virus spreads.

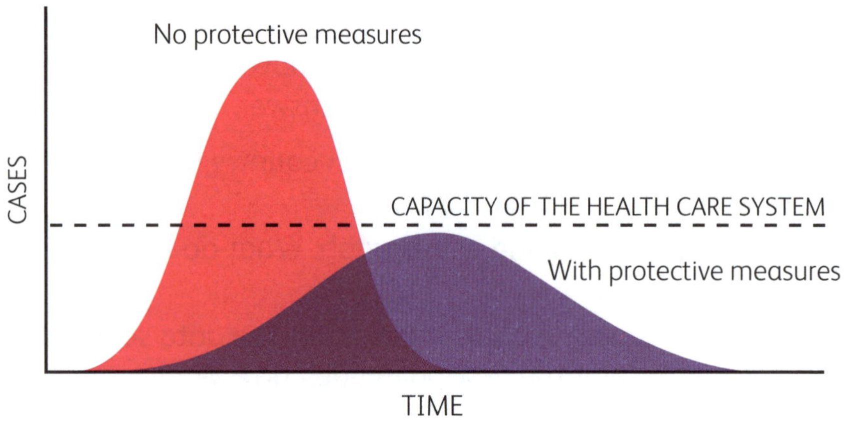

Protective measures
Personal: Wash hands, sanitise regularly
Community: Practise social distancing (1.5 m distance from others; staying home as much as possible)
Environment: Clean surfaces in public places regularly

a Why do the protective measures change the shape of the curve?

b Jamal asks: 'The area of the two curves is roughly the same anyway. So why does flattening the curve make a difference?' What would you say to Jamal?

Data handling

Key maths idea

Data is another word for information. We get data from many different **sources** – for example, books, the radio, news broadcasts and websites, and social media.

Data handling refers to the process of gathering, recording and presenting information in a way that is helpful to others – for instance in **graphs** or charts.

Data handling is sometimes called **statistics**.

Key words

data
source
data handling
graph
statistics

1 **a** What is your favourite TV programme?
b What do you think are your classmates' favourite TV programmes?
c How would you find out this information?

2 **a** Find out the favourite TV programme of each student in your class.
b Make a list of all the programmes mentioned and how many students named that programme as their favourite.
c Record your data in a table like this:

TV show	Number of students

3 **a** What is the most popular programme in your class?
b What is the least popular programme in your class?

4 How can you present your information in a way that is helpful to others, such as a graph or chart? Present your information using your chosen method. What are the labels, titles and colours that will make your findings clear?

Collecting information

Key maths idea

Researchers use many methods to collect information. They may conduct **interviews** – asking people questions. They may use **questionnaires** – lists of questions that people fill in and give back. They may use sources such as books, newspapers, records from governments or public institutions. Sometimes people perform telephone surveys or surveys online. Another approach is to make **direct observations** – watch something take place and record what you see. For some research, you could conduct experiments and see what happens.

Key words

interview
questionnaire
direct observation

Statistics and probability

1 Look at graphs A to E on the next page. Each of these presents data in a different way. Discuss with a partner:
a What type of graph is it?
b Is it possible to see what the graph is about? If the graph has information missing, what would you need in order to understand it?
c Try to guess how the researcher collected the data.

A How Grade 6 students travel to school

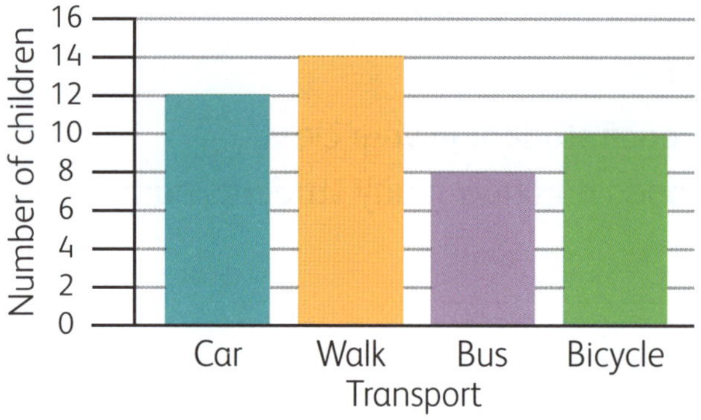

B

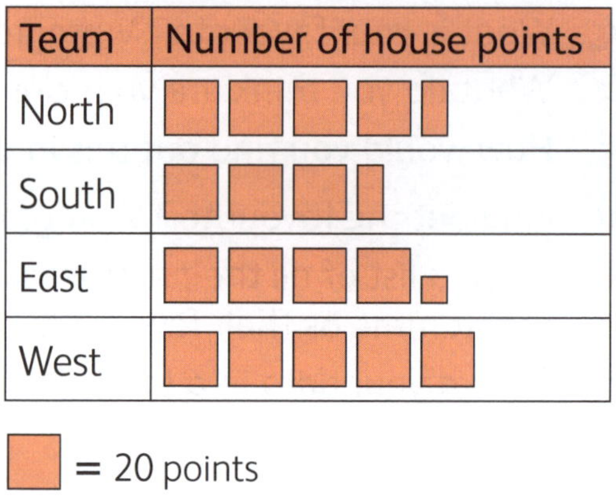

C Cars that drove past in an hour

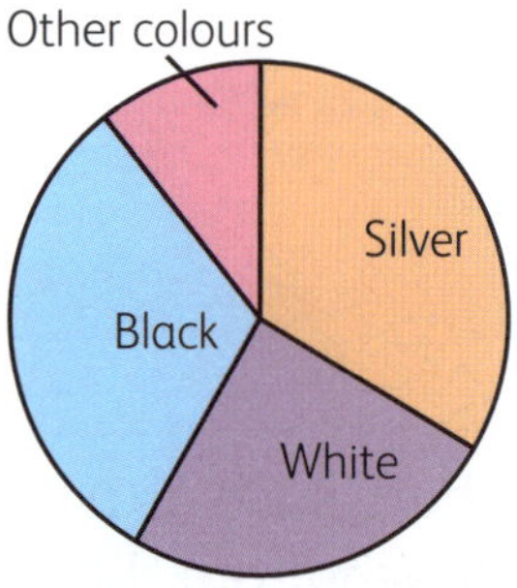

D

Name	Number of baskets
Jenny	𝍸
David	𝍸 ||
Michelle	||
Dinah	|||
Gareth	

E

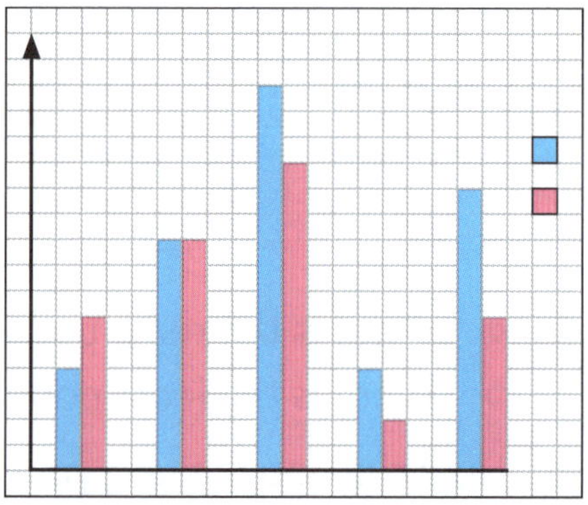

2 Mikayla did a survey to find out how many students had a birthday in each month. These are her results:

January March August November December June August June
June February June August January November September August February
July April May November September August January October
November October July April April May December

a How many people did she collect data from?

b Why is this data not easy to work with?

c How could you organise this data to make it easier to work with?

3 This table shows which foods students bought at the school tuckshop in one day.

sandwich cookies muffin chocolate cookies
fruit fruit sandwich muffin cookies
cookies muffin fruit chocolate sandwich
sandwich cookies muffin fruit muffin
sandwich cookies

a The students could choose between five options. What were they?

b Draw a tally chart with five rows. Fill in the tallies for how much was bought of each option.

c Use your tally table to make two statements about what students preferred at this tuckshop.

Presenting data in graphs

Key maths idea

We can show data in different types of graphs.
A **pictograph** is a simple table that uses symbols or pictures to represent information we can count. The **key** shows what each symbol represents.
A **bar graph** is also for data that we can count. We show the data in vertical or horizontal bars on a pair of **axes**. Double bar graphs split each bar into categories.
Circle graphs (**pie charts**) represent data as portions of a circle. The whole circle represents the whole data set, and the sectors (the wedges of the circle) show the size of each part in relation to the whole.
Line graphs show how a variable (such as temperature) changes over time.

Key words

pictograph
key
bar graph
axes
circle graph
pie chart
line graph

Look at these four graphs. For each graph, say which of the four types of graphs above it is, and how you know. Then explain what question the graph is answering, and what it shows.

a Mike's weekend activities over the last 2 months

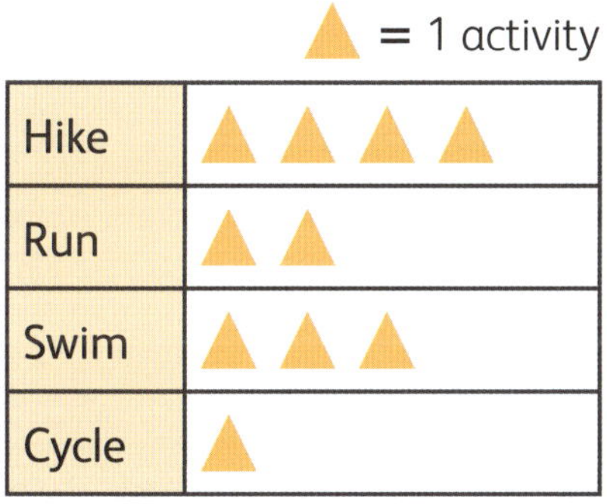

b How Ali spent their day

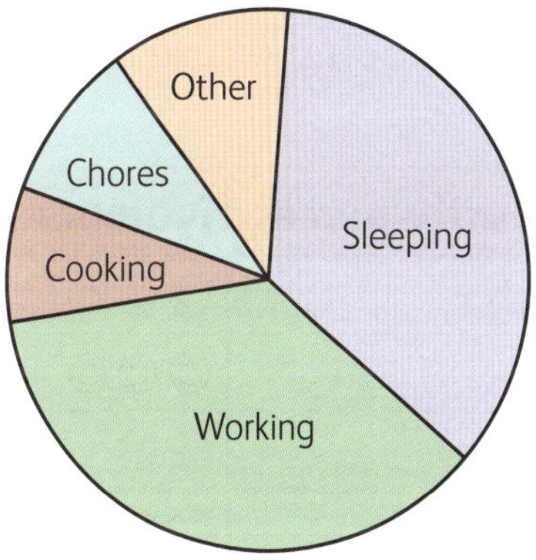

c Temperature of water in a pot

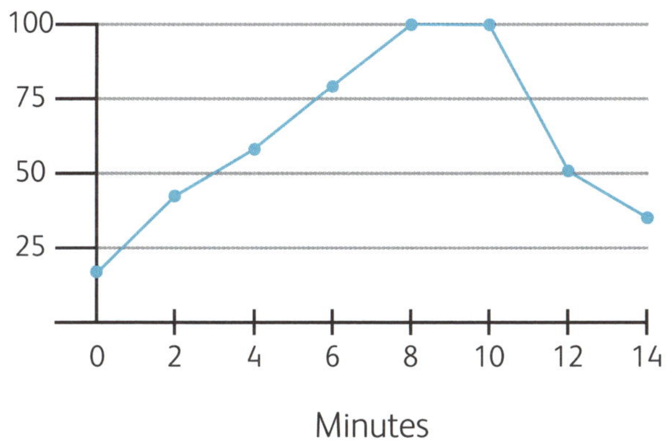

d Recycling collected by two classes over 3 weeks

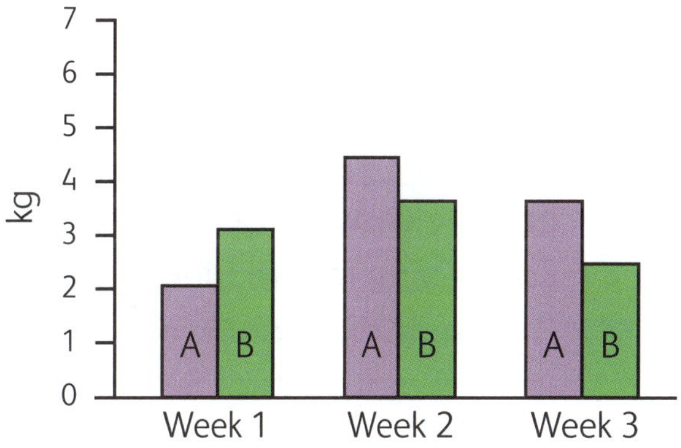

Revising tables, tallies and pictographs

Key maths idea

A **table** is a simple set of rows and columns. Sometimes we can arrange information in a table to make it easier to see or understand. A **tally chart** is a useful way to collect information. You draw **tally marks** for each item you count. Every fifth tally is drawn as a line through the previous four. This makes tallies easier to count.

Example

Candice asked her friends their favourite places to go on a weekend. She collected her results in a tally table. As each person answered, she drew a tally to show their choice.

Favourite place	Number of friends
Beach	~~////~~ //
Park	///
Cinema	//
Mall	~~////~~

Candice represents her results in a pictograph. A pictograph is a simple graph that uses pictures or symbols arranged in rows of a table.

Favourite place	Number
Beach	☺☺☺☺☺☺☺
Park	☺☺☺
Cinema	☺☺
Mall	☺☺☺☺☺

Key: ☺ = one person

Key words

table

tally chart

tally marks

1 What is similar and what is different about the tally chart and the pictograph?

2 Why is it important for a pictograph to have a key?

3 a Choose a question to ask your classmates. You might find out their favourite juice flavour, their favourite sport, their favourite type of ice-cream, or their favourite colour.

b Think about how to ask your question and how many options you will offer as possible answers. Do not use more than 5 possible answers.

c Collect your information using a tally chart.

d Represent your data in a pictograph. Make sure to include a title and a key.

Bar graphs

Key maths idea

Carla interviewed some students to find out their favourite sports. She gave them five sports to choose from. This table shows their answers:

Swimming	Cricket	Football	Netball	Athletics
8	10	6	7	4

Carla decided to show this information in a bar graph.

She followed these steps to draw the bar graph:

1 She determined the **range** of values.

> The lowest value is 4 and the highest is 10. The range is 6.

2 She decided on a **scale** for her graph.

> The y-axis will show the number of people, with $\frac{1}{2}$ cm representing 1. The bars will be vertical.

3 She drew the axes, and marked where she would place each bar.

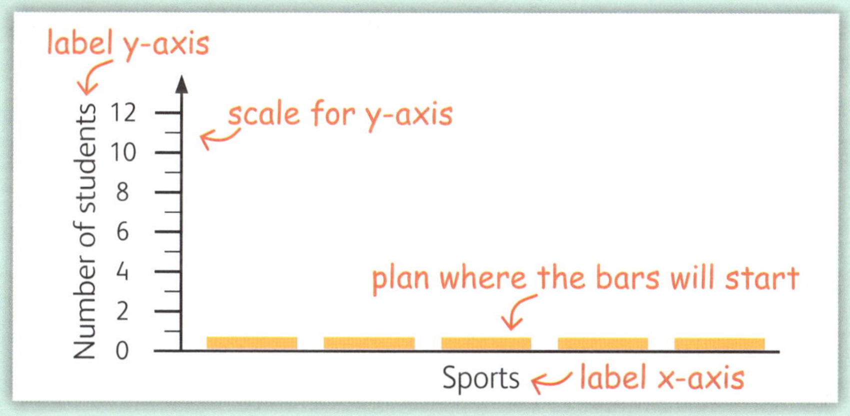

4 Last she used a ruler to help her draw each bar to the right height. She drew the bars and added a title.

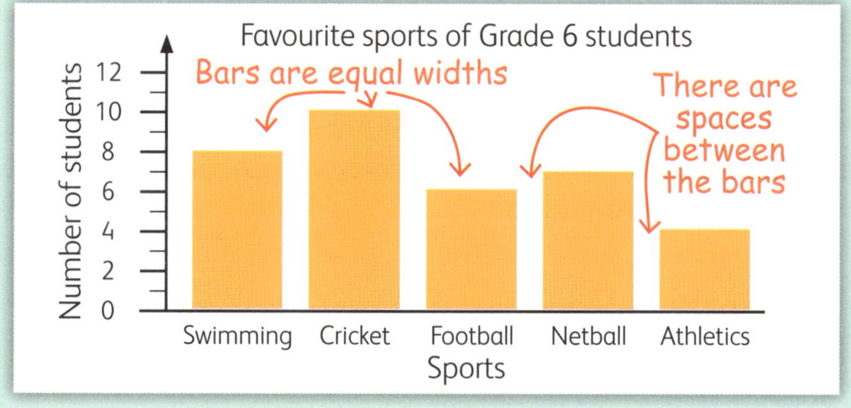

Key words
range
scale

Note that in a bar graph, there are spaces between the bars. This is because each bar shows a different category of information.

1 Here is some data about the number of tickets sold for a music festival.

Andre starts to create a bar graph to represent the number of festival tickets sold during the week.

Day	Number of tickets sold
Monday	55
Tuesday	30
Wednesday	45
Thursday	75
Friday	85

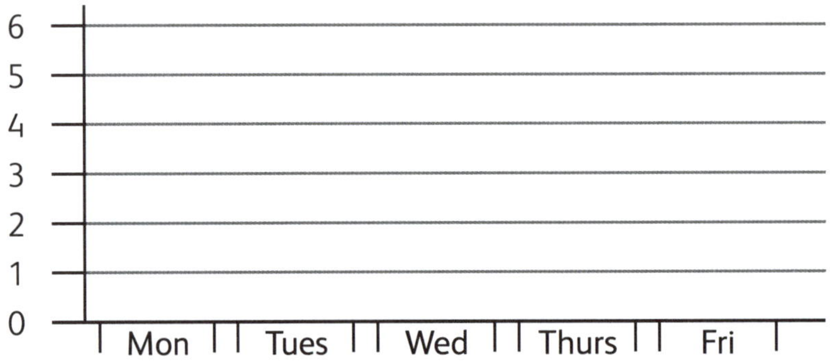

What advice would you give Andre about the scale he has chosen for his bar graph?
What would be a more suitable scale?
Is there anything else missing from the bar graph?

2 **a** Copy and complete the table. Use the data in the bar graph.

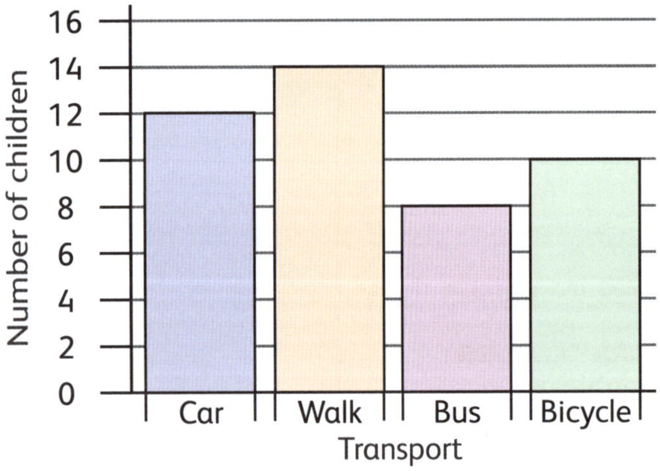

Transport	Number of children
Car	
Walk	
Bus	
Bicycle	

b What is the most popular way to travel to school?

c What is the least popular way to get to school?

d How many children are there in Grade 6?

e How many more children travel to school by cars than by bicycles?

f Produce your own table and bar graph showing how your class travels to school. Then represent the set of data in a bar graph.

3 Leon uses direct observation and creates a graph of buses passing Hagley Park Road in 1 hour.

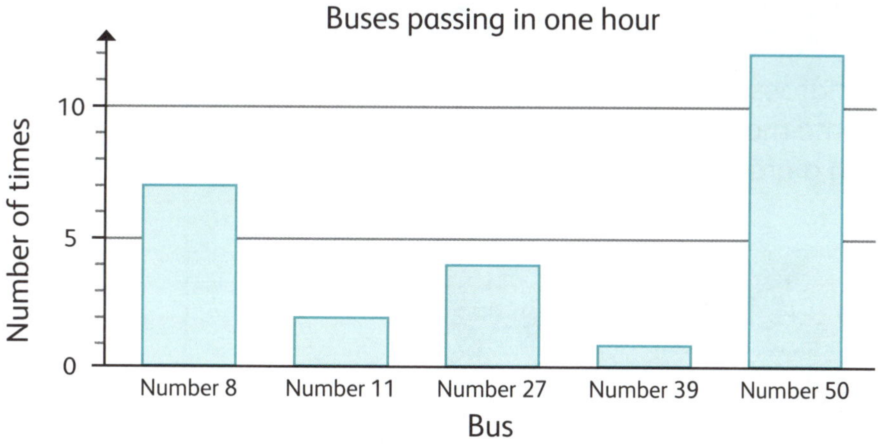

a How many times does a Number 27 bus pass in an hour?

b How many more Number 8 buses pass in the hour than Number 11 buses?

c Leon says: 'The Number 39 bus passes least often in the hour.' Explain how the graph shows this.

4 Classes 6A and 6B at a school are having a challenge to collect bottle caps for an art project. This graph shows the number collected each day over a week.

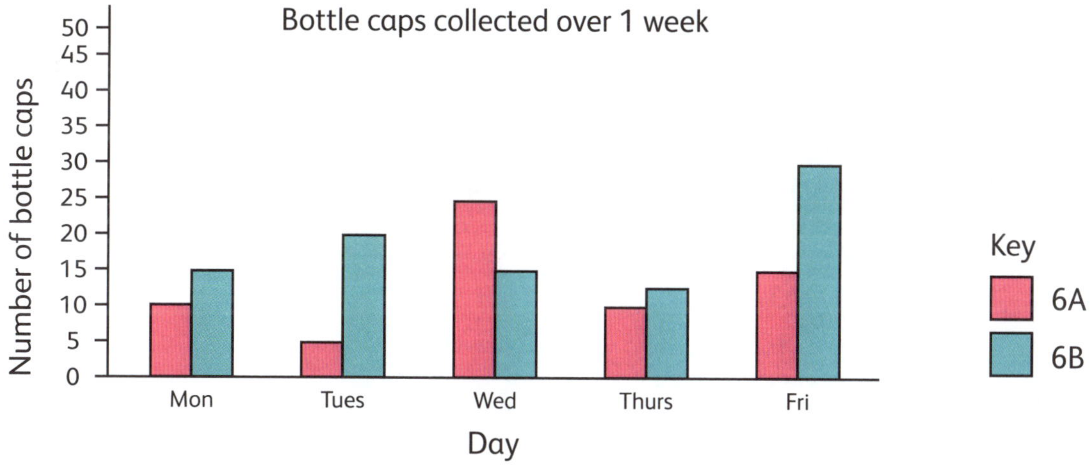

a Why do you think this kind of graph is called a double bar graph?

b What does the key tell us?

c On which day did 6A collect more?

d On which days did 6B collect more?

5 Leo decides to earn pocket money by offering dog walking services on weekends. He keeps a record of how many dogs he walks over four weeks.

	Saturday	Sunday
Weekend 1	4	5
Weekend 2	5	7
Weekend 3	6	6
Weekend 4	9	8

a What is the range of values for the number of dogs he walks each day?

b Draw a set of axes to make a bar graph. Each double bar should show the number of dogs walked on a weekend, split into Saturday and Sunday values.

Statistics and probability

Histograms

Key maths idea

Janice picks 35 tomatoes from her vegetable garden. She wants to measure the mass of each tomato and show how much they weigh in a graph. But each tomato is a slightly different mass.

She decides to use a histogram. A **histogram** is a bit like a bar graph, except that it groups data into **ranges** (or **class intervals**). This is a useful type of graph for **continuous data** such as height, mass, temperature or length, whereas a bar graph is more useful for **discrete data** – data that can be divided into separate categories.

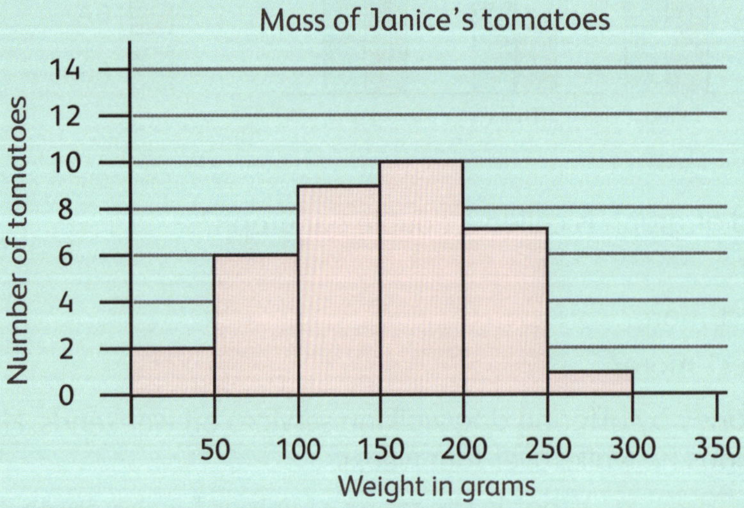

Key words

histogram
range
class interval
continuous data
discrete data

1 How many tomatoes had a mass between 0 and 50 g?

2 How many tomatoes had a mass between 200 and 250 g?

3 Janice divided the tomatoes into six class intervals. Each bar represents one of the intervals.

a Which class interval did most of the tomatoes fall into?

b Which class interval would a tomato weighing 297.8 g fall into?

4 Janice decides to do another graph to show the colours of the tomatoes. 18 were dark red, 6 were pink-red, 7 were yellow and 4 were green. Should she use a bar graph or a histogram? Explain your answer.

5 Janice plants a tray of beans. After one week, she measures the lengths of the shoots. These are the values:

12.1 cm	11.9 cm	8.5 cm	7.3 cm	13.8 cm	15.2 cm
10.5 cm	9.9 cm	10.1 cm	12.6 cm	13.3 cm	11.7 cm
11 cm	10.9 cm	11.5 cm	12.2 cm	13.1 cm	

a What are the greatest and smallest values in the data set?

b Copy and complete this table to group the data into class intervals:

Class interval	Values
Less than 8 cm	
Equal to or greater than 8 cm and less than 10 cm	
Equal to or greater than 10 cm and less than 12 cm	
Equal to or greater than 12 cm and less than 14 cm	
Equal to or greater than 14 cm and less than 16 cm	

c Draw a set of axes for a histogram that would show the length distribution of the seedlings.

d Use the data from your table to complete the histogram.

Drawing a line graph

Key maths idea

A line graph is also used for continuous data. When you draw a line graph, you first plan your axes and your scale, like you would for a bar graph. Then you plot the data points. After this, you join the points to make a line.

Robert filled a container with tap water and placed it in the freezer. He tested the temperature every 15 minutes. When the water was frozen, he took it out of the freezer, and continued to test the temperature.

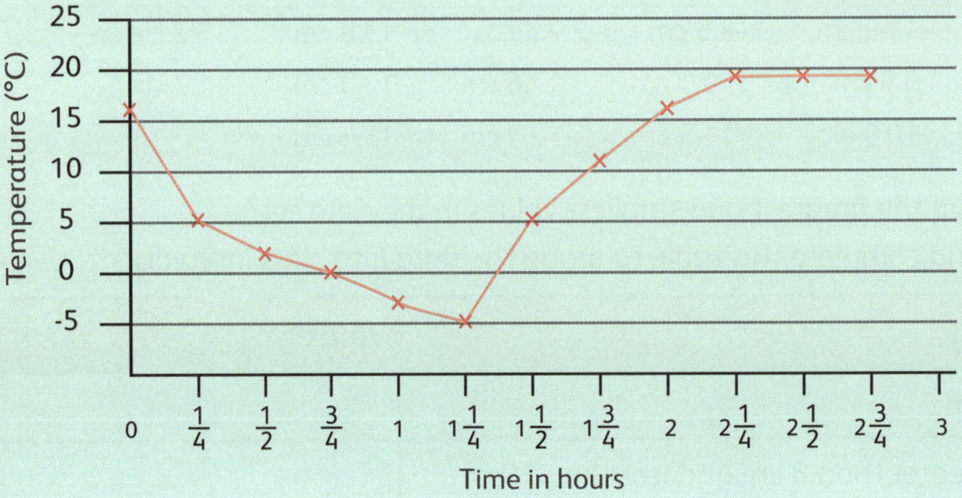

1 Answer these questions about the line graph.

- **a** What was the starting temperature of the water?
- **b** How long did it take to reach 0 °C?
- **c** After how long did Robert remove it from the freezer? How can you tell?
- **d** Robert decides to measure the temperature of a pot of water as he cooks some rice. How do you think the shape of that graph will differ from this one?

2 This table shows the mass of a kitten over the first six weeks of its life. Work with a partner.

- **a** Discuss how you could use it to draw a line graph showing the mass of the kitten.
- **b** Use graph paper. Draw the axes and decide on a suitable scale for each axis. Then plot the points and complete the line graph.

Age in weeks	Mass (g)
0 (at birth)	90
1	180
2	320
3	450
4	580
5	650
6	720

3 If you have a class thermometer, measure the temperature in the classroom each day for a week. Use your data to plot a line graph.

Pie charts

Key maths idea

Jess asks her classmates to choose their favourite sandwich fillings. She asks 24 classmates. This is the data she collects:

Peanut butter	~~////~~ /
Cheese and tomato	////
Chicken	~~////~~ ///
Tuna mayo	///
Butter and jam	///

She decides to show her data in a **pie chart** or **circle graph**. When you draw a pie graph, you first draw a **circle**, and then draw a **radius** (a line from the centre to the **circumference**). Then you need to draw **angles**.

Remember: we measure angles in degrees (°) and a full turn is 360°.

To work out the size of each angle in the pie chart, you calculate it as a fraction of 360°.

The total data set is 24. (There is a total of 24 tallies, meaning Jess asked 24 people their choices.)

6 out of 24 people chose peanut butter.

$\frac{6}{24} = \frac{1}{4}$

$\frac{1}{4}$ of 360° = 90° Do you remember that 90° is a right angle?

Discuss with your group how Jess worked out the other angles.

Favourite sandwich fillings

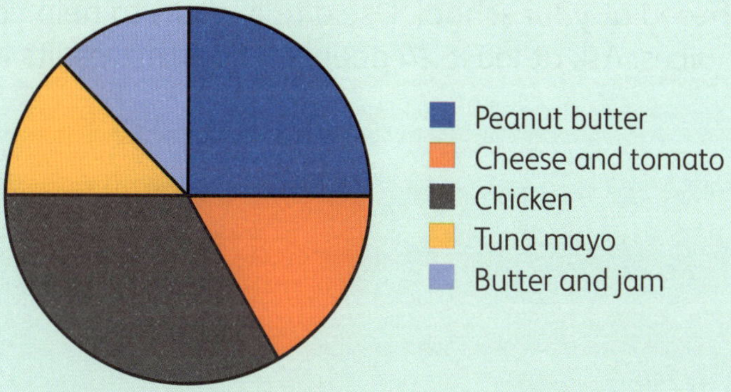

Key words

pie chart
circle graph
circle
radius
circumference
angle

Statistics and probability

1 A student made this pie chart to show how she used her free time each week:

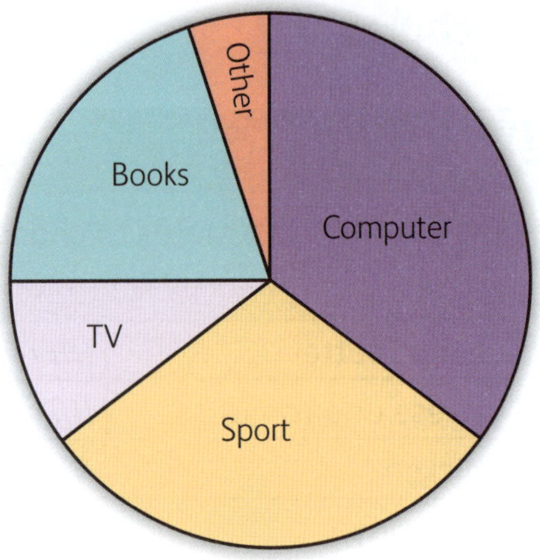

- **a** What does she spend most of her free time doing?
- **b** What does she spend the least time on?
- **c** Why is it not possible to tell how many hours she spends on each activity from this chart?
- **d** If her total free time is 40 hours per week, work out roughly how many hours she spends on each activity.

2 At a maths competition, there are 200 students. 150 are from Jamaica and the rest are from other islands. Which of these pie charts best shows the makeup of the students? Explain how you worked it out.

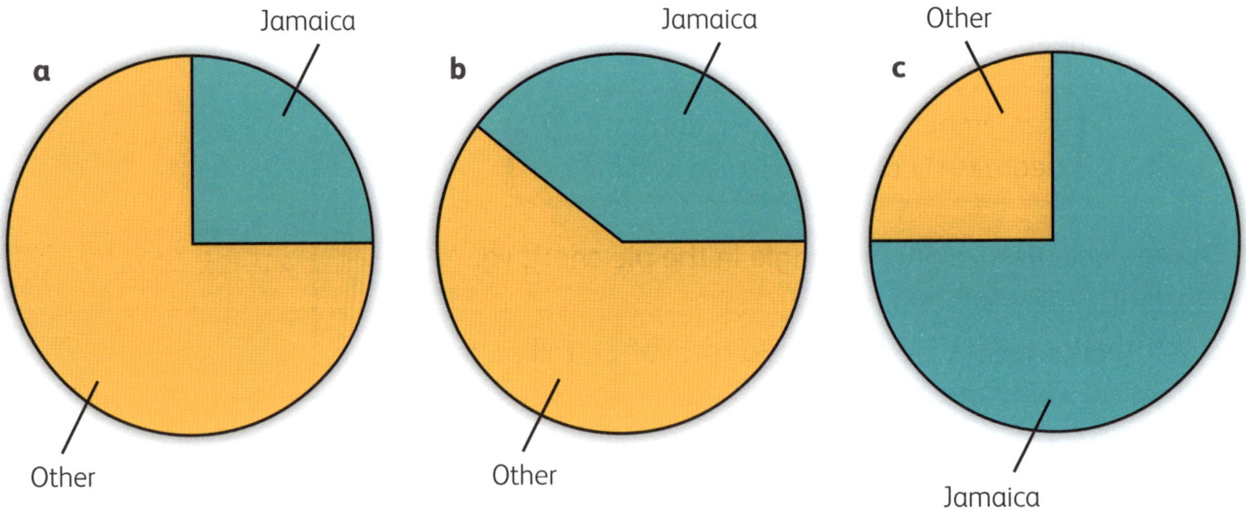

3 A fruit box contains 24 pieces of fruit. It includes 6 apples, 5 oranges, 3 bananas, 4 mangoes, 1 papaya, 2 pineapples and 3 ackee. Draw a pie chart showing the fruit amounts in a whole box.

4 Your school wants to encourage students to take part in a wider variety of sports. Select five sports that are not offered at your school. Use a tally chart to help you collect information about the choices. Ask at least 24 people. Show the results in a pie chart.

Stem-and-leaf plots

Key maths idea

A **stem-and-leaf plot** is a type of special table that splits each data value into two parts: the 'stem', which is made up of the first digit or digits, and the 'leaf', which is made of the last digit of each value. We group the values so that all values that share the same stem are listed in a row together. This is a way of showing the **distribution** of the values.

Example

Some Grade 6 students timed themselves to see how quickly they could tie a pair of shoelaces. These were the times in seconds:

2.8 3.4 5.5 4.7 3 3.4 3.1 2.9 3.8 4.5 2.9 2.6 4.8

First they arranged the times in order:

2.6 2.8 2.9 2.9 3 3.1 3.4 3.4 3.8 4.5 4.7 4.8 5.5

Next, they put the first digits in order in the Stem column, and the last digits in a row in the Leaf column:

Stem	Leaf
2	6 8 9 9
3	0 1 4 4 8
4	5 7 8
5	5

Key
2 | 6 means 2.6

Note that 'Stem 2 Leaf 6' means 2.6 in this table, because the leaf values are all decimals. This is not always the case.

Key words

stem-and-leaf plot
distribution

1 Look at the stem-and-leaf plot above and use it to answer these questions.

- **a** Why is there a 0 value in the second row?
- **b** What does the stem-and-leaf plot show us about the times?
- **c** The first two rows have a repeated value in the leaf column. What does this tell us?

2 These are the ages of the singers in a church choir:

23 28 34 58 65 54 43 43 29

78 63 69 74 70 65 57 49 50

- **a** What is the range of ages in the choir?
- **b** Work out the median age, and the mean age.

Remember that the mean is the sum of the values divided by the number of values. The median is the middle value, or the mean of the two middle values in an even set of values.

c Use the values to draw up a stem-and-leaf plot. Don't forget to write the values in order from least to greatest before you complete your table.

3 This table shows the scores of eight students in a test. Use this data set to draw a stem-and-leaf plot.

Name	**Score**
Leon	67
Jada	64
Jordan	43
Iris	77
Anthony	49
Joshua	54
Lamar	78
Sophia	73
Mikayla	54

4 Here are the prices of coco bread bought in different local stores.

$88 $75 $83 $79 $90 $96
$74 $90 $85 $76 $77 $81
$97 $82 $73 $95 $84 $72

a Represent this data in a stem-and-leaf plot. Don't forget to include a key.

Stem	**Leaf**

b Find the modal value of coco bread.

c Calculate the mean value.

d Calculate the median price of coco bread.

e Calculate the range in price of the coco bread.

5 This stem-and-leaf plot shows the times 10 students took to complete a jigsaw puzzle.

Stem	**Leaf**		
1	6	7	
2	4	6	8
3	2	4	4
4	0	3	

Key
2|4 means 24

a What was the median time taken to complete the jigsaw puzzle?
b What is the range of times?
c What was the modal time taken to complete the jigsaw puzzle?
d What is the mean time taken to complete the jigsaw puzzle?

6 This table shows how many minutes students spent watching television on Friday.

Student	Erica	Lola	Rosa	Dominic	Anthony	Mikayla	Joshua	Andre	Leon
Time spent watching television (minutes)	34	22	9	13	45	32	42	25	28

a Represent this data in a stem-and-leaf plot. Do not forget the key.

Stem	Leaf

b What is the range of times?
c What was the modal time spent watching television?
d What was the median time spent watching television?
e What was the mean time spent watching television?

Using data

Key maths idea

You have learned that:

- You **collect data** by conducting surveys or experiments. You may use tools such as interviews and questionnaires. During this step, you need to record the data, for example, using tables or tallies.
- You can **represent data** in different ways, for example, bar or line graphs, pie charts, histograms and stem-and-leaf plots. Some graphs are a combination of more than one of these.
- You **interpret** the graphs by asking questions and making observations about the trends they show.

Key words

collect data
represent data
interpret

1 Work in pairs. Suggest a topic that would be suitable to collect information about by using:

a a questionnaire   b an experiment

2 A school committee interviewed parents to find out whether they agree or disagree with plans to dig up one of the school sports fields to build a bigger cafeteria.

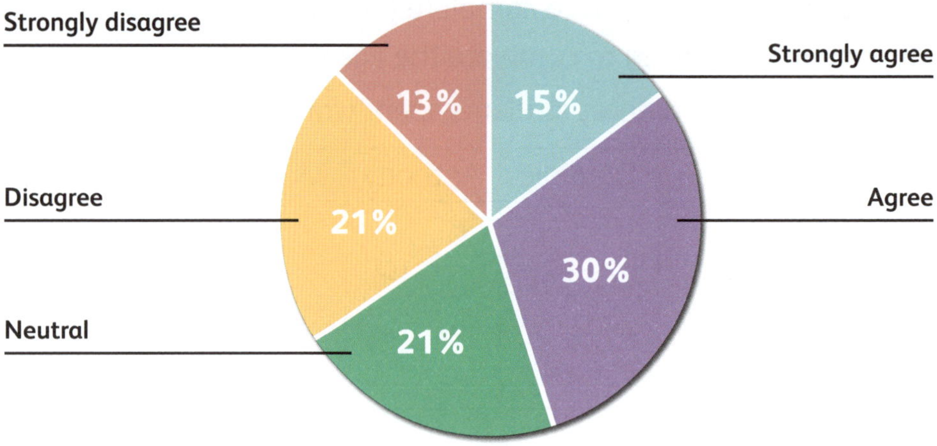

a Do you think the school should go ahead with the plans? Explain your ideas.

b What question do you think the school asked the parents? Write what you expect it looked like on the questionnaire.

c If you wanted to collect this data, would you have asked this question, or would you have done it differently? Give your own ideas.

3 Jada designed a questionnaire to learn about TV-watching habits. She intends to give this questionnaire to 30 of her classmates.

a Find two problems with this questionnaire and explain your answer.

b Redesign this questionnaire so that it is more likely to collect useful data.

4 A scientist is carrying out an experiment on climate change. She measures the noon temperature in the Blue Mountains over seven days. She records her findings in a table.

Day	Sunday	Monday	Tuesday	Wednesday	Thursday	Friday	Saturday
Noon temperature (°C)	25	28	26	22	19	23	27

a Create a line graph to show these readings.

b What was the lowest temperature recorded, and on what day did it occur?

c On what day was the noon temperature 26 °C?

d Why is a line graph the best data representation for temperature?

Real-world maths: Plan a picnic

Nisha and Leroy are planning a class outing to Dunn's Falls. They want to prepare packed lunches so they need to find out which sandwiches, sweets, fruit and juice are most popular. They want to offer a choice of 3 types of sandwiches, 2 types of juice, 2 different fruits and 3 different sweets as dessert.

1 Develop a short questionnaire to help them find the information they need. Decide which kinds of sandwiches, juice, fruit and sweets to include. Make sure you provide the correct number of choices for each type of item and make it very easy to answer the questions.

2 Use your questionnaire to do a survey in your class to find out what choices students would prefer.

3 Draw up a table to organise the data you will collect.

4 Use the results of the survey to draw up a menu for Nisha and Leroy's class outing.

Maths detective: Statisticians who solved real-life problems

The stem-and-leaf diagram was invented by a statistician called John Wilder Tukey. He also invented another statistical diagram called the box-plot.

Many of the ways we work with statistics and the graphs and other diagrams we draw to display data were developed by statisticians as they worked to solve real-life problems. In this maths detective you are going to use your research skills to find out more about statisticians and their work.

Case 1

Have you heard of Florence Nightingale?

Florence Nightingale is best known as a nurse. But she also collected data about injured soldiers during the Crimean War. She used the data to prove that 16 000 out of 18 000 deaths were due to bad hygiene and not battle wounds. She presented her work in a series of graphs and rose charts that were so clear she was able to convince the army that her data was correct and this saved many lives.

1 Find out more about rose charts. Draw a labelled sketch to explain how they work.

Hint: rose charts are often used to present wind data today.

2 Florence Nightingale David was also a statistician. (Her family named her after the famous nurse.) Do some research to find out about her and her work. Write a short biography about her for an online encyclopaedia.

Case 2

Work with another student. Choose any two of these famous statisticians:

- David Harold Blackwell
- Sir Ronald Aylmer Fisher
- Gertrude Mary Cox
- David R Brillinger
- William Playfair
- Dr John Snow

Do your own research to find out more about the statisticians you chose. Make an information poster to show what you find out. Include what each person is best known for.

Case 3

Have you ever wondered how we know these things?

- Smoking causes lung cancer.
- Breastfeeding is healthier for babies than bottle feeding.
- Women drivers have fewer accidents than men.

The answer, of course, is statistics.

1 Do some research to find out more about Dr Jerome Cornfield and his work on the effects of smoking.

2 Find out how Dr Janet Lane-Claypon proved that breastfeeding is healthier for babies.

3 The insurance industry makes use of statistics to work out how much people have to pay for insurance. Ask people in your community how this affects young people who have recently got a driving licence, people who have had more than one accident and people with very expensive cars.

Case 4

We live in the information age. Statistics and data play a huge role in decision making in many different fields, including weather forecasting, disaster management, ecology, environmental studies, human health and medicine.

Professor Cavell Brownie is a Jamaican-born statistician who developed a mathematical model to estimate bird populations. She used biometric methods and sampling in her work.

1 Why do you think it is important to know about wildlife populations?

2 What does 'biometric' mean? Give an example of how biometrics are used in daily life.

3 AmstatNews (https://magazine.amstat.org) is published by the American Statistics Association. Use the website to find out more about one of these modern statisticians and their work:

- Charlotte Baidoo
- Lorin Crawford
- Dionne Price
- Tommy Wright

4 Someone on the Amstat website wrote the comment 'Go data dawgs'. What do you think this means?

Case 5

The motto of the Statistical Institute of Jamaica (www.statinja.gov.ja) is 'We collect. We analyse. We publish.'

1 What do you think this means?

2 Each year, the Statistical Institute of Jamaica carries out several local surveys. Find information about the surveys carried out this year. List three of these and say how the data might be collected, how you think it would be presented and why it is useful to do this research.

Real-world maths: A litter campaign

Your town council wants to start a campaign to encourage people to become more environmentally aware. Your class wants to collect information that will help the town manage its litter problem. Design a research project to collect the most important information and decide how to present the results in an appropriate way.

1 Decide what kinds of information your class needs to collect (for example, type of litter, where it is found, where and how the town's rubbish is disposed of, what parts of the town are most affected, how much green space there is, people's attitudes towards litter, and so on).

2 Choose three topics to research from your list.

3 How would you gather data about each of the three topics? Would you use interviews, questionnaires, surveys, observations or written records?

4 Decide how you would present the data for each of the same three topics. Include at least two of these: pie chart, line graph, pictograph, bar graph.

5 If possible, divide into groups to research litter in your own town. Each group will gather data on one of the questions you formulated and present the data using different types of charts. Present your results to the class. If you have access to data presentation software, input your results and try to create different charts from your data.

What I have learned

1 In your own words, explain what statistics are.

2 Copy and complete this diagram.

STATISTICS

Organising data

List 2 ways of organising data.

1 ....

2 ....

Collecting data

List 5 ways of collecting data.

1 ....

2 ....

3 ....

4 ....

5 ....

Analysing data

List 4 ways of analysing data.

1 ....

2 ....

3 ....

4 ....

Representing data

List 6 ways of representing data.

1 ....

2 ....

3 ....

4 ....

5 ....

6 ....

Practice questions

1 Imagine that you want to find out the favourite subjects of students in your year group at school.

a Write a sentence to describe how you would collect this information.

b Which of the following types of graphs would you use to represent the data: stem-and-leaf plot, line graph, pictograph, histogram? Explain your answer.

2 This graph compares the numbers of followers that four famous sports stars had on a particular social media platform in 2021.

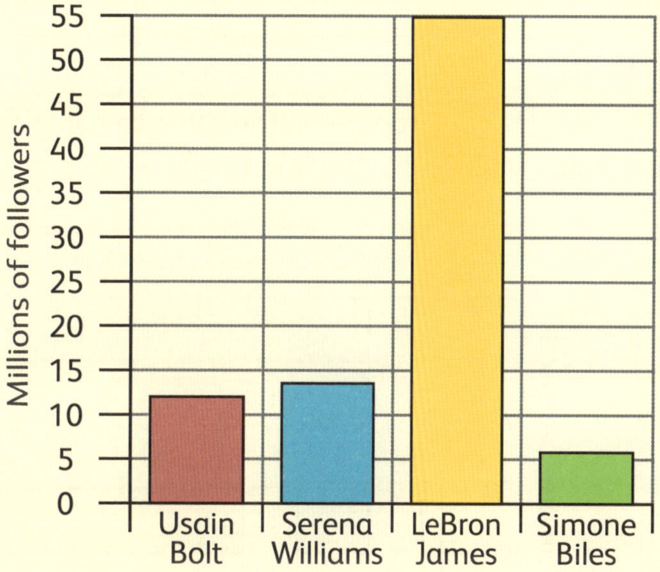

a How do you know that this is a bar graph and not a line graph or histogram?

b Zane says: ‘Usain Bolt only had 12 followers.’ What is Zane’s mistake?

c How many more followers would Serena Williams need to catch up to LeBron James on this platform?

d Do you think it is important for celebrities to have many followers on social media? Why or why not?

3 This circle graph shows the juice flavours that students in a class chose for a school outing.

a What was the total number of students in this data set?

b Which juice flavour was most popular?

c How can you use fractions to help you draw circle graphs?

4 Here are the heights (in centimetres) of 24 students.

151	152	162	143	128	151	144	165
160	166	159	158	125	124	171	139
167	157	141	138	132	135	147	144

Draw a stem-and-leaf plot for this data set.

5 This table shows the number of minutes Samuel and Anna exercised each day for five days.

	Samuel	**Anna**
Monday	60	45
Tuesday	30	35
Wednesday	20	60
Thursday	70	65
Friday	45	30

Use the information in the table to complete the double bar graph about Samuel and Anna's exercise times.

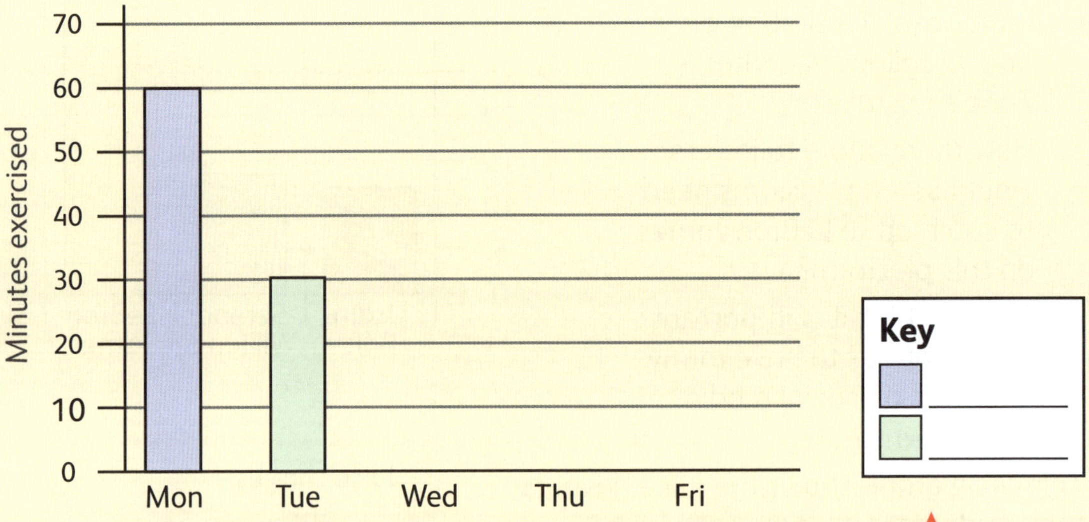

In a double bar graph, you can display two sets of data that are related. Remember that a double bar graph needs a key.

7 Self check

Check you can do this:	If you have difficulty, try this:
1 Collect data using direct observation, experiments, interviews and questionnaires.	Go back to pages 92–93. Make sure you understand the difference between these data collection methods. Remind yourself of the different ways you collected data in this chapter.
2 Identify patterns in data and interpret (make inferences from) these patterns.	Choose three of the graphs you answered questions about in this chapter. Think about what information is most clear when you look at the graph. Think about: - on line graphs: is the line going up or down? are there changes of direction, and what do they show? - on bar graphs: which bars are longest and which are shortest? - on histograms: what shape do the bars form? Do the bars get longer or shorter, or change direction? - on circle graphs: which sectors are larger and which are smaller? What are the angles of the sectors?
3 Represent data using a stem-and-leaf plot.	Go back to page 105. Make sure you know how to decide which part of the values make the stem part and the leaf part of the diagram. You also need to be able to look at a stem-and-leaf plot and list the data values by putting together each stem with the leaf digits.
4 Discuss the appropriate uses of various tables and graphs.	Look at the different types of graphs you learned about in this chapter on pages 96–103.
5 Solve problems in which data is given by means of a graph or diagram.	Look back at pages 107–109 at the other problem solving questions at the end of this chapter.

PEP task: A chocolate brownie fundraiser

This performance task has 4 parts. Your teacher will tell you when to do each part.

Part 1

Ella's youth group wants to raise funds to send some of their members to a climate summit. They decide to bake chocolate brownies to sell at a local farmer's market.

You will need:
- a ruler
- a pencil
- a calculator.
- If you have square grid paper it may be useful.

Ella found this recipe.

Chocolate brownies

Ingredients

185 g unsalted butter

195 g dark chocolate

85 g plain flour

40 g cocoa powder

45 g white chocolate

60 g milk chocolate

3 large eggs

275 g granulated sugar

Makes one 20 cm × 20 cm square pan.

1 Ella tried out the recipe at home using a 20 cm × 20 cm square pan. She cut the baked brownies into squares like this:

a What is the approximate size of each square? Show how you work this out.

b If the square brownies are 2.5 cm thick, what is the volume of each brownie?

c The cost of ingredients for this batch was $920. What is the cost per brownie?

d The recipe uses 300 grams of chocolate altogether. What percentage of the total chocolate are white, milk and dark chocolate?

2 The recipe says: 'Bake at 180 °C for 45 minutes.'

a Ella's oven is marked in Fahrenheit. Use the formula $F = C \times \frac{9}{5} + 32$ to convert 180 °C to °F.

b What should Ella set her oven at? Why?

c The baked brownies came out of the oven at 15:20. If it took Ella half an hour to prepare the brownie mix and 45 minutes to bake the brownies, at what time did she start making them?

Part 2

When Ella gets to the youth group kitchen she discovers they don't have a 20 cm square pan. They do have a 30 × 30 cm pan and a 20 cm × 15 cm pan. She decides they will make enough brownies to fill both those pans.

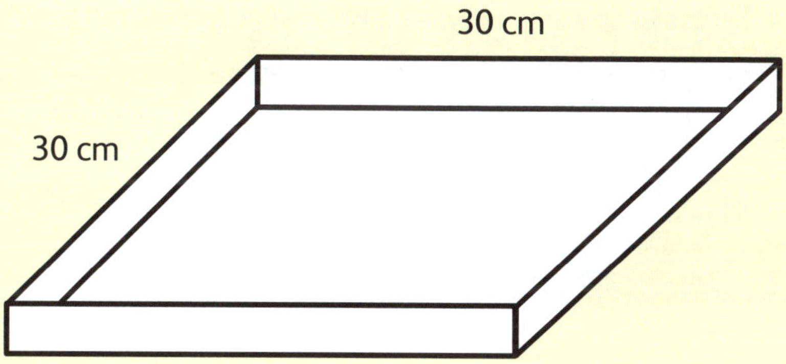

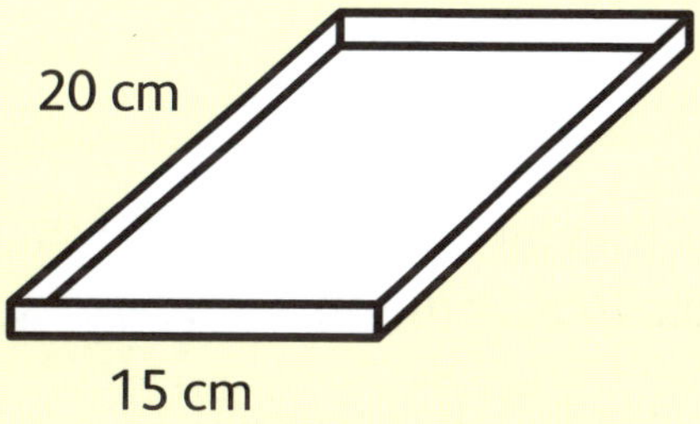

a Jerome says that the 30 cm × 30 cm pan is 10 cm longer than the 20 cm pan and because 10 is half of 20, they will need $1\frac{1}{2}$ times the ingredients. Draw a sketch and use what you know about area to explain why Jerome's thinking is incorrect.

b How will you need to scale up the recipe to fill both the 30 cm square pan and the 20 cm × 15 cm pan? Show how you work this out.

c Use your answer in (b) to calculate how much it will cost to buy the ingredients for these 2 batches of brownies.

d Sketch the base of both pans using a scale of 1 : 2. Draw lines on your scale drawings to show how you would cut the brownies in each pan so they are the same size as the ones Ella baked.

Part 3

The group decide to sell the brownies in boxes like these. One of their members will donate the flat boxes (nets) and the youth group will fold them up. Each box will hold four 5 cm square brownies.

a Sketch a net of the box. Add measurements to your net. Remember it will hold four brownies and that they are 2.5 cm high.

b Estimate the area of the plastic display window of the box. Show how you worked this out.

Part 4

The youth group did a survey to find how much money people usually spend on sweets at the farmer's market. They drew this histogram to show their findings.

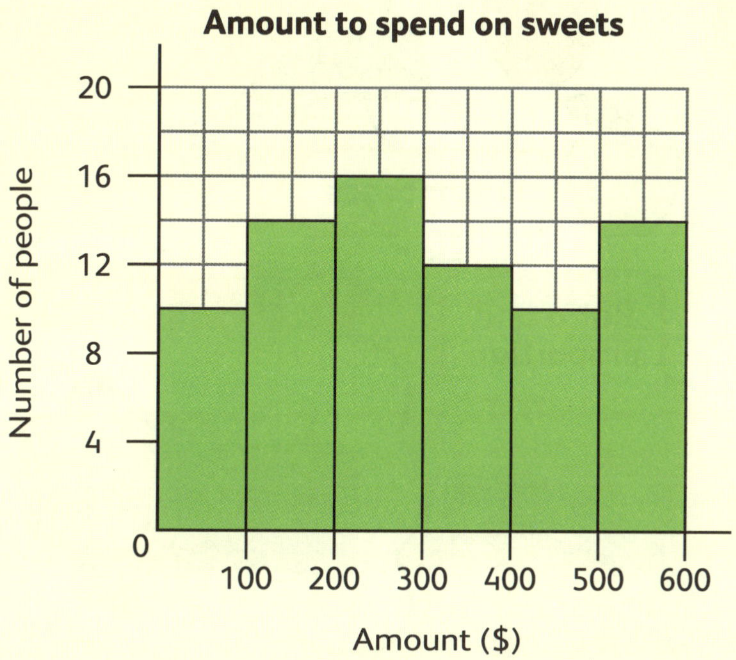

1 Answer these questions about the graph.

- **a** How many people spent between \$100 and \$300 dollars on sweets?
- **b** What is the modal amount spent on sweets? Give your answer as a range.
- **c** How many people took part in the survey?
- **d** Why is this information important to the youth group?

2 The youth group have to decide what price to charge for a pack of four brownies.

- They must pay for the ingredients but all other costs are donated.
- They want to make as much profit as they can so they can send more members to the climate summit.
- The price must not be so high that people won't buy the brownies.

- **a** Use the prices given and the other information you are given in this task to work out a selling price for the packs of brownies.
- **b** Write a few sentences explaining how you decided what the price should be.

TERM 2

UNIT 1: NUMBER
p121

Types of number

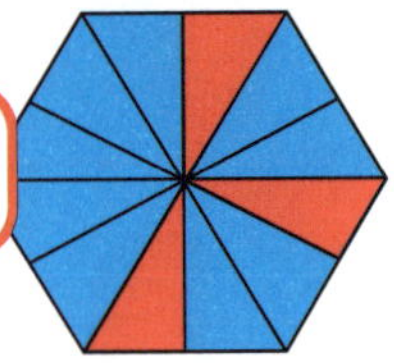

Fractions

Percentages

Decimals

Ratio and proportion

UNIT 2: MEASUREMENT
p207

Area and volume

Area and tessellations

UNIT 3: GEOMETRY
p237

Symmetry and congruence

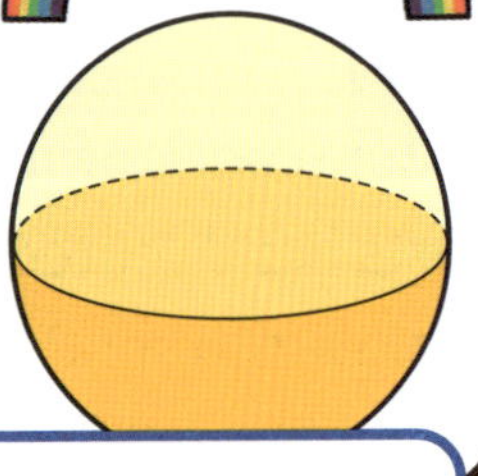

Coordinates and reflections

UNIT 4: ALGEBRA
p262

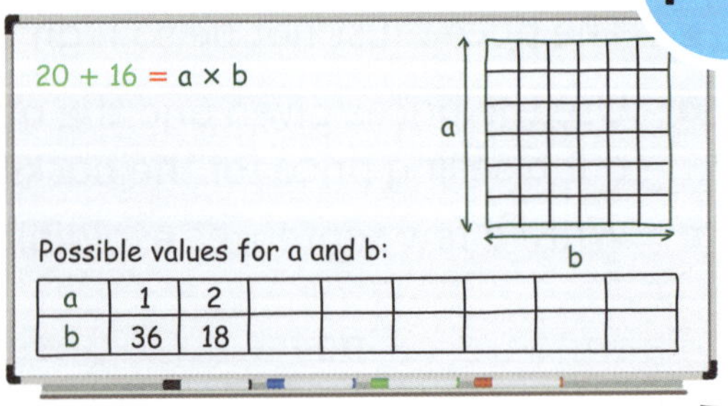

Inequalities, variables and number patterns

Unit 1: Number

Chapter 8 Types of number

In this chapter you will explore this focus question:

To answer this question, you will:

- show what you understand about numbers, types of numbers and their properties
- list all the prime factors of a given number
- write a composite number as a product of its prime factors
- write a composite number as a product of its primes in exponential form
- identify the highest common factor (HCF) of two numbers
- identify the lowest common multiple (LCM) of two or more numbers
- explain the difference between multiples and factors, and apply your understanding to different mathematical ideas.

Starting point

1 Look at this recipe for lentil soup.

The recipe is for 6 people.

To make the soup for different numbers of people, you will need to adapt the amount of each ingredient, but the proportions must stay the same. If you change the proportions, the soup might not taste right!

Re-write the recipe for the ingredients needed to make soup for:

- 3 people
- 9 people
- 1 person.

Talk through your method with a partner.

Recipe for 6 people

- 1 onion
- 60 g butter
- 180 g dried lentils
- 1.2 litres vegetable stock
- 480 ml tomato juice

2 Look at these numbers.

- **a** Which is the largest number you can find?
- **b** Which is the smallest number you can find?
- **c** Find an odd number, an even number, a prime number and a composite number.

Prime and composite numbers

Key maths idea

Prime numbers have only two factors.

These factors are 1 and the number itself.

A prime number can only be shown as an array in two ways – with one row or one column.

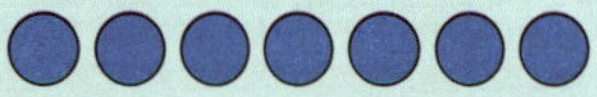

7 is a prime number because it can only be divided exactly by itself (7) or 1 without leaving any remainders:

$7 \div 7 = 1$ and $7 \div 1 = 7$

So the factors of 7 are 1 and 7.

Numbers that are not prime are called **composite numbers.**

A composite number can be shown as an array in more than two ways. Here is one of the arrays. Describe the other possible arrays to a partner.

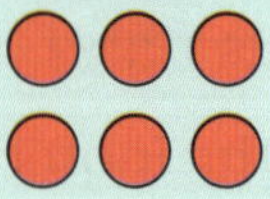

6 has four factors: 1, 2, 3 and 6.

The number 1 is different. It is neither prime nor composite. Try to explain why.

Key words

prime number
composite number
Hindu-Arabic number system

Do you remember from Chapter 1 that we use the **Hindu-Arabic number system** and that this system is based on 10? We use the 10 symbols 0, 1, 2, 3, 4, 5, 6, 7, 8 and 9 and our understanding of place value to write any number. We also say it is a decimal number system.

1 List each set of numbers.

- **a** {odd numbers less than 20}
- **b** {even numbers less than 20}
- **c** {prime numbers less than 20}
- **d** {composite numbers less than 20}

2 This is the beginning of a prime number sieve. Start by crossing out the number 1 because it is neither prime nor composite. 2 is a prime number, so circle it.
Next, cross out all multiples of 2, because they cannot be prime numbers.
Now look for the next prime number, which is 3. Circle it. All other multiples of 3 can now be crossed through, as these are not prime numbers.
Continue until you have circled every prime and crossed out every composite number up to 100.

~~1~~	(2)	3	~~4~~	5	~~6~~	7	~~8~~	9	~~10~~
11	~~12~~	13	~~14~~	15	~~16~~	17	~~18~~	19	~~20~~
21	~~22~~	23	~~24~~	25	~~26~~	27	~~28~~	29	~~30~~
31	~~32~~	33	~~34~~	35	~~36~~	37	~~38~~	39	~~40~~
41	~~42~~	43	~~44~~	45	~~46~~	47	~~48~~	49	~~50~~
51	~~52~~	53	~~54~~	55	~~56~~	57	~~58~~	59	~~60~~
61	~~62~~	63	~~64~~	65	~~66~~	67	~~68~~	69	~~70~~
71	~~72~~	73	~~74~~	75	~~76~~	77	~~78~~	79	~~80~~
81	~~82~~	83	~~84~~	85	~~86~~	87	~~88~~	89	~~90~~
91	~~92~~	93	~~94~~	95	~~96~~	97	~~98~~	99	~~100~~

- **a** Use a 100 square grid or software to complete the prime number sieve.
- **b** Make a list of all the prime numbers up to 100.

3 Leon says that all prime numbers are odd numbers. Candice says that all odd numbers are prime.

Both children are wrong. Explain why.

4

24 17 64 91 21 15 38

- **a** Write down all the odd numbers.
- **b** Write down all the even numbers.
- **c** Now write down all the prime numbers.
- **d** Which of the numbers are composite?

5 Tayla says she has found a rule for creating prime numbers.

Input → × 12 → − 7 → Output

- **a** Work with a partner. Try out Tayla's rule using the numbers 1 to 10.
- **b** Did you find that the output number is always a prime number?

Factors and prime factors

Key maths idea

A **factor** divides into a given number exactly, without leaving a remainder.

Example

The factors of 6 are 1, 2, 3 and 6.

A prime number is a number whose only factors are 1 and itself.

A **prime factor** is a combination of these two definitions: it is a factor that is also a prime number.

So the prime factors of 6 are 2 and 3.

To find the prime factors of larger numbers, we can use a prime factor tree. A factor tree is a handy way to factorise numbers into their prime factors. A factor tree starts at the root and grows upside down.

Let's write 90 as a product of its prime factors.

Write the number as a product of any two of its factors.

Circle the primes.

Write any non-prime factors as a product of any two of its factors.

Circle the primes.

Continue until all factors are prime.

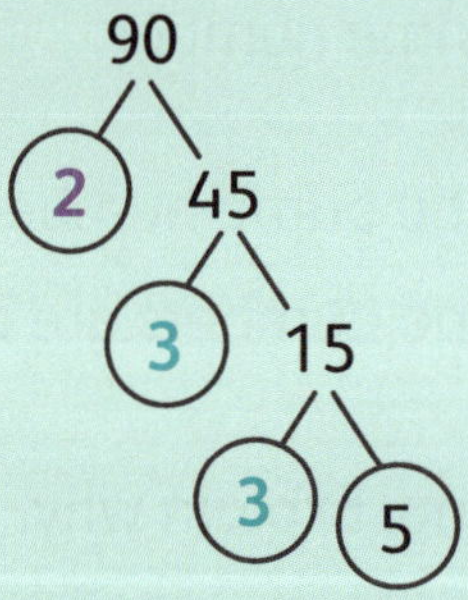

You can write the product either showing all of the individual factors:

$2 \times 3 \times 3 \times 5 = 90$

or in **exponential form** where, in this example, the **exponent** shows that the number 3 is squared:

$2 \times 3^2 \times 5 = 90$

Key words

factor

prime factor

exponential form

exponent

1 Find all the factors of:

a 32 **b** 25 **c** 81 **d** 100 **e** 48

2 Now circle the prime factors for each number in Exercise 1.

3 Copy and complete the table.

Number	Factors	Prime factors
16	1, 2, 4, 8, 16	2
20		
30		
42		

4 Draw a prime factor tree for each number. Express the number as a product of its prime factors.

a 24 **b** 18 **c** 28 **d** 36 **e** 50 **f** 120 **g** 225

5 Write each of these numbers as a product of its prime factors in exponential form. Draw a prime factor tree if you need to.

a 40 **b** 236 **c** 498 **d** 56 **e** 624 **f** 2 000 **g** 500

The highest common factor (HCF)

Key maths idea

The **highest common factor (HCF)** is the largest factor shared by two or more numbers. The HCF is sometimes called the GCF (greatest common factor).

Example

Find the HCF of 12 and 20.

List all the factors of 12 and 20.
Circle the largest number that appears in both factor lists.
The HCF of 12 and 20 is 4.

Factors of 12	Factors of 20
1	1
2	2
3	(4)
(4)	5
6	10
12	20

Key word

highest common factor (HCF)

1 What is the highest common factor of:

a 6 and 15 **b** 45 and 105 **c** 24 and 40

d 27 and 126 **e** 12, 18 and 30

2 Copy and complete this Venn diagram.

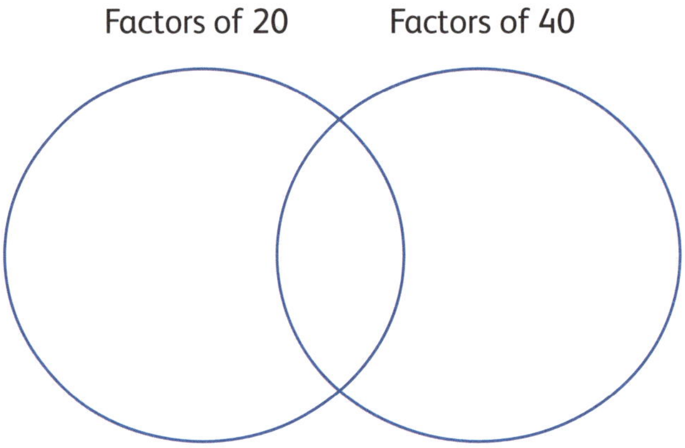

a Use the Venn diagram to help identify the common factors of 20 and 40. List them.

b What is the highest common factor (HCF) of 20 and 40? Explain how you know.

c Where would the number 3 go on this Venn diagram? Why?

3 **a** What is the largest number that divides into 42 and 48 without leaving a remainder?

b List all the common factors of 42 and 48.

c What is the largest number that divides into 32 and 96 without leaving a remainder?

d List all the common factors of 32 and 96.

Real-world maths: Equal shares

1 Rolls of wire are cut into a number of equal pieces to be sold. The length of each piece is in whole metres. There is no wire left over.

a What are the possible lengths of each piece for a 72 metre roll?

b What are the possible lengths of each piece for a 45 metre roll?

c Could a customer buy the same length piece of wire from either roll? Explain your answer.

2 A teacher has 60 markers and 330 coloured pencils. She shares these among the students in her class so that the students all get an equal number of markers and an equal number of pens. What is the maximum number of students in the class?

The lowest common multiple (LCM)

Key maths idea

A **multiple** of a number is the result of multiplying this number by another.

Example

$1 \times 2 = 2$, $2 \times 2 = 4$, $3 \times 2 = 6$, $4 \times 2 = 8$,
$5 \times 2 = 10$, $6 \times 2 = 12$.

So the multiples of 2 are: 2, 4, 6, 8, 10, 12, …

A common multiple is a shared multiple of two or more numbers.

Example

10 and 20 are common multiples of 2 and 5.

12 is a common multiple of 2, 3, 4 and 6.

The **lowest common multiple (LCM)** of a set of numbers is the smallest number that is exactly divisible by each number in the set. We can also say that it is the first or the lowest multiple that is shared by all the numbers.

Example

Multiples of 2: 2, 4, 6, 8, 10, …
Multiple of 3: 3, 6, 9, 12, 15, …

The lowest multiple that appears in both lists is 6.
So 6 is the lowest common multiple (LCM) for the two sets.

Key words
multiple
lowest common multiple (LCM)

1 List the first ten multiples of:
a 4 **b** 8 **c** 7 **d** 9 **e** 11

2 List three numbers that are common multiples of:
a 2 and 3 **b** 5 and 6 **c** 4 and 5 **d** 9 and 12

3 What is the LCM of each set of numbers?
a 4 and 6 **b** 3, 4 and 8 **c** 12 and 8 **d** 2, 5 and 10

4 Copy the Venn diagram.
a Fill in the missing multiples in each of the given sets.
b Where are the common multiples in this diagram?
c Put a circle around the lowest common multiple.

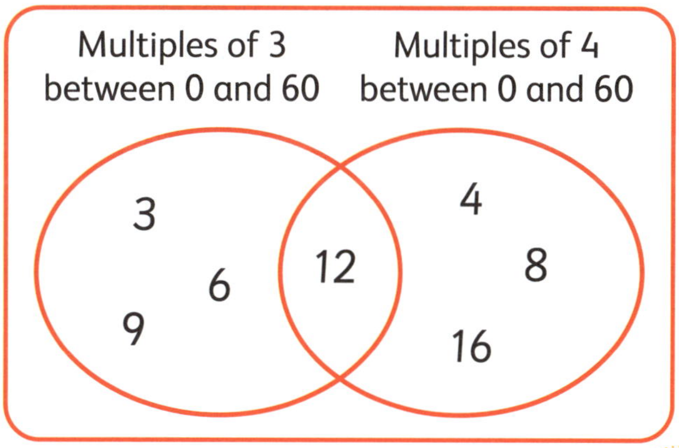

5 Find the LCM of 10 and 11.

Real-world maths: Harbour lights

At a harbour entry there are two warning lights.

1 One light flashes every 20 seconds and the other flashes every 30 seconds. If the lights flashed together at 23:01, what is the next time they will flash at the same time?

2 An engineer changes the timing on the lights so that one flashes every 25 seconds and the other flashes every 35 seconds. If they flash at the same time at 12:00 noon, when is the next time they will flash at the same time?

Maths detective: Trains and timetables

Use what you know about the HCF and the LCM of two or more numbers to solve these cases.

Case 1

Kaleem has two electric train sets, A and B. Each train is on its own circular track. He starts both trains at the same time. Train A returns to its starting point every 12 seconds. Train B returns to its starting point every 9 seconds.

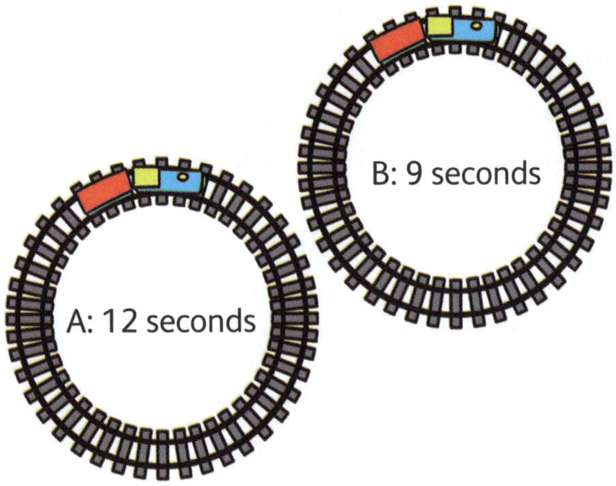

1 If the trains continue travelling, what is the least amount of time, in seconds, that both trains will arrive at the starting points at the same time?

2 It is now 11:20 a.m. Three stoplights on the train track show a green light. Stoplight A changes after 20 seconds, stoplight B changes after 45 seconds, and stoplight C changes after 30 seconds. In how many seconds will the three stoplights be on green light at the same time?

3 At what time will the three stoplights be on a green light at the same time?

Case 2

Kaleem now wants to design a bus timetable for a bus station alongside his train track. The buses begin their routes at 6:00 a.m. He bases the schedule for his two bus lines on these time intervals:

- Bus A has a long route and leaves the station every 75 minutes.
- Bus B has a short route and leaves the station every 15 minutes.

What are the next four times that Bus A and Bus B will leave the station at the same time?

Case 3

Kaleem has 48 pieces of train track and 48 joiners. He wants to group these so that there is an equal number of tracks and an equal number of joiners in each group. Kaleem wonders what is the smallest number of groups he can make.

Kaleem uses prime factors to solve this problem:

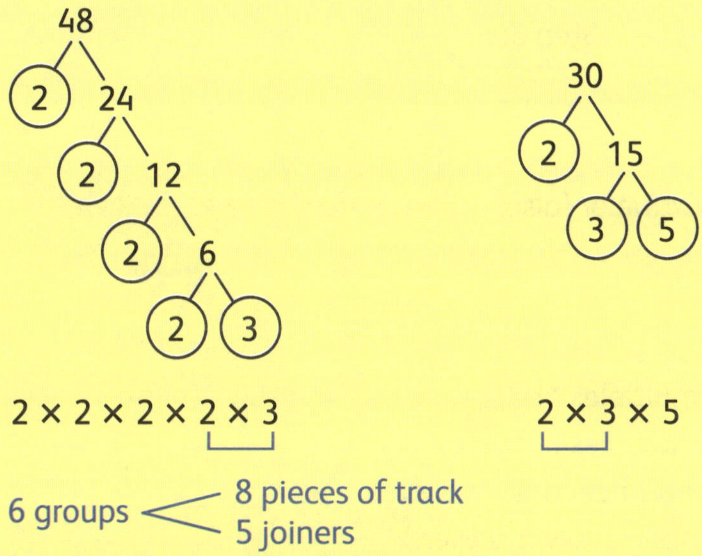

1 Discuss Kaleem's method with a partner. What is the rule for using prime factors to find the HCF of two (or more) numbers?

2 Use prime factors and the rule you have worked out to find the HCF of 60, 72 and 96.

What I have learned

Copy the flow diagram and write down the steps you need to use to write 72 as a product of its primes in exponential form.

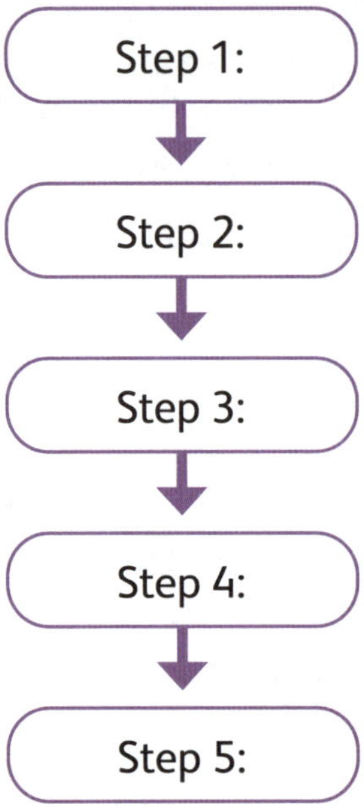

Practice questions

1 Say whether each statement is true or false.
  a 4 is a factor of 25.
  b 6 is a factor of 84.
  c 7 is a factor of 56 and 84.
  d All multiples of 6 are also multiples of 3.
  e All multiples of 4 are also multiples of 8.
  f The product of two prime numbers is always even.
  g The sum of two prime numbers is always even.

2 a Find the LCM of 2, 3, 4 and 6.
  b What is the HCF of 2, 3, 4 and 6? Why?

3 Express the following as a product of prime factors.

a 14	b 32	c 40	d 36	e 100
f 156	g 225	h 80	i 24	j 1 000

4 List:
  a all the factors of 36
  b the prime factors of 36
  c the four lowest multiples of 36
  d the factors of 36 that are square numbers.

5 Find the difference between the sum of the prime numbers less than 10 and the sum of the prime numbers between 10 and 20.

6 Kevin wants to draw a Venn diagram to show information about the factors of two numbers.

P = {factors of 32}

Q = {factors of 56}

a List the members of $P \cap Q$.

b Describe in words the region represented by $P \cap Q$.

c What is the HCF of 32 and 56?

7 Copy and complete this diagram.

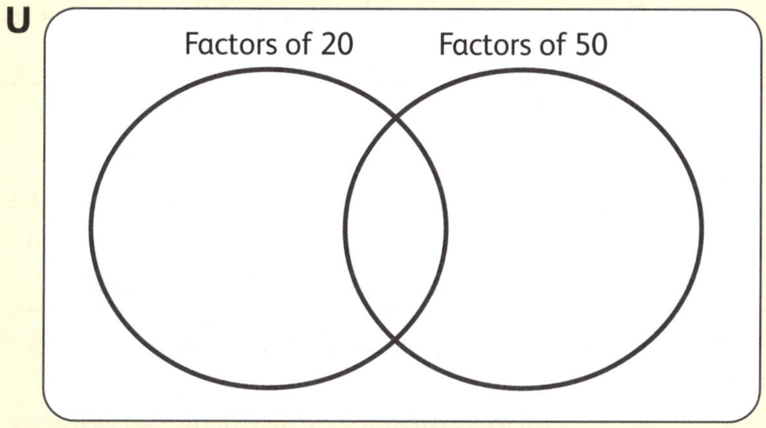

Use it to find:

a the HCF of 20 and 50.

b List the members of the universal set.

c List the common factors of 20 and 50.

8 **a** After a lesson on prime factors, Gynelle went home and said to her mom, 'I learned that 168 is the product of the prime factors $2 \times 3 \times 4 \times 7$'. Is she correct? Use words, numbers, tables or a diagram to justify your answer.

b After a lesson on writing numbers as a product of its primes in exponential form, Samuel wrote that 392 is equal to $3^2 \times 2^7$. Is he correct in his thinking? Use words, numbers, tables or a diagram to justify your answer.

9 Sage is taking two types of medicine.

- She takes one medicine every 6 hours.
- She takes the other medicine every 4 hours.
- She takes both medicines at 9:00 a.m.

At what time will Sage take both medicines together again?

10 Sandra has 24 eggs she needs to package for distribution. Help Sandra design trays for packaging the eggs.

a Show different ways the trays can be designed.

b What size tray is most suitable for packaging the eggs? Explain why this size is most suited.

8 Self check

	Check if you can do this:	If you have difficulty, try this:
1	Explain the difference between prime and composite numbers, and explain why the number 1 is neither prime nor composite.	Go back to page 122. Remind yourself that a prime number only has itself and 1 as a factor. Composites have other factors. Think about why 1 does not fall into either group.
2	Identify and list the prime factors of a number.	Write all the factors of the number and then circle all the factors that are prime numbers.
3	Write a number as a product of its prime factors.	Use a factor tree to help break down a number into its prime factors. Go back to page 124 if you need to check how a factor tree works.
4	Write a number as a product of its prime factors in exponential form.	Practise writing examples like 3×3 and $2 \times 2 \times 2$ in exponential form. Remember that multiplication is commutative so you can reorder the factors to group like ones together.
5	Explain the difference between multiples and factors.	We can write a multiplication sentence as factor x factor = multiple. We can see that a multiple is the result of multiplying two numbers together so $3 \times 2 = 6$, where 6 is a multiple of 3 and of 2. A factor is a number that divides into another number exactly, so 2 and 3 must both be factors of 6.
6	Identify the highest common factor (HCF) of two numbers.	Remind yourself that the HCF is the largest factor shared by two or more numbers. Try to list the factors for both numbers first and then find the highest factor that is shared by both numbers.
7	Identify the lowest common multiple (LCM) of two or more numbers.	Remind yourself that the LCM is the lowest multiple shared by two or more numbers. List the multiples for each number, for example, 2, 4, 6, 8 for multiples of 2. Then find the first multiple that you find that is shared by both numbers.

Chapter 9 Fractions

In this chapter you will explore this focus question:

To answer this question, you will:

- find the reciprocal of a whole number or fractional number
- use the four basic number operations: addition, subtraction, multiplication and division
- quickly and accurately carry out calculations involving fractions, including examples in real-life problem-solving contexts
- divide a fraction or mixed number by a whole number
- divide a whole number by any fractional number.

Starting point

1 **a** Estimate what fraction of each flag is green.

b Tell your partner how you worked out each answer.

c Which flag was easiest to estimate? Why?

d Which was most difficult to estimate? Why?

2 **a** Sort these numbers into the correct groups.

$\frac{7}{5}$ $5\frac{3}{10}$ $\frac{9}{9}$ $\frac{7}{10}$ $\frac{5}{7}$ $2\frac{3}{4}$ $\frac{1}{3}$

Proper fractions	Improper fractions	Mixed numbers

b Which of the fractional numbers is equivalent to 1?

c Which fraction is also an example of a unit fraction? Explain why.

d What is $\frac{7}{5}$ as a mixed number?

e What is $2\frac{3}{4}$ as an improper fraction?

3 What is the same and what is different about these fractional representations? Discuss your ideas with a partner.

a

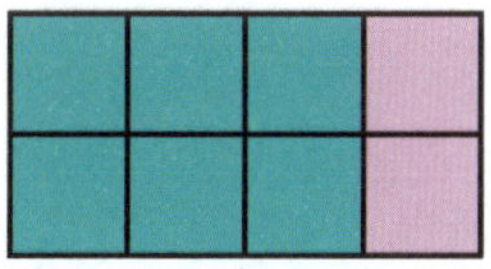

b

c

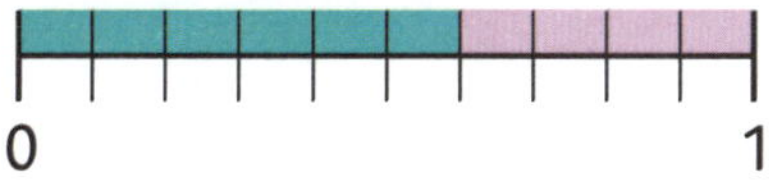

d

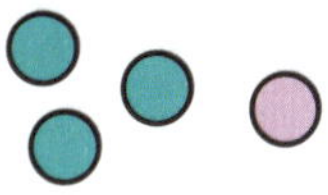

Skills to help you calculate with fractions

Key maths idea

The students in the picture are talking about some of the skills you need when you work and calculate with fractions. Why do you think they are important?

I need to find a **common denominator** to help me add $\frac{2}{3}$ and $\frac{2}{5}$.

Leon

I needed to convert $1\frac{3}{8}$ to an **improper fraction** to help me find $\frac{13}{8}$ subtract $\frac{5}{8}$.

Rhianna

I need to use what I know about **common factors** to help me simplify the fraction $\frac{12}{15}$.

Malcolm

I can use what I know about **equivalent fractions** to help find $\frac{4}{5} - \frac{3}{10}$.

Candice

I'm going to use what I know about **lowest common multiples (LCM)** to help find the **lowest common denominator (LCD)** so I can order the set of fractions $\frac{1}{2}$, $\frac{1}{3}$ and $\frac{5}{6}$.

Hattie

Jayden

I need to round **mixed numbers** to the nearest whole number to help estimate $6\frac{7}{8} + 4\frac{1}{6}$.

Key words

common denominator
improper fraction
common factors
equivalent fractions
lowest common multiple (LCM)
lowest common denominator (LCD)
mixed number

Talk about

Talk about some other examples where these skills will be useful.

1 Order each set of values in ascending order. Which skills will you use?

a $\frac{4}{5}$ $\frac{7}{10}$ $\frac{2}{3}$ **b** $3\frac{2}{3}$ $\frac{10}{3}$ $\frac{9}{3}$ **c** $1\frac{3}{5}$ $\frac{13}{10}$ $1\frac{1}{2}$

2 Which number is the odd one out? Which skills will you use?

a $\frac{4}{5}$ $\frac{16}{20}$ $\frac{13}{15}$ **b** $\frac{9}{12}$ $\frac{26}{36}$ $\frac{6}{8}$ **c** $\frac{16}{8}$ $\frac{12}{6}$ $\frac{10}{5}$ $\frac{7}{2}$

3 True or false? Correct any that are false.

a The fraction $\frac{21}{30}$ can be simplified as $\frac{7}{10}$.

b The mixed number $5\frac{20}{30}$ can be simplified as $5\frac{1}{3}$.

c $12\frac{5}{9}$ rounded to the nearest whole number is 12.

d The improper fraction $\frac{18}{4}$ can be written as the mixed number $4\frac{2}{4}$ and simplified as $5\frac{1}{2}$.

4 Complete these calculations. Which skills will you use?

a $\frac{3}{5} + \frac{3}{4}$

b $1\frac{2}{3} - \frac{5}{6}$

c $\frac{5}{8} + \frac{2}{3} + \frac{1}{6}$

d $2\frac{3}{8} + \frac{11}{8}$

e $2 - \frac{5}{9}$

f $10\frac{1}{5} - 2\frac{4}{5}$

Estimate the answers first to make sure your answer is sensible!

Real-world maths: Pizza fractions

Jayden and Femi have made pizzas. Each pizza is cut into 8 equal slices. Altogether they have 24 equal slices. How many whole pizzas do they have?

Adding fractions and mixed numbers

Key maths idea

It is easy to add fractions with the same denominator.

Example $\frac{1}{8} + \frac{3}{8} = \frac{4}{8}$

When you add fractions with different denominators, you need to find and use a common denominator. Try to find the lowest common denominator (LCD) each time. This way you can avoid tricky simplifications of the answers.

Example

$\frac{4}{5} + \frac{1}{2}$ The LCD is 10

$= \frac{8}{10} + \frac{5}{10}$

$= \frac{13}{10}$

$= 1\frac{3}{10}$

$\frac{2}{3} + \frac{1}{6}$ The LCD is 6

$= \frac{4}{6} + \frac{1}{6}$

$= \frac{5}{6}$

Explain to a partner why the LCD is not found by multiplying the denominators 3 and 6 in the second example.

When you add mixed numbers, add the whole numbers together first, then work with the fractions. Combine the whole number and fraction to find the answer.

Example

$2\frac{1}{4} + 1\frac{1}{6}$

$= 2 + 1 + \frac{1}{4} + \frac{1}{6}$ The LCD is 12

$= 3 + \frac{3}{12} + \frac{2}{12}$

$= 3 + \frac{5}{12}$

$= 3\frac{5}{12}$

1 Add each of these fractions. Estimate first whether the total will be greater or less than 1.

a $\frac{3}{8} + \frac{3}{4}$ **b** $\frac{2}{3} + \frac{5}{6}$ **c** $\frac{3}{4} + \frac{9}{16}$ **d** $\frac{5}{8} + \frac{13}{16}$

e $\frac{9}{16} + \frac{1}{2}$ **f** $\frac{3}{4} + \frac{5}{6} + \frac{1}{2}$ **g** $\frac{8}{9} + \frac{7}{9} + \frac{2}{9}$ **h** $\frac{3}{10} + \frac{2}{5} + \frac{4}{15}$

2 Mrs Henry has baked a cake.

She gives $\frac{1}{3}$ of the cake to Jayden, $\frac{2}{5}$ to Iris and $\frac{1}{4}$ to Joshua.

a Who received the largest piece of cake?

b Who received the smallest piece of cake?

c What fraction of the cake was left?

3 Find the sum of each of these mixed numbers and fractions.

a $1\frac{1}{3}$ and $\frac{2}{15}$ **b** $4\frac{5}{8}$ and $\frac{3}{4}$

c $3\frac{7}{8}$ and $\frac{5}{8}$ **d** $2\frac{2}{5}$ and $\frac{2}{3}$

4 Find the sum of these mixed numbers.

a $3\frac{4}{5}$ and $1\frac{1}{6}$ **b** $3\frac{1}{7}$ and $4\frac{2}{3}$

c $4\frac{3}{10}$ and $1\frac{5}{12}$ **d** $7\frac{1}{2}$ and $3\frac{4}{5}$

5 Anthony is baking a chocolate cake. He mixes $3\frac{1}{2}$ cups of flour, $1\frac{2}{3}$ cups of butter, $\frac{2}{3}$ of a cup of cocoa and $\frac{1}{8}$ of a cup of sugar. How many cups is this altogether?

Real-world maths: Passenger problems

A group of passengers travelled by bus. $\frac{7}{15}$ of the passengers got off the bus at Plaza A.

$\frac{1}{6}$ of the passengers got off at the post office and $\frac{1}{5}$ of the passengers get off at the supermarket.

1 What fraction of the passengers in total went to these three destinations?

2 What fraction of the passengers went to other destinations?

3 What is the minimum number of seats the bus could have?

Subtracting fractions and mixed numbers

Key maths idea

Just like adding fractions, it is easy to carry out a subtraction with fractions that share the same denominator.

Example

$\frac{3}{7} - \frac{1}{7} = \frac{2}{7}$ or $4\frac{8}{9} - 1\frac{1}{9} = 3\frac{7}{9}$

If the fractions have different denominators, we must find the LCD first.

Example

$\frac{3}{4} - \frac{2}{3}$
$= \frac{9}{12} - \frac{8}{12}$
$= \frac{1}{12}$

$1\frac{3}{4} - 1\frac{1}{6}$
$= \frac{7}{4} - \frac{7}{6}$
$= \frac{21}{12} - \frac{14}{12}$
$= \frac{7}{12}$

$3 - \frac{5}{8}$
$= \frac{24}{8} - \frac{5}{8}$
$= \frac{19}{8}$
$= 2\frac{3}{8}$

1 Calculate.

a $\frac{10}{16} - \frac{4}{16}$ **b** $\frac{5}{9} - \frac{1}{4}$ **c** $\frac{13}{16} - \frac{3}{4}$

d $\frac{3}{8} - \frac{1}{16}$ **e** $\frac{5}{7} - \frac{5}{14}$ **f** $\frac{17}{20} - \frac{3}{5}$

2 Calculate.

a $8\frac{3}{6} - 3\frac{5}{6}$ **b** $3\frac{3}{4} - 2\frac{1}{10}$ **c** $3\frac{2}{7} - 1\frac{5}{7}$

d $1 - \frac{3}{8}$ **e** $2 - \frac{11}{15}$ **f** $4 - \frac{5}{3}$

3 Andrea gives Leon $\frac{2}{5}$ of a bag of cookies.
What fraction of the bag of cookies was left?
If Andrea's bag contained 20 cookies, how many cookies did Andrea give to Leon?

4 Mikayla gave $\frac{3}{8}$ of a box of plums to her father and $\frac{1}{5}$ to her friend, Brianna. $\frac{3}{10}$ of the box were bad, so she threw them away.
Mikayla ate the rest!

a What fraction of the plums did Mikayla give away?

b What fraction did she eat?

5 A recipe for whole-wheat cookies uses $2\frac{3}{4}$ kg of sugar. The recipe for oatmeal cookies uses $1\frac{1}{2}$ kg of sugar. How much more sugar is needed for whole-wheat cookies?

6 Gabrielle swam $12\frac{1}{3}$ lengths of the pool and Jada swam $14\frac{3}{4}$ lengths of the pool. How much further did Jada swim?

Multiplying fractions

Key maths idea

Look at the dimensions of the different sets of tiles that Hattie wants to use in her house.

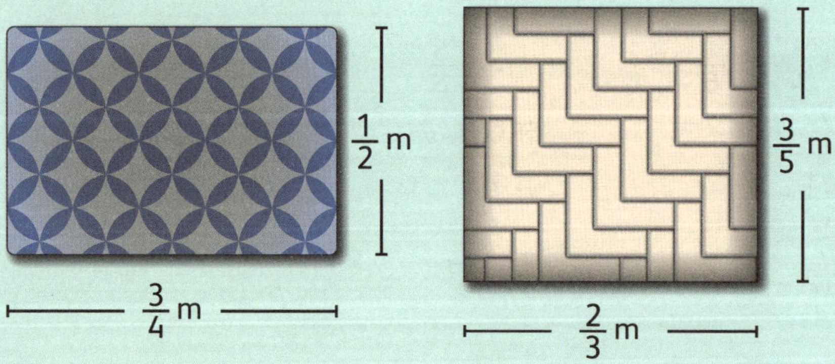

Which do you think has the largest area? How could we find out?

Let's look at methods to multiply fractions.

$$\frac{3}{4} \times \frac{1}{2}$$
$$= \frac{(3 \times 1)}{(4 \times 2)}$$ Multiply the **numerators** together
Multiply the **denominators** together
$$= \frac{3}{8}$$

$$\frac{2}{3} \times \frac{3}{5}$$
$$= \frac{(2 \times 3)}{(3 \times 5)}$$
$$= \frac{6}{15}$$
$$= \frac{2}{5}$$

Key words
numerator
denominator

How can you explain why $\frac{2}{5}$ m² is a larger area than $\frac{3}{8}$ m²?

When multiplying mixed numbers, it is easier to convert them to improper fractions first:

$1\frac{3}{4} \times 2\frac{1}{5}$

$= \frac{7}{4} \times \frac{11}{5}$ Convert the mixed numbers into improper fractions

$= \frac{(7 \times 11)}{(4 \times 5)}$ Multiply the numerators together
Multiply the denominators together

$= \frac{77}{20}$

$= 3\frac{17}{20}$ Convert back to a mixed number

1 Calculate.

a $\frac{1}{2} \times \frac{3}{4}$ **b** $\frac{3}{4} \times \frac{5}{6}$ **c** $\frac{7}{8} \times \frac{7}{8}$ **d** $\frac{3}{5} \times \frac{2}{3}$ **e** $\frac{2}{3} \times \frac{5}{6}$

2 Find the product.

a $5\frac{2}{3} \times \frac{3}{4}$ **b** $2\frac{1}{3} \times 4\frac{1}{5}$ **c** $1\frac{1}{3} \times \frac{3}{4}$ **d** $3 \times \frac{5}{6}$ **e** $7 \times \frac{8}{9}$

3 Grandma has baked a pie. She gives $\frac{3}{4}$ of the pie to her friend, Annie.
Annie gives her daughter $\frac{1}{5}$ of the pie.
What fraction of the whole pie does Annie's daughter get?

Sketch diagrams to help you! Remember that $\frac{1}{5}$ of an amount is the same as multiplying the amount by $\frac{1}{5}$.

4 In a survey $\frac{3}{8}$ of the people interviewed preferred to go shopping than to the beach.
Of the people who preferred to go shopping, $\frac{1}{2}$ liked to shop for clothes. What fraction of the whole group liked to shop for clothes?

5 A pool has a length of $10\frac{1}{2}$ metres and a width of $6\frac{3}{4}$ metres. What is the area of the pool?

Making calculating easier

Key maths idea

When multiplying fractions, it is useful to see if we can make the calculation easier before we complete it. Let's look at the following example:

$$\frac{3}{\cancel{4}_{2}} \times \frac{\cancel{2}^{1}}{5}$$

Note how the highest common factor (HCF) 2 has been used to simplify the denominator in $\frac{3}{4}$ and the numerator in $\frac{2}{5}$.

What do you notice about the way the numerators and denominators are simplified in these two multiplications?

$$\frac{3}{4} \times \frac{\cancel{2}^{1}}{\cancel{10}_{2}} \qquad \frac{\cancel{3}^{1}}{\cancel{4}_{2}} \times \frac{\cancel{2}^{1}}{\cancel{9}_{3}}$$

Note that you can only **simplify** a numerator and a denominator, otherwise you change the proportions you are multiplying.

Key word
simplify

1 Simplify these fractions. Remember to find the highest common factor (HCF) each time:

a $\frac{4}{8}$ **b** $\frac{6}{9}$ **c** $\frac{8}{12}$ **d** $\frac{4}{10}$ **e** $\frac{15}{15}$

2 How will you simplify these multiplications? Solve them.

a $\frac{4}{8} \times \frac{5}{6}$ **b** $\frac{5}{6} \times \frac{3}{10}$ **c** $\frac{8}{12} \times \frac{6}{7}$ **d** $\frac{9}{10} \times \frac{5}{9}$ **e** $\frac{5}{12} \times \frac{6}{9}$

3 Look at these multiplications. How can you simplify these calculations first?

a $1\frac{4}{5} \times \frac{2}{3}$ **b** $2\frac{3}{4} \times 5$ **c** $3\frac{5}{8} \times \frac{4}{7}$

4 Malcolm is multiplying pairs of fractions. He uses the highest common factor (HCF) 3 each time to simplify the calculation.
What could the missing numbers in the calculations be? Try to make them different.

$\frac{\square}{7} \times \frac{5}{\square}$ $\frac{\square}{\square} \times \frac{4}{5}$

Dividing a fraction by a whole number

Key maths idea

These students are discussing how they will solve these divisions:

$\frac{3}{4} \div 2$ $\frac{3}{4} \div 3$ $\frac{3}{4} \div 4$

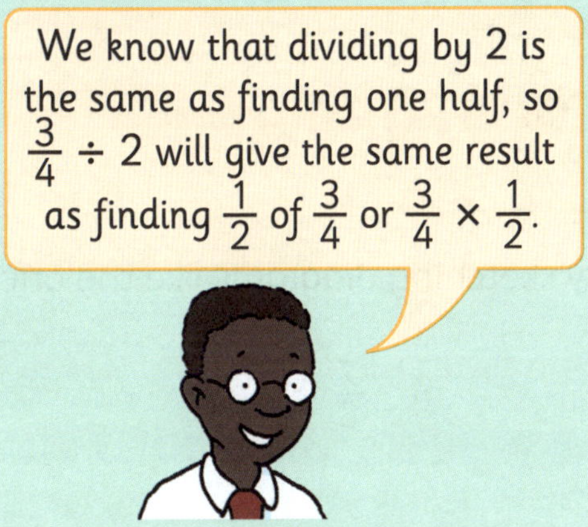

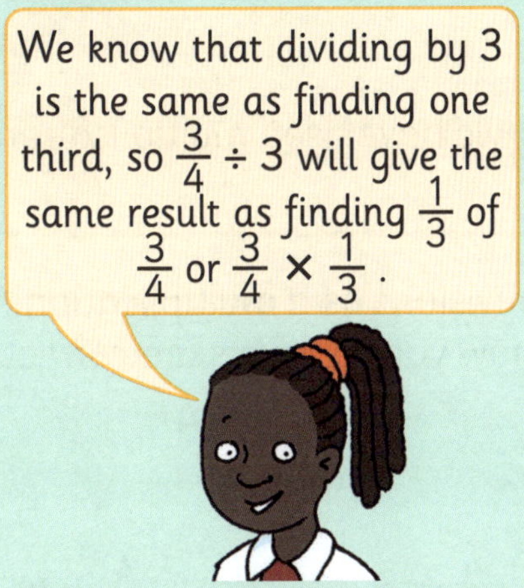

Explain to your partner how the diagrams show that $\frac{3}{4} \div 2 = \frac{3}{4} \times \frac{1}{2}$

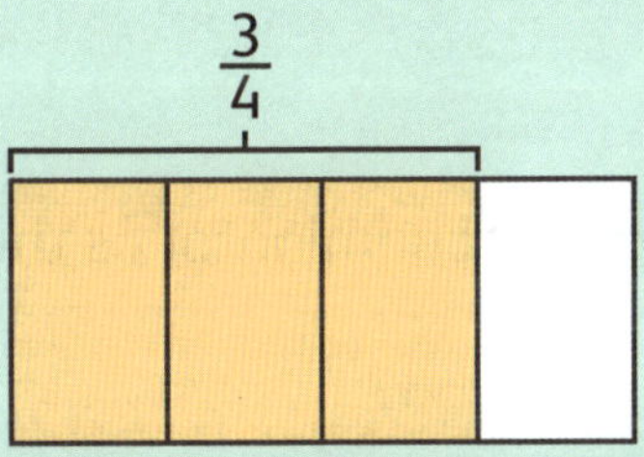

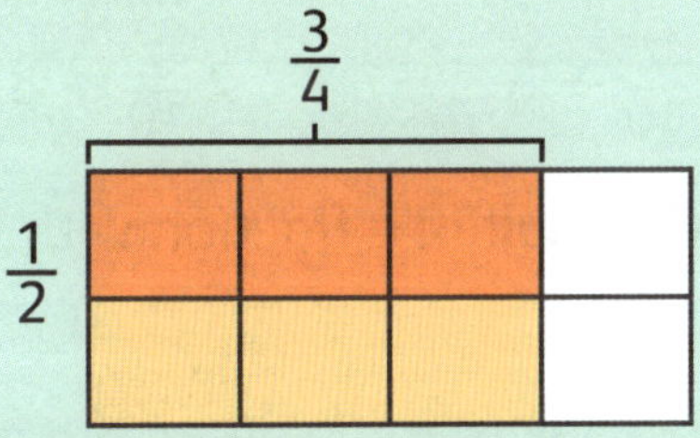

What can you say about dividing by 4?

Use reasoning to decide which division will have the largest answer.

How can you explain why the answer to $\frac{3}{4} \div 3$ must be $\frac{1}{4}$?

Now let's calculate:

$\frac{3}{4} \div 2$	$\frac{3}{4} \div 3$	$\frac{3}{4} \div 4$
$\frac{3}{4} \times \frac{1}{2}$	$= \frac{3}{4} \times \frac{1}{3}$	$= \frac{3}{4} \times \frac{1}{4}$
$= \frac{(3 \times 1)}{(4 \times 2)}$	$= \frac{(3 \times 1)}{(4 \times 3)}$	$= \frac{(3 \times 1)}{(4 \times 4)}$
$= \frac{3}{8}$	$= \frac{3}{12}$	$= \frac{3}{16}$

Key word

inverse

We can also divide mixed numbers by whole numbers in a similar way.

Example

$2\frac{4}{5} \div 3$

$= 2\frac{4}{5} \times \frac{1}{3}$ Dividing by 3 is the same as multiplying by $\frac{1}{3}$

$= \frac{14}{5} \times \frac{1}{3}$ Convert the mixed number to an improper fraction

$= \frac{(14 \times 1)}{(5 \times 3)}$

$= \frac{14}{15}$

In all of these examples, we have used multiplication – the **inverse** of division.

1 Rewrite each of these divisions as a multiplication. Try sketching diagrams like the ones in key maths idea to show your answers are correct.

a $\frac{2}{3} \div 5$ **b** $\frac{9}{10} \div 4$ **c** $\frac{7}{8} \div 3$

2 Calculate.

a $\frac{4}{5} \div 2$ $\frac{4}{5} \div 3$ $\frac{4}{5} \div 4$

b $\frac{5}{8} \div 5$ $\frac{5}{8} \div 6$ $\frac{5}{8} \div 7$

c $2\frac{7}{10} \div 4$ $2\frac{7}{10} \div 6$ $2\frac{7}{10} \div 8$

3 Which of these divisions will result in a fraction with the same denominator as the one being divided?

$\frac{6}{10} \div 3$ $\frac{4}{5} \div 5$ $\frac{8}{9} \div 3$ $\frac{8}{9} \div 4$ $2\frac{3}{5} \div 3$

What will you sketch or write to help explain your decisions? What do you notice? Now give an example of your own.

4 Solve these problems.

a $3\frac{1}{2}$ cakes are shared equally among 6 children. What fraction of a piece does each child get?

b $\frac{3}{4}$ of a bread loaf must be shared among 5 people. What fraction of the loaf will each person get?

c Gabrielle has $\frac{7}{8}$ kg of coconut drops. She divides them equally into 6 packets. What fraction of a kilogram of coconut drops is in each packet?

d Daniel has $2\frac{1}{2}$ litres of pineapple juice. He divides the juice equally among 8 glasses. What fraction of a litre of juice is in each glass?

Reciprocals

Key word
reciprocal

Key maths idea

Look at these multiplications. What do you notice about each of the answers?

$\frac{1}{3} \times \frac{3}{1}$ $\quad \frac{4}{1} \times \frac{1}{4}$ $\quad \frac{1}{5} \times \frac{5}{1}$ $\quad \frac{2}{5} \times \frac{5}{2}$ $\quad 5 \times \frac{1}{5}$

We say that the **reciprocal** of a number is the number you must multiply it by to get the product 1. The reciprocal of any whole number x is $\frac{1}{x}$, which is the same as 1 divided by x.

We can also write any whole number as a fraction with denominator 1.

So 3 can be written as $\frac{3}{1}$.

This will help you find the reciprocal of a fraction.

Let's use two diagrams to show this.

$\frac{1}{3} \times \frac{3}{1} = \frac{3}{3}$ or 1

Remember that $\frac{3}{1}$ is the whole number 3, so $\frac{1}{3} \times 3$ or $3 \times \frac{1}{3} = 1$

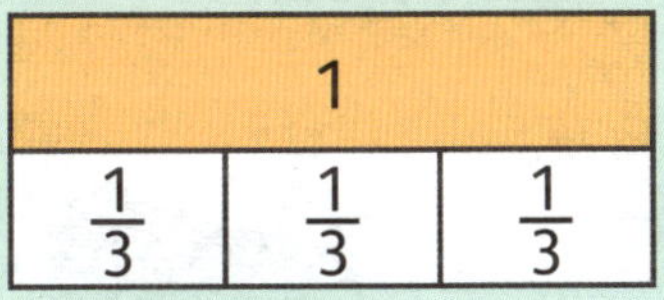

$\frac{4}{1} \times \frac{1}{4} = \frac{4}{4}$ or 1

Remember that $\frac{4}{1}$ is the whole number 4, and so $\frac{1}{4}$ of 4 or $4 \times \frac{1}{4} = 1$

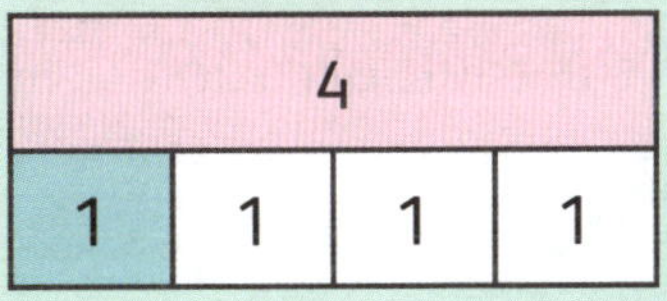

$\frac{2}{5} \times \frac{5}{2}$ is a little different because neither fraction is equivalent to a whole number, but the product is still 1 because

$\frac{(2 \times 5)}{(5 \times 2)} = \frac{10}{10} = 1$

The reciprocal of $\frac{2}{5}$ is $\frac{5}{2}$ and the reciprocal of $\frac{5}{2}$ is $\frac{2}{5}$.

1 Write down the reciprocal of each whole number.

a 6 **b** 10 **c** 24

d 17 **e** 9 **f** 12

2 Write the reciprocals of these fractions.

a $\frac{1}{2}$ **b** $\frac{1}{6}$ **c** $\frac{2}{7}$

d $\frac{5}{8}$ **e** $\frac{11}{12}$ **f** $\frac{17}{24}$

3 Write each of these mixed numbers as improper fractions first, then find their reciprocals.

a $1\frac{1}{2}$ **b** $4\frac{4}{5}$ **c** $9\frac{9}{10}$ **d** $11\frac{2}{3}$ **e** $6\frac{7}{9}$ **f** $5\frac{2}{7}$

Using reciprocals for division

Key maths idea

We can use what we know about reciprocals to help with division.

Remember that dividing by a whole number, for example, 3, is the same as finding one third or multiplying by $\frac{1}{3}$.

We can write:

$\frac{5}{8} \div 3 = \frac{5}{8} \times \frac{1}{3}$

Note how the reciprocal of 3 is used and $\frac{5}{8}$ is now multiplied by $\frac{1}{3}$.

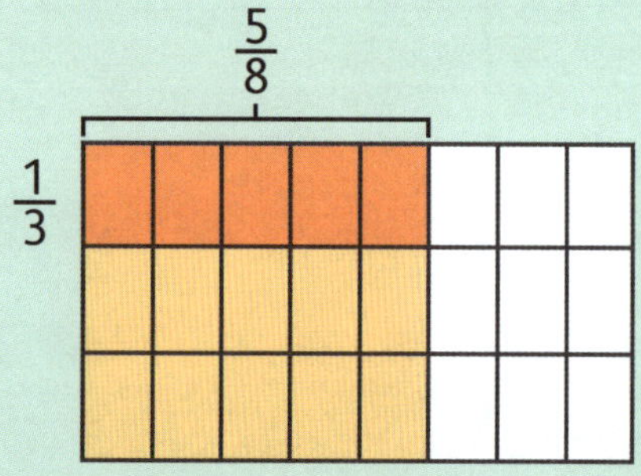

$\frac{5}{8} \times \frac{1}{3} = \frac{(5 \times 1)}{(8 \times 3)}$

$= \frac{5}{24}$

Explain why $\frac{5}{24}$ is in its simplest form.

We can also use reciprocals to help to divide whole numbers by a fraction.

So $5 \div \frac{2}{3}$ can be re-written as the multiplication $5 \times \frac{3}{2}$, because $\frac{3}{2}$ is the reciprocal of $\frac{2}{3}$.

$5 \times \frac{3}{2} = \frac{(5 \times 3)}{(1 \times 2)}$

$= \frac{15}{2}$

$= 7\frac{1}{2}$

Talk about

Try solving this division:

$2\frac{5}{8} \div 3$

What is the same and what is different about it? Remember to convert the mixed number to an improper fraction first.

1 Solve each division calculation. Remember to think about using reciprocals.

a $\frac{3}{4} \div 2$ **b** $\frac{7}{10} \div 3$ **c** $\frac{2}{3} \div 5$

d $\frac{5}{6} \div 4$ **e** $1\frac{2}{3} \div 5$ **f** $2\frac{5}{6} \div 4$

Choose one of these divisions and sketch a diagram (like the one in the key maths idea) to show you are correct.

2 Solve these divisions.

a $\frac{3}{4} \div 3$ **b** $\frac{5}{8} \div 8$ **c** $\frac{2}{5} \div 2$

d $\frac{9}{10} \div 9$ **e** $\frac{5}{12} \div 5$ **f** $3\frac{1}{2} \div 7$

3 **a** What do you notice about the answers in Exercise 2? Explain why.

b Make up a similar example of your own.

4 Divide.

a $2 \div \frac{1}{3}$ b $5 \div \frac{1}{10}$ c $8 \div \frac{1}{4}$

d $12 \div \frac{5}{9}$ e $4 \div \frac{9}{10}$ f $10 \div \frac{2}{3}$

5 Copy and complete. How will you find the missing numbers?

a $\square \div 6 = \frac{1}{2}$ b $\square \div 9 = \frac{1}{3}$ c $\square \div 8 = \frac{3}{4}$

d $\square \div 12 = \frac{2}{3}$ e $\square \div 3 = \frac{2}{3}$ f $\square \div 16 = \frac{3}{8}$

6 Leon says that when you divide a whole number by a proper fraction, the result is larger than the original whole number.
Is this always true, sometimes true or never true?
Use your own examples to investigate.

7 A soup recipe for 6 people uses 2 litres of water.
How many litres of water are used in a soup recipe for:

a 2 people b 3 people

c 4 people d 5 people?

Maths detective: Home maintenance and décor

Use your knowledge of dividing and multiplying fractions and using reciprocals to solve these cases.

Case 1

For an event, Peter has to make some scatter cushion covers. He designs each cover using three different colour materials. He needs one-and-a-half metres of pink fabric, one-and-a-quarter metres of blue fabric and one eighth metre of white fabric.

1 Peter has to make one dozen covers. How much material of each colour does he need? Give your answer in metres.

a Pink fabric

b Blue fabric

c White fabric

2 Peter purchased eleven metres of pink fabric at $516.00 per metre, ten metres of blue fabric at $608.10 per metre and one metre of white fabric at $479.50 per metre.

a How many covers can Peter make?

b Which type of fabric does Peter use up first? Explain your answer.

c What was the total cost of the fabrics Peter purchased?

Case 2

Peter also has a piece of board that is 2 metres long, which is to be cut into segments that are $\frac{5}{6}$ metres long. How many segments can he make? What fractional part will be left?

Case 3

Peter then needs to repaint some bookcases. He has $5\frac{1}{4}$ litres of paint. All the bookcases are the same size and each one requires $\frac{3}{4}$ litre of paint.

1 How many bookcases will he be able to repaint with the paint he has?

2 He is paid $3 725.00 to repaint each bookcase. How much pay does he receive for the painting?

Case 4

1 Peter has $1 200 and spends $\frac{1}{5}$ of this amount on buying his lunch. How much money does he have left?

2 Peter's assistant buys some juice. He spends the same amount of money on the juice as Peter spent on his lunch. The assistant now has only $\frac{1}{3}$ of the amount that he had before he bought the juice. How much money did he have before he bought the juice?

Real-world maths: Dividing a pie

Candice bakes 4 pies. She puts $\frac{2}{5}$ of a pie onto each plate until she has used up all the pies. How many plates does she use altogether?

What I have learned

Read the information in the table. Then rewrite the sentences that have words missing. Fill in the missing words in each sentence.

Example	Method
$\frac{4}{5} + \frac{1}{2}$	Is one denominator a multiple of the other? No.
$= \frac{8}{10} + \frac{5}{10}$	Express both fractions using the lowest common denominator (LCD). Fifths and halves can become tenths. The LCD is _____ .
$= \frac{13}{10}$	Combine the tenths: 8 + 5 = 13
$= 1\frac{3}{10}$	Thirteen tenths can be regrouped as a whole with three tenths remaining.
$2\frac{1}{4} - 1\frac{5}{6}$	Rewrite mixed numbers into _____ .
$= \frac{9}{4} - \frac{11}{6}$	Is one denominator a multiple of the other? No. Express both fractions using the _____ .
$= \frac{27}{12} - \frac{22}{12}$	Fourths and sixths can become twelfths. The LCD is 12.
$= \frac{5}{12}$	Subtract the twelfths: 27 – 22 = 5
$3\frac{3}{4} \times 1\frac{3}{7}$	Rewrite _____ into improper fractions.
$= \frac{15}{4} \times \frac{10}{7}$	Divide the numerator and denominator by the _____ . $4 \div 2 = 2$ and $10 \div 2 = 5$
$= \frac{15}{2} \times \frac{5}{7}$	Multiply the numerators: $15 \times 5 = 75$ Multiply the denominators $2 \times 7 = 14$
$= \frac{75}{14}$	
$= 5\frac{5}{14}$	Rewrite the improper fraction as a _____ .
$1\frac{3}{4} \div 3$	Rewrite the _____ as an _____ .
$= \frac{7}{4} \div 3$	
$= \frac{7}{4} \times \frac{1}{3}$	Multiply by the _____ of the divisor.
$= \frac{7 \times 1}{4 \times 3}$	Multiply the _____ : $7 \times 1 = 7$ Multiply the _____ : $4 \times 3 = 12$
$= \frac{7}{12}$	

Use each method in the table to do these calculations.

1 $\frac{5}{9} + \frac{3}{4}$

2 $3\frac{1}{6} - 1\frac{2}{9}$

3 $2\frac{2}{3} \times 1\frac{4}{5}$

4 $3\frac{3}{5} \div 4$

Practice questions

1 Order each set of fractions from largest to smallest.

a $\frac{2}{5}$ $\frac{1}{3}$ $\frac{3}{4}$ $\frac{4}{5}$ $\frac{1}{2}$

b $\frac{1}{12}$ $\frac{3}{5}$ $\frac{1}{6}$ $\frac{2}{3}$ $\frac{1}{10}$

c $\frac{5}{6}$ $\frac{9}{10}$ $\frac{11}{12}$ $\frac{4}{5}$ $\frac{2}{3}$

2 Bailey has a bag of guavas. $\frac{2}{3}$ of the guavas are rotten. If 36 guavas are rotten, how many guavas does Bailey have altogether?

3 A recipe for a large batch of cookies for a cake sale needs $2\frac{1}{3}$ kilograms of white sugar and $1\frac{3}{8}$ kilograms of brown sugar. How much sugar is needed?

4 On a flight, $\frac{2}{3}$ of the passengers are from Cuba. If there are 192 passengers altogether, how many are from Cuba?

5 Add these fractions. Give your answers as proper fractions or mixed numbers.

a $\frac{1}{4}+\frac{1}{5}$ b $\frac{1}{3}+\frac{11}{12}$ c $\frac{1}{7}+\frac{2}{3}$

6 Subtract.

a $1\frac{2}{3}-\frac{3}{4}$ b $5\frac{1}{6}-2\frac{7}{8}$ c $3\frac{7}{8}-1\frac{3}{4}$

7 Multiply.

a $2\frac{5}{8}\times 1\frac{3}{7}$ b $1\frac{1}{9}\times 7\frac{1}{5}$ c $6\frac{1}{2}\times 1\frac{1}{5}$

8 Divide.

a $\frac{1}{2}\div 3$ b $\frac{2}{3}\div 5$ c $\frac{3}{5}\div 3$

d $6\div\frac{5}{8}$ e $4\div\frac{2}{3}$ f $5\div\frac{7}{10}$

9 Self check

Check if you can do this:	If you have difficulty, try this:
1 Find the lowest common denominator (LCD) for 2 or more fractions.	Remind yourself that to find the LCD, you need to find the lowest common multiple. You could check the multiples for each number on a multiplication grid and find the first multiple that appears in each multiplication table.
2 Simplify a fraction by finding the highest common factor (HCF) to help.	Remind yourself that to find the HCF, you need to think about the factors for both numbers and then find the highest one that is common to both.
3 Convert a mixed number to an improper fraction, and vice versa.	Remind yourself that an improper fraction is greater than or equal to 1. Try splitting up an improper fraction into parts that equal 1, for example, $\frac{7}{2}$ as $\frac{2}{2} + \frac{2}{2} + \frac{2}{2} + \frac{1}{2} = 3\frac{1}{2}$. You can do the opposite to convert a mixed number to an improper fraction, for example, $2\frac{3}{4} = \frac{4}{4} + \frac{4}{4} + \frac{3}{4} = \frac{11}{4}$. You can also multiply the whole number by 2 to get $\frac{8}{4}$ and add on the extra $\frac{3}{4}$.
4 Add and subtract fractions quickly and accurately.	Go back to pages 135–137. Make sure you know how to use the LCD to help convert fractions to fractions with the same denominator.
5 Multiply fractions quickly and accurately.	Go back to pages 138–139. Remind yourself that you multiply the numerators together and then the denominators.
6 Identify the reciprocal of a whole number and a fractional number.	Go back to page 143. Remind yourself that the reciprocal of a number is the number you must multiply it by to get the product 1.
7 Use reciprocals for division with fractions and mixed numbers.	Go back to page 144. Make sure you know how to find a reciprocal of a whole number or a fraction.
8 Solve problems involving all four operations.	Remember that you can draw diagrams to help you make sense of problems with fractions or mixed numbers.
9 Make calculating easier by simplifying fractions.	Use what you know about highest common factors (HCF) to help simplify a fraction or mixed number before calculating.

Chapter 10 Decimals

In this chapter you will explore this focus question:

To answer this question, you will:

- divide a decimal fraction by a whole number
- divide a decimal fraction by a power of ten
- divide a decimal fraction by another decimal fraction with answers written to two or three decimal places
- use the four operations to calculate with decimal fractions
- use a variety of strategies to solve problems involving decimals and justify your decisions
- use mathematical tools to solve problems involving decimals and understand how to use the displayed information correctly in the context of measurement.

Starting point

1 Four students compared how far they walked. Here are their smart watch readings.

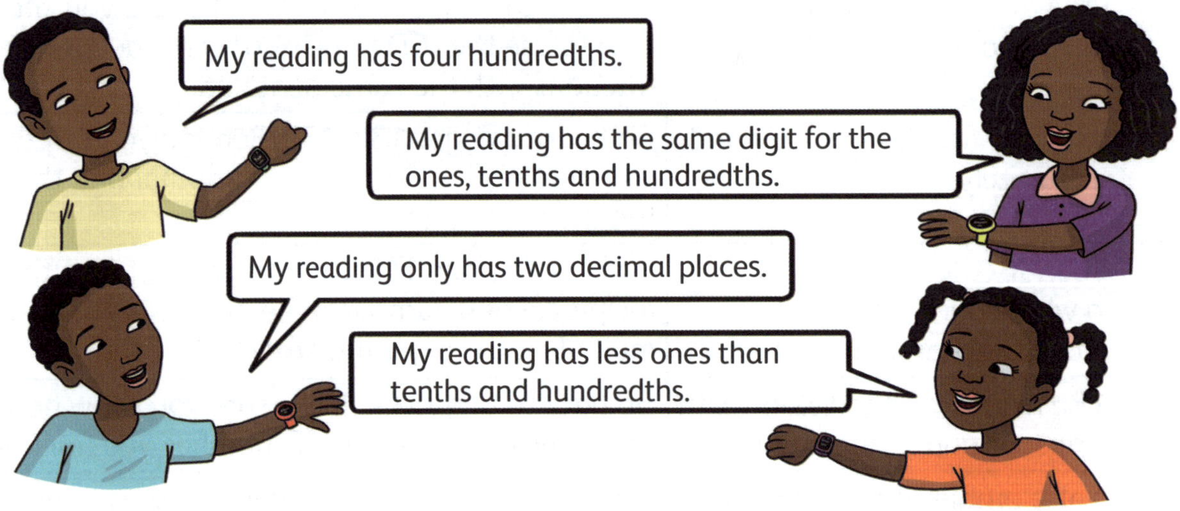

3.454 | 4.445 | 4.345 | 3.54

Match each number to the correct student.

2 Work with a partner and write out these numbers in expanded notation, for example, 4.91 = 4 + 0.9 + 0.01.

23.7 2.37 9.08 0.98

3 Copy and complete.

a 3.4 × 10 = □ **b** 0.1 × 100 = □ **c** 0.06 × 1 000 = □

d 3.01 × □ 3100 **e** 9.45 × 100 = □ **f** 9.321 × 100 = □

g 4.32 × 1 000 = □ **h** 1.2 × 1 000 = □ **i** 0.8 × 1 000 = □

4 A carpenter needs 100 m of timber at $962.68 per metre. How much will it cost him?

Remember:

- When a number is multiplied by 10, the value of each part of the number is made 10 times as large. So the digits all move 1 place to the left.
- When a number is multiplied by 100, the value of each part of the number is made 100 times as large. So the digits all move 2 places to the left.

Skills to help work and calculate with decimals

Key maths idea

When working and calculating with decimals, there are many skills we need to apply.
The children in the picture are talking about some of these skills.
Why do you think they are important?

I can use what I know about **place value** to help multiply and **divide** whole numbers and decimal numbers by 10, 100, 1 000, and so on.

Hattie

I need to **convert** the fraction $\frac{3}{4}$ to a **decimal fraction** to help me find the total mass of 2.5 kilograms and $\frac{3}{4}$ kilograms.

Candice

I can use what I know about the relationship between tenths and hundredths to help me add 3.5 and 2.34.

Rhianna

I'm going to use place value to help **compare** and **order** decimal numbers.

Leon

I will apply number facts to help me calculate with decimals. I can use the fact 6×4 to work out 0.6×4, because 6 tenths $\times 4 = 24$ tenths or 2.4.

Jayden

Malcolm

I need to **round** decimal numbers to help me estimate answers.

Key words

place value
divide
convert
decimal fraction
compare
order
round
equivalent

Talk about

Talk about some other examples where these skills will be useful.

There are many fraction and decimal equivalents that you should already know. Here are a few:

$\frac{1}{2} = 0.5$ $\frac{1}{4} = 0.25$ $\frac{3}{4} = 0.75$

$\frac{1}{5} = 0.2$ $\frac{2}{5} = 0.4$ $\frac{1}{10} = 0.1$

Remember that any fraction can be written as a division, so $\frac{5}{8}$ can be written as the division 5 ÷ 8. Therefore, any equivalences that you do not know can be found by solving the division on a calculator or as short division:

$$\begin{array}{r|l} & 0.\,6\,2\,5 \\ 8 & 5.{}^{5}0{}^{2}0{}^{4}0 \end{array}$$

Fractions like $\frac{7}{25}$ can easily be converted without a calculator because the **equivalent** fraction $\frac{28}{100}$ can be found first.

$$\frac{7}{25} \xrightarrow{\times 4} \frac{28}{100}$$

$$\frac{7}{25} = \frac{28}{100}$$

(× 4 on the numerator and × 4 on the denominator)

Then we can use understanding of place value to divide 28 by 100.

T	O	t	h
2	8 •		
	0 •	2	8

So $\frac{7}{25}$ is equivalent to the decimal fraction 0.28.

1 Order each set of values in ascending order. Which skills will you use?

a 1.25 3.67 0.35 4.328

b 1.5 0.6 0.62 1.55

c 0.4 $\frac{7}{10}$ $\frac{2}{5}$ 0.5

d 3.2 20.5 12.45 5.36

> Remember to line up the decimal points when comparing decimals so the place values are correct.

2 Which is the odd one out? Which skills will you use?

a $\frac{4}{5}$ 0.8 $\frac{80}{100}$ 0.08

b 0.5 0.50 $\frac{1}{5}$ $\frac{1}{2}$

c 0.07 $\frac{7}{10}$ 7 ÷ 10 0.7

> Always look to see if you can use an equivalent fraction with a denominator that is a power of 10.

3 True or false? Correct all the statements that are false.

a $3.45 > 3.5$ **b** $1\frac{3}{4} < 1.8$

c 13.6 rounded to the nearest whole number is 13

d 1.43 rounded to the nearest tenth is 1.4

e 4.306 rounded to the nearest hundredth is 4.3

4 Complete these calculations. Which skills will you use?

a 0.6 + 0.4 **b** 1.5 – 0.5

c 0.4 × 3 **d** 1.5 ÷ 3

e 0.45 + 0.55 **f** 2.5 – 1.55

5 Anthony says that $\frac{83}{100}$ is less than 0.85.
Is Anthony correct? Explain your answer.

Real-world maths: A javelin competition

In a javelin competition, each person gets six throws. Only the three best results are counted.

These are Orrin's six throws in a competition:

84.58 m	84.39 m
85.28 m	85.38 m
84.38 m	82.67 m

1 Arrange the distances thrown in ascending order.

2 Which three results are counted in the competition?

3 Use the top three throws to calculate the mean score for this event.

4 The Olympic record for men's javelin is 90.57 m. How much further would Orrin's best throw have to be to beat this record by $\frac{1}{100}$th of a metre?

5 Convert the lengths of the six throws to centimetres (cm).

Adding decimals

Key maths idea

When we carry out addition with decimals we must add 'like' place value parts, so ones with ones, tenths with tenths, and so on.

It is important to line up the decimal points correctly. This way we know we are adding 'like' parts.

Sometimes we have to add decimals with a different number of decimal places. We can use knowledge of place value equivalence to write these with the same number of decimal places.

Always remember to make an estimate first.

Example

Calculate 11.055 + 2.1 + 9 + 0.6

Estimate: 11 + 2 + 9 + 1 = 23

Line up decimal points

```
  11.055
   2.100
   9.000
 + 0.600
 -------
  22.755
```

Remember that 9 = 9.0 = 9.000

Note how zeros have been used as place holders, so all the decimals are written to the same number of decimal places

We can also use the facts we know to help solve additions mentally:

- What is the connection between 7 ones + 2 ones and 7 tenths + 2 tenths?

So what is 0.7 + 0.2?

- What is the connection between 32 ones + 24 ones and 32 hundredths + 24 hundredths?

So what is 0.32 + 0.24?

1 Add each set of decimals using mental methods. Write the answers only.

a 0.7 + 0.2 **b** 0.8 + 0.2
c 0.8 + 0.3 **d** 2.7 + 1.2
e 6.8 + 3.2 **f** 0.23 + 0.65
g 0.65 + 0.26 **h** 0.35 + 0.65
i 0.37 + 0.65 **j** 0.58 + 0.58

2 Use the column method to complete these additions. Make estimates first. Check your answer against your estimate and with a calculator.

a 1.49 + 0.99 **b** 2.34 + 0.07
c 5.32 + 5.48 **d** 4.09 + 2.8
e 6.3 + 5.25 **f** 1.235 + 2.143
g 2.763 + 5.015 **h** 2.76 + 5.015

3 Estimate first, then calculate.

a 5.99 + 15.32 + 231.09
b 214.6 + 87.99 + 234.9
c 612.5 + 132.09 + 99.5

4 At the hardware shop, many things are measured in metres and centimetres.

- **a** What is the combined length of the two shortest wooden planks?
- **b** What is the total length of the three rope coils?
- **c** If you need 18 m of rope, will there be enough?
- **d** Add up the lengths of the two shortest poles.
- **e** How much wire is there altogether?

5 A plane flies 123.62 km on Friday, 132.085 km on Saturday and $115\frac{3}{4}$ km on Sunday. How far did the plane fly in total?

Real-world maths: Route distances

David uses a mobile phone app to map the route he followed on a day hike.

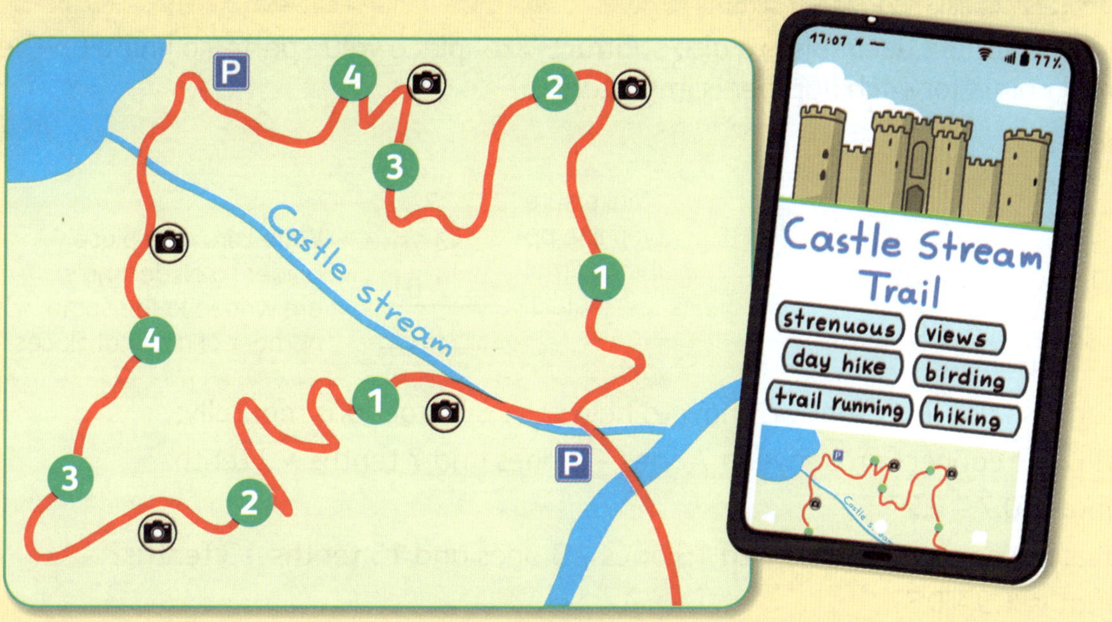

David made the following note of the rest stops along the way:

	Rest stop 1 (RS1)	Rest stop 2 (RS2)	Rest stop 3 (RS3)	Rest stop 4 (RS4)	End
Distance	3.6 km from start	4.35 km from RS1	5.128 km from RS2	$1\frac{3}{4}$ km from RS3	2.3 km from RS4

1 How far had David hiked by the time he reached rest stop 3?

2 How far was David's hike in total?

Candice followed the other route on her hike that day. She stopped at the following rest points:

	Rest stop 1 (RS1)	Rest stop 2 (RS2)	Rest stop 3 (RS3)	Rest stop 4 (RS4)	End
Distance	4.3 km from start	6.32 km from RS1	4.325 km from RS2	1.7 km from RS3	2.04 km from RS4

3 Who hiked the shorter route?

4 How much shorter was the route?

Subtracting decimals

Key maths idea

When we subtract decimals, we also subtract ‘like’ place value parts, so lining up the decimal points for each number is important.

Example

54.8 – 7.01

First, make an estimate:

55 – 7 = 48

Make sure you line up the decimal points.

```
  54.70
–  7.01
  47.69
```

Write zero as a place holder so all decimals are written to the same number of decimal places.

We can also use facts that we know to help solve subtractions mentally:
What is the connection between 7 ones – 2 ones and 7 tenths – 2 tenths?
So what is 0.7 – 0.2?
What is the connection between 15 ones – 3 ones and 15 tenths – 3 tenths?
So what is 1.5 – 0.3?

1 Try to complete these subtractions mentally.

a 0.8 – 0.3 **b** 0.9 – 0.6
c 0.8 – 0.1 **d** 1.7 – 0.6
e 2.6 – 0.9 **f** 0.08 – 0.03
g 0.09 – 0.06 **h** 0.08 – 0.01
i 0.17 – 0.06 **j** 0.26 – 0.12

2 Use the column method to complete these subtractions. Make estimates first. Check your answer against your estimate and with a calculator.

a 12.4 – 3.6 **b** 12.4 – 3.62
c 14.36 – 3.09 **d** 14.36 – 3.095
e 8.172 – 3.533 **f** 8.17 – 3.533
g 7.4 – 2.039

3 Estimate and then calculate.

a 325 – 124.865 **b** 243 – 124.55 **c** 412 – 128.8
d 100 – 45.087 **e** 210 – 78.77 **f** 345 – 55.988

4 Work out how much fabric is left on each roll after the pieces in front have been cut off. All the lengths are in metres.

5 The table shows the masses of five children.

Child	Mass in kg
May	36.06
Paul	38.11
Kit	$41\frac{1}{2}$
Bob	44
Joan	35.92

What is the difference in mass between:

a Bob and May
b Kit and Joan
c Paul and May
d Bob and Paul
e Bob and Joan
f Kit and May?
g Between which two children is there the smallest difference in mass?
h What is this difference in mass?

6 Gabrielle gets a restaurant bill for \$2 235.70 and pays with a \$5 000 note. How much change will Gabrielle receive?

Real-world maths: Recipe calculations

Here are some of the ingredient for a recipe for a sweet potato curry:

Ingredients

2 onions
0.05 kg ginger
0.4 kg chopped tomatoes
1 kg sweet potato
0.4 kg chickpeas
0.45 kg roasted red pepper

1 If the average mass of a large onion is 0.15 kg, what is the total mass of 2 of these onions?

2 What is the difference in mass between the chopped tomatoes and roasted red pepper?

3 How many more kilograms of chickpeas are needed than ginger?

4 How many more kilograms of sweet potatoes are needed than red pepper?

5 This is a recipe for 10 people. Write the ingredients needed for a party of 20 people.

Multiplying decimals and whole numbers

Key maths idea

We can multiply decimal numbers by whole numbers or by other decimal numbers. In either case, it is important to think carefully about place value and use **estimates** to give you an idea of the size of the answer.

Let's look at multiplying by a whole number.

Examples

2.7 × 6

The **product** will be less than 18 because we know that 3 × 6 = 18.

We can solve this by expanding 2.7 and then multiplying each part by 6:

$$\begin{array}{rcr} 2 \times 6 & = & 12.0 \\ 0.7 \times 6 & = & 4.2\ + \\ \hline & & 16.2 \\ \hline \end{array}$$

Remember that we can describe 0.7 × 6 as 7 tenths × 6 = 42 tenths or 4.2.

6.1 × 12

The product will be a bit more than 72 because 6 × 12 = 72.

Let's use a column method:

$$\begin{array}{rl} 6.1 & \\ \times\ 12 & \\ \hline 12.2 & \text{1 tenth × 2 and then 6 × 2} \\ +\ 61.0 & \text{1 tenth × 10 and then 6 × 10} \\ \hline 73.2 & \end{array}$$

Note how the number of **decimal places** for all parts of the calculation match the number of places in 6.1.

2.35 × 24

The product will be more than 48 because 2 × 24 = 48

$$\begin{array}{rl} {}^{1}2.{}^{1}3{}^{2}5 & \\ \times\quad 24 & \\ \hline 9.40 & \text{5 hundredths × 4, 3 tenths x 4 and then 2 × 4} \\ +\ 47.00 & \text{5 hundredths × 20, 3 tenths × 20 and then 2 × 20} \\ \hline 56.40 & \end{array}$$

Note how the number of decimal places for all parts of the calculation match the number of places in 2.35. However, because there are no hundredths in the answer 56.40, the product is written as 56.4.

Key words

- estimate
- product
- decimal places

1 Find the product. Make estimates first. Check your answers with a calculator.

a 0.9 × 5 **b** 0.09 × 5

c 0.009 × 5 **d** 1.3 × 15

e 2.6 × 15 **f** 14.83 × 13

g 23.62 × 32 **h** 1.194 × 23

2 Calculate. Estimate each answer first.

a How many litres will 5 cans hold?

b How many litres will 20 mugs hold?

c How many litres do 10 buckets hold?

d How many litres do 3.5 kettles hold?

3 A jar of sweets has a mass of 1.213 kg. What is the mass of 4 jars?

Real-world maths: Comparing prices

At a garage, the cost of diesel is $132.50 per litre. A second garage charges $131.95 per litre.

A customer buys 45 litres of diesel. How much will they save if they buy it from the second garage?

Another customer fills their 70 litre tank with diesel at the first garage. They have a voucher that gives them 75 cents off per litre. How much does it cost to fill the tank?

Multiplying decimals together

Key maths idea

When multiplying decimal numbers together, we must also think very carefully about place value and make estimates first.

Look at these examples:

2.7×0.6

The **product** will be less than 2.7 because we know that $2.7 \times 1 = 2.7$.

We can solve this by expanding 2.7 and then multiplying each part by 0.6:

$$
\begin{aligned}
2 \times 0.6 &= 1.20 \\
0.7 \times 0.6 &= \underline{0.42} + \\
& \underline{1.62}
\end{aligned}
$$

We can describe 0.7×0.6 as 7 tenths $\times$ 6 tenths = 42 hundredths or 0.42.

6.1×1.2

The product will be a bit more than 6.1 because $6.1 \times 1 = 6.1$.

Let's use a column method:

$$
\begin{array}{r l}
6.1 & \\
\times\ 1.2 & \\
\hline
1.22 & \text{1 tenth} \times \text{2 tenths and then } 6 \times \text{2 tenths} \\
+\ 6.10 & \text{1 tenth} \times 1 \text{ and then } 6 \times 1 \\
\hline
7.32 &
\end{array}
$$

Note how numbers are written to 2 decimal places because tenths $\times$ tenths = hundredths.

2.35×2.4

The product will be more than 4.8 because $2 \times 2.4 = 4.8$

$$
\begin{array}{r l}
{}^{1}2.{}^{1}3{}^{2}5 & \\
\times\ \ 2.4 & \\
\hline
0.940 & \text{5 hundredths} \times \text{4 tenths, 3 tenths} \times \text{4 tenths and then } 2 \times \text{4 tenths} \\
+4.700 & \text{5 hundredths x 2, 3 tenths x 2 and then 2 x 2} \\
\hline
5.640 &
\end{array}
$$

Note how numbers are written to 3 decimal places because hundredths $\times$ tenths = thousandths. However, because there are no thousandths in the answer 5.640, the product is written as 5.64.

Key word

product

1 Calculate. Make estimates first.

a 0.9×0.1 b 0.3×0.8

c 1.2×0.6 d 0.05×6

2 Make estimates first and then calculate. Check your answers with a calculator.

a 4.5×4 b 2.4×8

c 25.4×3.1 d 9.13×2.5

e 5.37×4.6 f 5.75×6.4

g 3.176×2.3 h 3.176×2.5

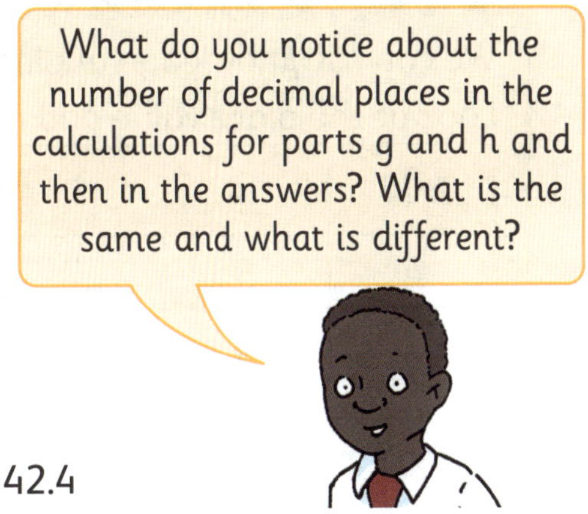

3 True or false? Correct all the equations that are false.

a $5 \times 0.05 = 25$ b $3 \times 2.7 = 8.1$

c $0.9 \times 0.2 = 0.18$ d $10.6 \times 0.4 = 42.4$

e $3.2 \times 4.5 = 14.04$ f $11.7 \times 1.2 = 140.4$

4 A kilogram of sweet potatoes cost $195.75. Use the column method or a calculator to find the cost of:

a 2.5 kilograms b 5 kilograms

c 1.75 kilograms d 4.35 kilograms

e 1.86 kilograms f 2.32 kilograms.

5 Malcolm used the multiplication 3.45×2.4 to solve a word problem in the context of measurement.

a What could the problem be? Make up at least two different examples.

b What is the answer to each of your problems?

Real-world maths: Map distances

1 The distance from Sea Point to Ocean Point is 1.5 times as long as the distance from Sandy Hill to Rocky Hill. The distance from Sandy Hill to Rocky Hill is 40.83 km. What is the distance from Sea Point to Ocean Point?

2 Use a ruler to measure the distances on the map in millimetres. Use the scale to calculate the real distances. Write the real distances in kilometres.

a Rocky Hill to Sea Point b Ocean Point to the lighthouse.

Dividing decimals by powers of 10

Key maths idea

We can use knowledge of place value to help divide decimals by powers of 10.
Look at the place value grid here:

100	10	1	0.1	0.01	0.001	
4	5	7				
	4	5	7			475 ÷ 10
		4	5	7		475 ÷ 100
		0	4	5	7	475 ÷ 1 000

Remember that division is the inverse of multiplication, so the digits behave in the opposite way as the values become 10, 100 and 1 000 times as small.

When a number is divided by 10, the value of each part of the number is made 10 times as small or $\frac{1}{10}$ of the size. As a result, the digits all move 1 place to the right.

When a number is divided by 100, the value of each part of the number is made 100 times as small or $\frac{1}{100}$ of the size. As a result, the digits all move 2 places to the right.

Tell your partner what happens when a number is divided by 1 000.

1 Divide each set of numbers as shown. Use a place value grid to help you.

- **a** Divide each of these numbers by 10. 5 15 115 1 150
- **b** Divide each number by 100. 3 23 232 2 325
- **c** Divide each number by 1 000. 3 5 15 23 115 232

2 Copy and complete.

- **a** 3.9 ÷ ☐ = 0.39
- **b** ☐ ÷ 1 000 = 5.04
- **c** ☐ ÷ 10 = 0.19
- **d** ☐ ÷ 10 = 0.78

3 Each coil of rope must be divided into equal lengths.

- **a** Work out the length of the pieces if each coil is divided into 10 equal pieces.
- **b** Work out the length of the pieces if each coil is divided into 100 equal pieces.
- **c** Another coil is 34 metres long. Work out the length of a piece if the coil is divided into 1 000 equal pieces.
- **d** What do you notice about your answer to part c and one of the answers you found for part b. Explain why this has happened.

4 Solve this riddle.

> *The hippopotamus is 10 times as heavy as the motorbike.*
> *The dog is one tenth of the mass of the motorbike.*
> *The feather is 1 000 times as light as the dog.*
> *The dog has a mass of between 17 kg and 18 kg.*

What could the masses of the dog, feather, motorbike and hippopotamus be?
Find at least five solutions.
Write the calculations you use each time. Record your solutions in a table.

Dividing a decimal number by a whole number

Key word
division

Key maths idea

André made some toto cake.
He cut it into ten equal parts (tenths).
André kept $\frac{2}{10}$ or 0.2 of the whole cake for himself.

He shared the remaining $\frac{8}{10}$ or 0.8 with his friends.
We can use multiplication facts we know to help us.
8 ones ÷ 4 = 2 ones so 8 tenths ÷ 4 = 2 tenths or 0.2.

So each person received $\frac{2}{10}$ or 0.2 of the whole cake.

Let's look at how the number sentence and diagram represent how André divided the remaining cake with his 4 friends:

$0.8 \div 4 = 0.2$

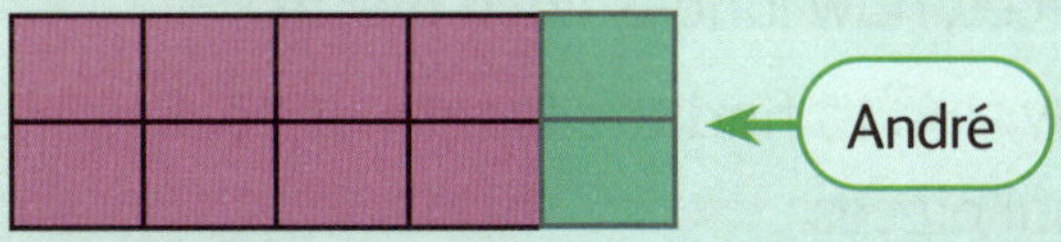

Let's look at another problem.
Malcolm spends \$0.75 on 3 sweets. What is the cost of each sweet?
We can solve this problem by finding 0.75 ÷ 3.
We can use known facts to help us:
75 ones ÷ 3 = 25 ones so 75 hundredths ÷ 3 = 25 hundredths or 0.25
Each sweet costs \$0.25.

Short **division** can also be used as a method to solve these division problems and those with more decimal places.

```
   0. 2 5          2. 2 1 3
3 | 0. 7¹5      4 | 8. 4 5¹2
```

1 Draw a diagram to show each division. Show your working and answers.

a 0.6 ÷ 3 **b** 0.8 ÷ 2 **c** 0.9 ÷ 3 **d** 0.5 ÷ 5

2 Mr Foster and Mr Nelson share six tenths (0.6) of a whole cake equally between them. How much of the whole cake do they each get?

3 Complete these divisions problems.

a 0.24 ÷ 6 **b** 0.48 ÷ 6 **c** 0.72 ÷ 3

d 0.72 ÷ 6 **e** 0.72 ÷ 9

4 Use the short division method to calculate each problem.

a 12.8 ÷ 2 **b** 19.6 ÷ 4 **c** 6.75 ÷ 3 **d** 30.75 ÷ 5

e 57.61 ÷ 7 **f** 12.369 ÷ 3 **g** 20.656 ÷ 8 **h** 4.172 ÷ 4

5 Each roll of cloth is divided into pieces of equal length.

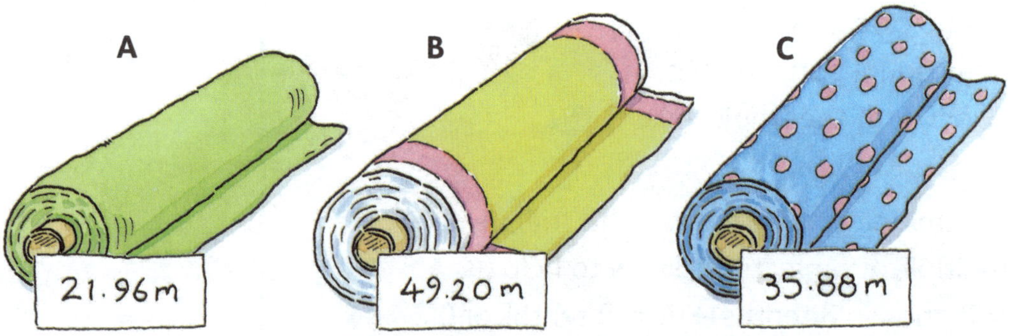

a If roll A is divided into 3 pieces, how long is each piece?

b If roll B is divided into 5 pieces, how long is each piece?

c If roll C is divided into 8 pieces, how long is each piece?

Now use a calculator or long division to find the length of each piece when:

d roll A is divided into 12 equal pieces

e roll B is divided into 15 equal pieces

f roll C is divided into 13 equal pieces.

6 Yohan runs the length of the park in 78.54 seconds.

Candice cycles the length of the park in one third of the time. How many seconds does it take Candice?

Dividing decimals by decimals

Key maths idea

Remember that we have used equivalences to make calculating with fractions easier, for example, finding an equivalent fraction for $\frac{4}{5}$ to help us add $\frac{4}{5}$ and $\frac{3}{10}$.

We can also use equivalences to help make division with decimals easier.

Remember that any fraction can be written as a division, and vice versa.

Let's look at the division 4.8 ÷ 0.3. To make an **equivalent** division, we must do the same to each number so the relationship between them stays the same – just like we do for numerators and denominators.

It is much easier to calculate with whole numbers, so let's multiply both numbers by a power of 10, until we reach a whole number **divisor**.

We can multiply 4.8 × 10 and also 0.3 × 10 to make the whole number divisor, 3.

So, the division 4.8 ÷ 0.3 and 48 ÷ 3 are equivalent divisions, each with the answer 16. Check this with a calculator!

Let's look at another example:

1.15 ÷ 0.5

11.5 ÷ 5 Multiply by 10 to get the whole number divisor 5.

Calculate 11.5 ÷ 5 either as a short division or with a calculator.

So, 11.5 ÷ 5 = 2.3 and 1.15 ÷ 0.5 = 2.3.

Key words

equivalent

divisor

1 Divide. Write the equivalent divisions you use.

a	1.8 ÷ 0.6	**b**	5.6 ÷ 0.8
c	3.5 ÷ 0.7	**d**	7.2 ÷ 0.9
e	3.6 ÷ 0.9	**f**	2.4 ÷ 0.6
g	1.2 ÷ 0.3	**h**	6.3 ÷ 0.7

Check all your answers with a calculator for both the original division and the equivalent division to make sure they are the same.

2 Calculate. Write the equivalent division you use in each case.

a $3.6 \div 0.3$ **b** $0.24 \div 0.4$ **c** $0.8 \div 0.02$

d $5.25 \div 0.5$ **e** $9.36 \div 0.3$

3 Solve these division problems. Use long division to calculate, as in the example.
Example: $9.1 \div 2.6$ can be calculated as $91 \div 26$.

```
     3.5
 26)91
   -78
    13.0
   -13.0
    00.0
```

a $12.4 \div 1.6$ **b** $34.8 \div 2.4$ **c** $44.8 \div 3.2$

d $1.24 \div 1.6$ **e** $3.48 \div 2.4$ **f** $4.48 \div 3.2$

4 Divide. Round each answer to three decimal places.

a $0.5 \div 0.6$ **b** $0.7 \div 0.8$ **c** $0.5 \div 0.7$

d $0.2 \div 0.9$ **e** $0.6 \div 0.7$ **f** $0.4 \div 0.6$

5 Divide. Round each answer to two decimal places.

a $0.3 \div 0.8$ **b** $0.4 \div 0.9$ **c** $0.1 \div 0.7$

d $0.5 \div 0.8$ **e** $0.7 \div 0.4$ **f** $0.3 \div 0.2$

Talk about

Rhianna notices that the division $0.8 \div 0.2$ can also be re-written using fractions:
$\frac{8}{10} \div \frac{2}{10}$ or $\frac{4}{5} \div \frac{1}{5}$
She then uses what she knows about reciprocals to find
$\frac{8}{10} \times \frac{10}{2}$ and $\frac{4}{5} \times \frac{5}{1}$.
Rhianna says that this will give the same answer as $0.8 \div 0.2$.

- Do you agree? Explore her idea and explain your decision.
- Now try her method for $0.6 \div 0.5$ and $0.9 \div 0.2$.

Maths detective: Number puzzles

Use what you know about decimal numbers to solve these number puzzles.

Case 1

Lira told her cousin Joshua that she could get some numbers under 10.00 to always add up to 10.89. Joshua said that there must be exceptions to this rule. Lira said there are no exceptions. Who is correct?

This is the procedure that Lira followed:

Steps	Example
Step 1: Write down a decimal number that is less than 10.00.	5.48
Step 2: Write down the digits in the reverse, with the decimal after the first digit.	8.45
Step 3: Find the difference of the two numbers.	8.45 – 5.48 = 2.97
Step 4: Write the reverse of your answer to Step 3.	7.92
Step 5: Add the answers in Steps 3 and 4.	2.97 + 7.92 = 10.89

Lira challenged Joshua to follow the procedure and find a set of numbers that does not add up to 10.89.

1 Help Joshua by performing the procedure with 15 decimal numbers less than 10.00. Write down all the answers for the procedure each time.

2 Which numbers add up to 10.89?

3 What pattern do you see? Try to explain the pattern in your own words. Look at all the tenths columns for the answers to the subtraction. Explain what they have in common. Do the same for the hundredths column.

> Hint: try starting with a decimal number with the three same digits. Try to find a pattern when you use a decimal number in which the digits are all the same.

4 Which amounts do not add up to 10.89?

5 Who is correct, Lira or Joshua?

Case 2

Do this decimal number investigation. Write the examples in your notebook.

Steps	Example
Step 1: Use a calculator to divide any two-digit number by 7. If your answer is a whole number, divide it by 7.	a
Step 2: Prediction: After the decimal point, the number 1 is followed by number 4.	b
Step 3: Add the six numbers after the decimal point. The total is 27.	c
Step 4: Repeat Steps 1 to 3 with a three-digit number. The prediction is that the total of the 6 digits after the decimal point is still 27.	d
Step 5: Why do you think the total of the first 6 digits after the decimal point is always 27 when the original number is divided by 7?	e

Real-world maths: More bang for your buck

1 Salim Holdings are planning a staff day. The managing director wants to calculate which of four restaurants offers the best price for the chicken buckets he would like to order. From which restaurant should Mr Salim purchase the chicken if he wants to pay as little as possible?

Restaurant	Price per bucket	Number of pieces of chicken per bucket
Restaurant A	$895.25	25
Restaurant B	$1 028.90	22
Restaurant C	$918.62	24
Restaurant D	$1 104.75	21

2 Serena will go to the swimming pool on 18 different days this month.

- A one-day pass costs $637.24.
- A monthly pass costs $9 648.85.

How much money will Serena save if she buys a monthly pass?

3 This table shows the average number of kilometres per litre for four popular model cars in Jamaica. Use the information in the table to work out the answers to the questions on the next page.

Car model	Daily rental	Kilometres per litre
Honda	$2 300	9
Nissan	$1 900	8
Toyota	$2 000	6
BMW	$2 200	10

The distance from Kingston to Montego Bay is 201.6 Kilometres. Gasoline costs $158.30 per litre.

- **a** Amanda has to rent a car to make a trip from Kingston to Montego Bay. Select a car model. Find the total number of litres of gasoline needed to drive the car from Kingston to Montego Bay.
- **b** Add the daily rental to calculate the total cost of driving the car from Kingston to Montego Bay.
- **c** Now calculate the cost of the journey for each of the other cars and add the daily rental amount.
- **d** Which car will be cheapest to rent for the journey?
- **e** Which one will be most expensive?

What I have learned

1. Explain the concept of decimals and give three examples of decimal numbers.
2. How can you order a list of decimals?
3. Make a list of five examples of decimals you come across in your daily life.
4. Find these numbers.
   - **a** If you divide me by 36 you get 1.15.
   - **b** If you multiply me by 9 you get 32.4.
   - **c** I am 2.6 less than 107.1 ÷ 3.
   - **d** I am 15 more than 3.3 multiplied by 7.1.

Practice questions

1 Here are the heights of trees in a botanical garden.

- **a** Write down the letters for the three tallest trees.
- **b** What is the difference in height between the tallest and smallest tree?
- **c** Which tree is taller, B or C?
- **d** Which two trees are closest in height?

2 Calculate.

a 4.8 + 6.1

b 7.5 + 8.46 + 0.9

c 2.148 + 4.25 + 17.6

d 18.8 – 9.251

3 Sage buys a piece of green ribbon that is 1.62 metres long and a piece of red ribbon that is 0.825 metres long. What is the total length of ribbon Sage buys?

4 A shopkeeper buys 50 m of rope. He sells 13.5 m and 8.35 m. How much rope does he have left?

5 Sophia weighed a bag of cashew nuts. She found that the bag had a mass of 5.72 kg. She took out 1.925 kg and weighed the cashew nuts again. How much did she have left?

6 Mia is baking cookies for a school bake sale. She has 4.5 kg of flour. The recipe requires 0.75 kg of flour for one batch of cookies. How many batches of cookies can Mia make?

7 Copy and complete this table.

	÷10	÷100	÷1 000
30			
3 kg			
	0.9		
			9.0
		9.09	

8 Multiply.

a 1.5 × 6

b 0.12 × 7

c 0.2 × 0.9

d 1.2 × 0.8

9 Divide.

a 2.46 ÷ 3

b 9.12 ÷ 3

c 0.255 ÷ 3

d 30.69 ÷ 3.1

10 A space shuttle travels at 88.5 m per second. How far will it travel in 25.5 seconds?

10 Self check

	Check if you can do this:	If you have difficulty, try this:
1	Add and subtract decimal numbers quickly and accurately.	Go back to pages 154–158. Remind yourself that when you add or subtract, you should make sure that each quantity is written to the same number of decimal places.
2	Divide a decimal number by a power of 10.	Go back to page 164. Remind yourself that when a number is divided by 10, the value of each part of the number is made 10 times as small or $\frac{1}{10}$ of the size. As a result, the digits all move 1 place to the right. Think about what then happens when a number is divided by 100 or 1 000.
3	Multiply decimals quickly and accurately.	Go back to page 160. Remind yourself that you can use multiplication facts and understanding of place value to help you.
4	Divide a decimal number by a whole number.	Remember to make use of division facts you know and place value, for example, relating $0.9 \div 3$ to 9 tenths $\div 3 =$ 3 tenths or 0.3. You can also use short division or a calculator.
5	Divide a decimal number by another decimal number with answers written to two or three decimal places.	Go back to page 167. Remind yourself that you can find an equivalent division to make calculating easier. Try to find an equivalent division with a whole number divisor.
6	Solve problems involving all four operations.	Remember that you can draw diagrams to help you make sense of problems with decimal numbers.
7	Use a calculator to help solve problems with decimals and read and use the displayed information correctly in the context of measurement.	Remember that calculators are useful mathematical tools but it is important to interpret the display in the correct way. Check to see if the question requires you to round your answer. Be particularly careful with money, as values should be given to only 2 decimal places.

Chapter 11 Ratio and proportion

In this chapter you will explore this focus question:

To answer this question, you will:

- use ratio in problem-solving and decision-making
- use ratio to compare quantities
- write a ratio to compare the numbers of items in two sets or two parts of a single set
- write ratios in different ways
- write equivalent ratios for a given ratio
- solve problems using equivalent ratios.

Starting point

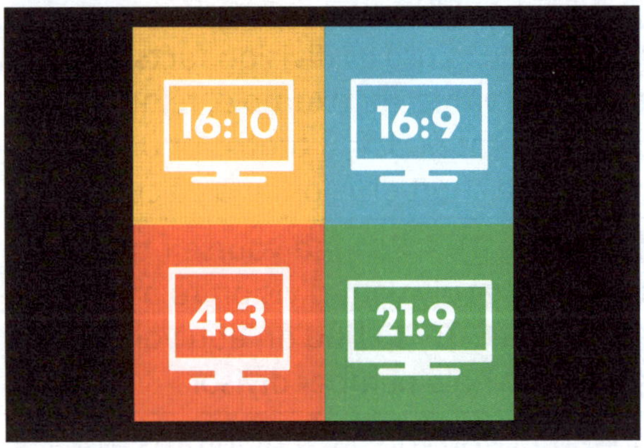

1 These diagrams show the aspect ratio of different kinds of computer and TV screens.

- What do you think the aspect ratio compares?
- If a tablet screen has an aspect ratio of 16 : 10 (length : width) and the screen is 20 cm wide, how long is the screen?
- Measure the length and width of a phone, computer or television screen. Does it match any of these aspect ratios?

2 Look at the bunting. Complete the sentences.

a For each piece with a Jamaican flag there are _____ pieces without a Jamaican flag.

b Every five rectangles there are _____ triangles and _____ pentagons.

c We can say the ratio of triangles to rectangles is 2 to _____ .

d The ratio of rectangles to triangle is _____ to _____ .

e If there were six Jamaican flags, there would be _____ striped flags.

Reading, writing and representing ratios

Key maths idea

A **ratio** shows the relationship between two or more quantities.

We can **compare** the number of items in two sets or two parts of a single set.

In this set of apples, we can compare the number of green apples to red apples. There are four green apples and one red apple.

The ratio of green to red apples is 4 to 1, or 4 : 1.

The ratio of red to green apples is 1 to 4. We can write this as 1 : 4.

It is important to write the ratio in the same order as the quantities we are comparing.

We often use the words 'for every' to describe a ratio. So we can say that

- *for every* 1 red apple there are 4 green apples, or
- *for every* 4 green apples, there is 1 red apple.

For the ratio 1 : 4, the number of red apples (1) is a quarter of the number of green apples (4).

We can also show this relationship as a **fraction**:

$\frac{1}{4}$ $\frac{\text{red apples}}{\text{green apples}}$

Try to explain why we can also show the relationship between the number of green apples and the number of red apples as $\frac{4}{1}$.

Key words

ratio
compare
fraction

Talk about

Identify some examples of ratios in your environment with a partner. Try to describe the ratios to each other as accurately as possible.

1 Look at the pictures. Describe the ratios you see.

a 

b 

c 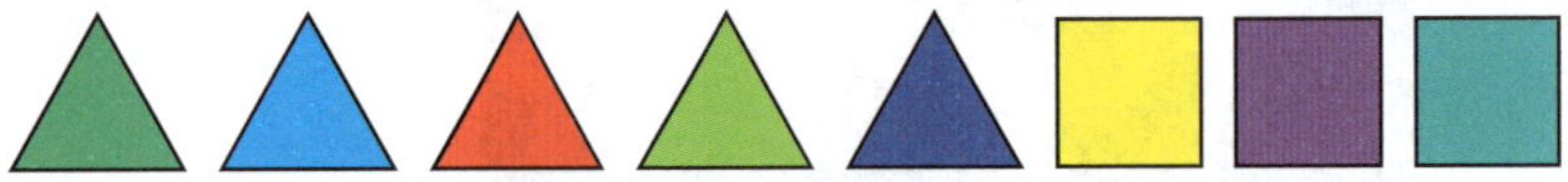

2 Draw diagrams to represent the ratios.

a When Jada has 2 mangoes, Joshua has 3 mangoes.

b When Femi has 1 ice-cream, Anthony has 2 ice-creams.

c When Lamar has 3 biscuits, Sophia has 4 biscuits.

d Now use ratio notation (:) to write the ratios next to each diagram in parts a to c.

3 True or false? Correct any that are false.

a 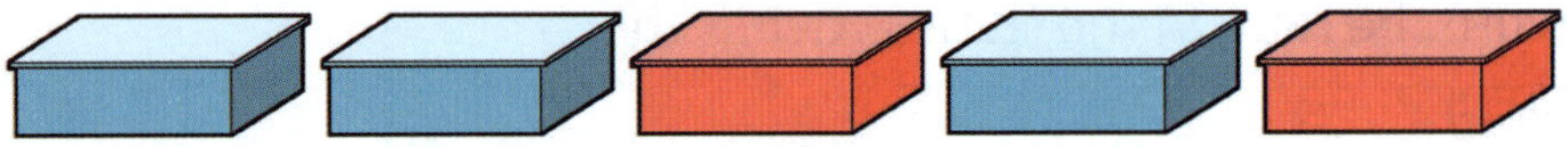

The ratio of blue counters to red counters is 2 : 3.

b 

The ratio of crabs to fish is 2 : 6.

c 

The ratio of paintbrushes to pens is 1 : 5.

4 Malcolm compares the number of ducks to ducklings on a pond.

For every 2 ducks, there are 5 ducklings.

a Sketch a diagram to show the number of ducks as a fraction of the number of ducklings.

b Now write the fraction: $\frac{\square}{\square}$ ducks / ducklings

Equivalent ratios

Key maths idea

The diagram represents the ratio of cars to motorbikes in a car park.

For every 1 motorbike, there are 3 cars. This ratio is 1 : 3.
For every 3 cars, there is 1 motorbike. This ratio is 3 : 1.

Let's think about what we need to do when we double the number of motorbikes:

Note how we have made a copy of the ratio.

To keep the relationship between the number of motorbikes and cars the same, we must also double the number of cars.
For every 2 motorbikes, there are 6 cars. The ratio is 2 : 6.
Now look at 3 times as many motorbikes.
Explain to your partner why there are now 9 cars.

We say that the ratios 1 : 3, 2 : 6 and 3 : 9 are **equivalent**.
We can also use what we know about equivalent fractions to help find equivalent ratios. What is the same and what is different about the equivalent fractions for $\frac{1}{3}$ ($\frac{1}{3} = \frac{2}{6} = \frac{3}{9}$, and so on) and the equivalent ratios for 1 : 3?

Key word
equivalent

1 Look at this diagram.

Then read each statement. Is it true or false?

- **a** For every red cube, there are 8 blue cubes.
- **b** For every 4 blue cubes, there is one red cube.
- **c** For every 3 red cubes, there will be 12 blue cubes.
- **d** For every 16 cubes, 4 will be red and 12 would be blue.
- **e** For every 20 cubes, 4 will be red and 16 would be blue.
- **f** Now write the corrected statements in ratio notation (:).

> You can use counters, cubes or coloured beads to help you make copies of the ratio.

2 Represent the ratio 3 : 2. You can colour squares in your notebook or use counters. Make copies of the ratio to help you find equivalent ratios. Record the ratios you find. Find at least five equivalent ratios.

3 In a maths test, there are four questions about calculating for every one question about shape and measure. How many questions about calculating are there when there are:

- **a** 2 questions about shape and measure
- **b** 4 questions about shape and measure
- **c** 6 questions about shape and measure?
- **d** If there are 16 questions about calculating, how many shape and measure questions are there?
- **e** If there are 24 questions about calculating, how many shape and measure questions are there?

4 Leon says that the ratio 6 : 15 is equivalent to the ratio 3 : 5. Use equivalent fractions to explain why Leon has made a mistake.

More about equivalent ratios

Key maths idea

Remember we say that a ratio is equivalent when the relationship between the quantities remains the same.

Look at the way that a set of **equivalent ratios** are represented here:

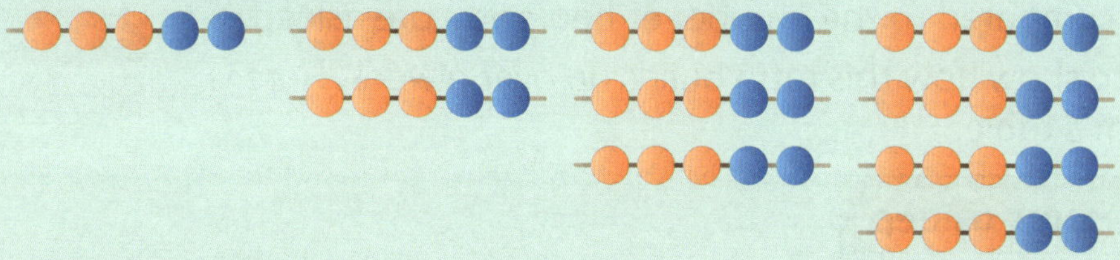

Describe the ratios as orange beads to blue beads and then as blue to orange beads.

Use the words 'for every' to help you.

It can be useful to represent ratios in a table to help you identify patterns and solve problems:

Orange beads	Blue beads
3	2
6	4
9	6
12	8

How would the table continue so we know:

- how many orange beads there are when there are 12 blue beads
- how many beads there are in total when there are 15 orange beads?

Like fractions, ratios are usually written in their simplest form.

To write a ratio in its simplest form, divide the values in the ratio by the highest common factor (HCF).

For example, 12 : 4 is not in its simplest form.

The HCF of 12 and 4 is 4.

Divide both values by 4.

The ratio in its simplest form is 3 : 1.

Key word

equivalent ratios

1 Joshua and Jordan share mangoes in the ratio 5:4.

If Joshua has 20 mangoes, how many does Jordan have?

2 Write equivalent ratios for each set of data.

a To make jam, use 6 kilograms of sugar for every 3 kilograms of fruit.

b For every 15 minutes I sit down, I spend 60 minutes walking.

3 a Make up a ratio example about the number of hours each day you are awake compared to the number of hours you are asleep.

b Draw a ratio table to show how this ratio builds up over a week. Here is the start of Malcolm's table.

Hours awake	Hours asleep
16	8
32	16

4 Write these ratios in their simplest form.

a 4:2 b 5:15 c 2:8

d 20:10 e 12:3 f 4:20

5 The ratio 3:8 is in its simplest form. Explain why.

Real-world maths: Equivalent ratios

At an outing, there are 2 girls for every 1 boy.

1 How many girls are at the outing if there are:

a 10 boys b 20 boys?

2 If there are 10 girls, how many boys will there be?

3 How many girls are there if there are 24 children?

4 What is the number of boys if the group consists of 30 children?

Ratio and proportion

Key maths idea

We know that a ratio compares two or more quantities. We can also say that it compares parts. For example, the ratio of red to green apples is 1 : 4.

Like fractions, a **proportion** compares a part to the whole.

In this example, we give the proportion of red apples as 1 out of 5 apples or $\frac{1}{5}$. So 1 out of every 5 apples is red.
Tell your partner why the proportion of red apples can also be expressed as 20%.

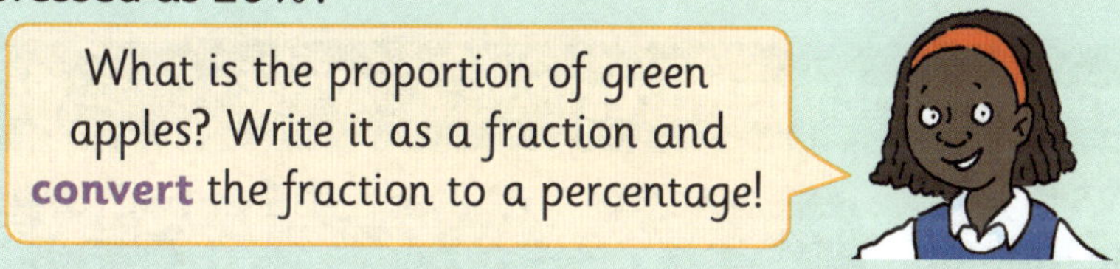

Key words

proportion
convert

1 Look at each set of pictures.

A

B

C

D

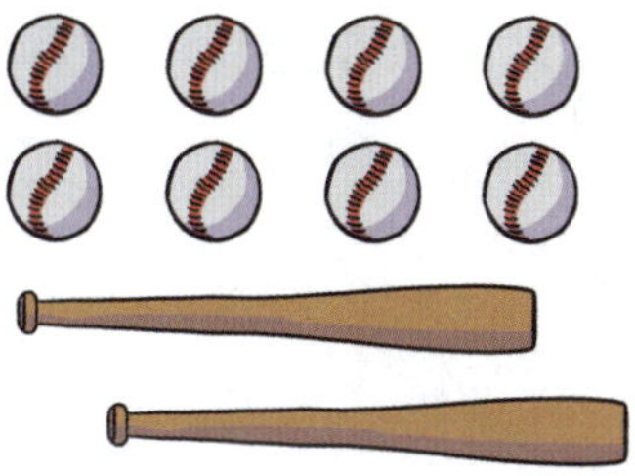

a Use ratio notation (:) to describe each picture as a ratio.

b Now describe each part as a proportion of the whole.

For example: $\frac{\square}{\square}$ of the insects are butterflies and $\frac{\square}{\square}$ of the insects are ants.

2 Draw and then write the ratio of triangles to circles each time.

- **a** $\frac{1}{4}$ of the shapes are circles. $\frac{3}{4}$ of the shapes are triangles.
- **b** $\frac{3}{5}$ of the shapes are circles. $\frac{2}{5}$ of the shapes are triangles.
- **c** $\frac{5}{8}$ of the shapes are circles. The rest are triangles.
- **d** $\frac{7}{12}$ of the shapes are circles. The rest are triangles.

3 Think about the alphabet.

- **a** Write the ratio of vowels to consonants. Check to make sure the ratio is in its simplest form.
- **b** Write the vowels as a proportion of the whole alphabet.
- **c** Write the consonants as a proportion of the whole alphabet.

4 Candice planted some seeds. 3 out of every 10 seeds she planted grew. The rest did not grow. Candice planted 100 seeds.

- **a** How many seeds did grow?
- **b** How many seeds did not grow?
- **c** Now describe these proportions as percentages:

  _____% did grow _____% did not grow

Real-world maths: An investment in shares

Two investors, Mr Harris and Mrs Bryant, hold shares in a company in the ratio of 2 : 3.

1 Investigate the possible total number of shares in the company.

2 Determine how many shares Mr Harris and Mrs Bryant each hold if the company has a total of 100 shares.

3 Write the proportion of the total shares Mr Harris holds as a fraction.

4 Now write the proportion of the total shares Mrs Bryant holds as a percentage.

Applying our understanding of ratio

Key maths idea

Ratios do not have units. So a ratio of 1 : 5 could mean 1 cup to 5 cups, 1 sweet to 5 sweets, 1 gram to 5 grams.

Example

Dana's mom has a bottle of orange squash.

The label on the bottle states that the squash should be mixed with water in the ratio 1 : 4.

This means she needs to mix 1 cup of squash with 4 cups of water.

Always check in which order you write the quantities in a ratio.

The amount of squash (1 cup) must come first, so you write 1 : 4. If you wrote it as 4 : 1 it would mean 4 cups of squash for each cup of water, and that would be very sweet!

We can also compare more than two quantities as a ratio.

For a fruit cocktail, bananas, apples and kiwis are used in the ratio 2 : 3 : 5.

When 4 bananas are used, we must use 6 apples and 10 kiwis to keep the ratio the same.

Use the ratios above to answer Exercises 1 and 2.

1 How much water would Dana's mom need to add to the following amounts of squash?

a 1 bottle **b** 3 cups **c** 1 tall glass **d** 2 capfuls

2 Solve these problems about making a fruit cocktail.

a How many kiwis are needed when 12 apples are used?

b How many bananas are needed when 25 kiwis are used?

c How many apples are needed when 10 bananas are used?

d How many of each piece of fruit are used when 50 pieces are used in total?

> You can use a ratio table to help you with these problems.

3 Write these measures as ratios.

a 10 cm to 35 cm **b** 200 km to 250 km

c $1 000 to $1 500 **d** 3 goals to 5 goals

4 A decorator wants to paint the walls of a room pink.
She mixes white and red paint in the ratio 3 : 1.
She paints one wall, but the pink colour is not bright enough for her.
Which of these ratios of paint mix would she use to make the walls brighter pink: 1 : 5 or 2 : 1?

5 Rhianna spends her pocket money on snacks, magazines and going out with friends. For every \$40 she spends on snacks, she spends \$70 on magazines and \$90 on going out with friends.

a Use ratio notation (:) to write the amount she spends as snacks : magazines : going out with friends.

b Use a ratio table to help solve the following problems. Here is the start of one:

Snacks	Magazines	Going out with friends
\$40	\$70	\$90

How much does Rhianna spend on magazines when she spends \$160 on snacks?

c How much does spend on snacks when she spends \$450 going out with friends?

d How much does she spend on going out with friends when she spends a total of \$800?

e What proportion of pocket money does Rhianna spend on magazines?

f Write your answer to part e as a percentage.

Maths detective: Ratio, scale and proportion

Use what you know about ratio, scale and proportion to solve these cases.

Case 1

Look at these three shapes.

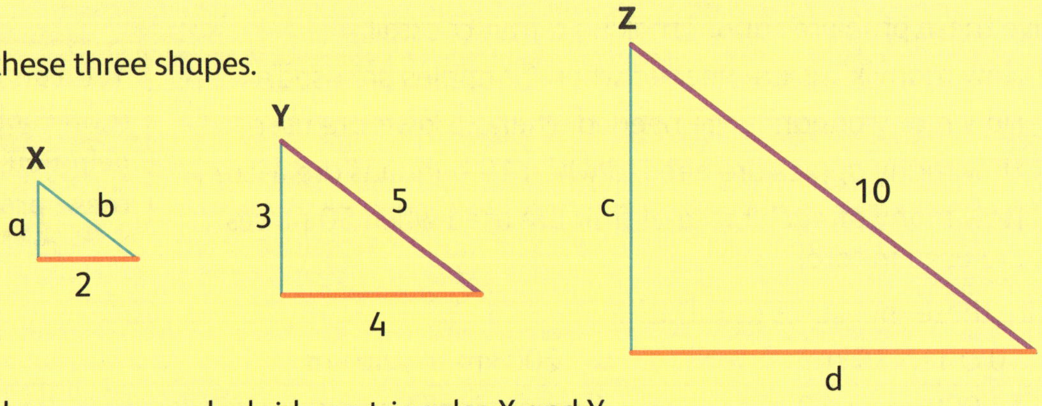

Look at the orange marked side on triangles X and Y.

This side is 2 units long on X and 4 units long on Y. The ratio of these sides is 2 : 4, simplified as 1 : 2. This tells you that each side on triangle Y is twice as long as each side on triangle X.

1 If the side lengths are given in centimetres, how long are sides *a* and *b*?
2 What is the ratio of the sides marked in purple on triangles Y and Z?
3 What does this tell you?
4 Use the ratio to work out the lengths of *c* and *d* in centimetres.

Case 2

This heart design has been drawn on a grid of squares.

Draw your own grid or use squared paper to make an enlarged copy of this design. The ratio of the length of lines on this grid to your copy should be 1 : 2.

Case 3

1 Look at this number sequence:

It is called the Fibonacci sequence. How do you think this number sequence will continue? You can research for information on the internet.

2 The Fibonacci sequence can also be expressed as the value 1.618… This ratio is found in nature in many places. Carry out some research about where the Fibonacci sequence is found in nature. You could create a class poster with pictures of all the examples you found.

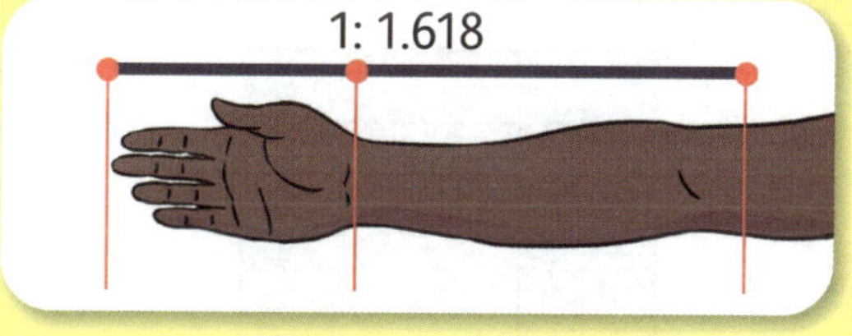

3 The ratio 1 : 1.618 has also been used over many centuries by artists, scientists and musicians, and is regarded as a code to beauty. It is called the golden ratio. Have a look at this illustration.

Now check whether your own arm measurements also represent the golden ratio. Measure the length of your hand. Then measure the distance from your wrist to your elbow. What do you find?

4 Draw up a table like this one for your class.

Name	1st measurement	2nd measurement	Ratio

What do you notice about the ratios in the right-hand column?

5 Now do the same, but measure the distance from the ground to your belly button. Then measure the distance from your belly button to the top of your head. Draw up another table for your class like the one above. What do you notice about your new set of ratios?

What I have learned

Make a mind map to show the following information about ratio:

- What is a ratio? Give some examples.
- In what ways can you write or describe a ratio?
- What do you know about equivalent ratios? How do you make an equivalent ratio?
- How can you simplify a ratio? How do you know a ratio is in its simplest form?
- What is the difference between ratio and proportion? Give examples.
- Explain how you can share set of objects in a given ratio, for example, sweets in the ratio 4 : 3.

Practice questions

1 Write a ratio to compare the quantities in each statement.

a In Jamaica there is one teacher for every 35 students.

b To cook rice, you use 3 cups of water for each cup of rice.

c Jayden is twice as tall as his younger brother.

2 Look at the shapes from A to D.

a What fraction of each shape is blue?

b What fraction of each shape is red?

c What is the ratio of blue parts to red parts in each shape?

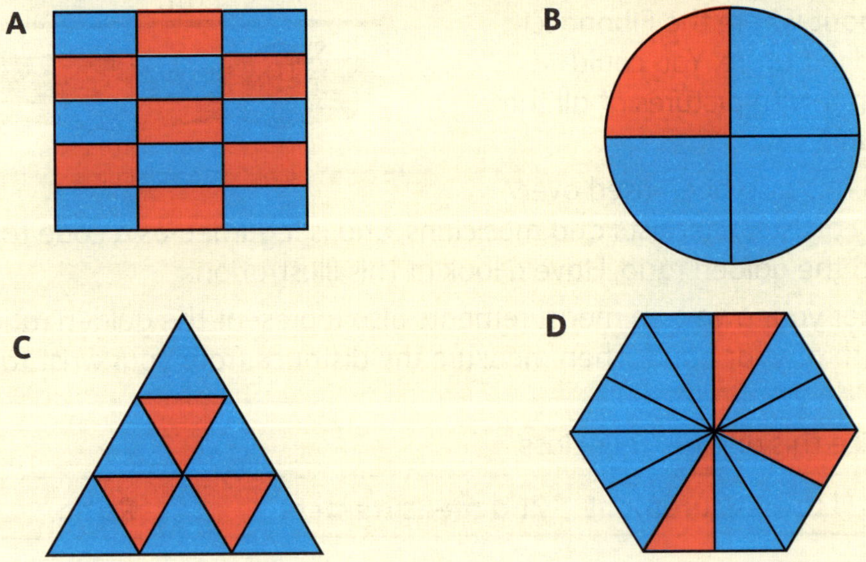

3 Simplify these ratios.

a 2 : 50 b 10 : 100 : 50

4 Look at the results of a survey on what drink people prefer in the morning.

Juice	Tea
39	18

Say if each statement is true or false.

a The total number of people surveyed was 39.

b 6 out of 13 people drink tea on the morning.

c The ratio of people who drink tea to people who drink juice is 6 : 13.

5 A recipe for 20 muffins uses 2 eggs. How many eggs are needed for:

a 40 muffins b 10 muffins c 50 muffins?

6 If it takes Elaine 20 seconds to run 200 metres in training, how many seconds will it take her to run 160 metres at the same speed?

7 A man makes 10 models a day. If he continues at the same rate, how long will it take him to make 35 models?

8 Mrs Nelson can sew 9 dresses in two days. How many days will she take to sew 36 dresses if she works at the same rate?

9 Jacqueline charges $275.00 per hour to babysit. Copy and complete this table to show how much she earns for different numbers of hours.

Hours worked	Amount earned
1	
3	
5	
7	
9	
11	

10 At a store, Susan selected a pumpkin that weighed 1 425.98 grams. Pumpkins cost $180.00 per kilogram. How much did Susan's pumpkin cost? Express your answer in dollars and cents.

Remember that there are 1 000 g in 1 kg. !

11 The ratio of Savion's oranges to Stephen's oranges is 11 : 5. If Savion gives 36 oranges to Stephen, they each have the same number of oranges. How many oranges do they have altogether?

11 Self check

Check if you can do this:	If you have difficulty, try this:
1 Use the language of ratio to compare quantities.	Sketch the quantities you are comparing. Use the language, ‘for every’ to help you compare them.
2 Use ratio notation (:) to compare two or more quantities.	Remember that the order you write the ratio is important. Use the symbol : to replace the language ‘for every’ when you are describing the ratio in words.
3 Explain the connection between a ratio and a fraction, for example, 2 : 3 and $\frac{2}{3}$, by giving one quantity as a fraction of the other.	Draw diagrams to help you identify the relationship. For the ratio 2 : 3 we can draw the following to show the relationship between the two quantities. In this example, a pattern uses tiles in the ratio 2 orange for every 3 blue. 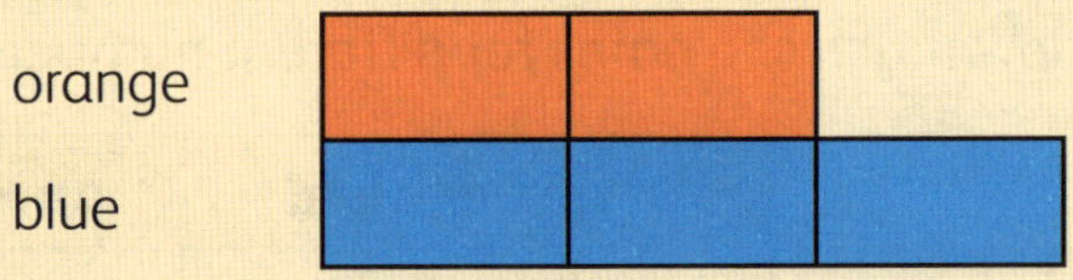 The number of orange tiles used is $\frac{2}{3}$ of the number of blue tiles used.
4 Write equivalent ratios for a given ratio.	Go back to page 179. Remind yourself that the relationship between the two quantities must stay the same. You can use what you know about equivalent fractions to help you.
5 Solve problems using equivalent ratios.	Remember that you can sketch diagrams or use small objects to help make copies of the ratio. You can do this until you reach the answer you need to solve the problem. You can also draw a ratio table.
6 Explain the difference between a ratio and a proportion.	Go back to page 181. Remind yourself what you are comparing when describing a ratio, and what you are comparing when describing a proportion.
7 Give a proportion as a fraction or as a percentage of a whole.	Remember to use what you know about the equivalence between fractions and percentages to help you here. Remind yourself that percentages can be written as a fraction with denominator 100.

Chapter 12 Percentages

In this chapter you will explore this focus question:

To answer this question, you will:

- show how you can apply your understanding of percentages to real-world contexts
- apply your understanding of ratio to percentages and use the symbol % correctly
- identify a given percentage of a set or of an object
- write a percentage as a fraction with denominator 100 or in its simplest form
- write a percentage as a decimal
- convert between fractions and percentages to help solve problems
- recognise that 100 % is a whole
- express a given number as a percentage of another given number (multiple of 10)
- calculate a given percentage of a number, amount of money, measure of mass, capacity, etc.
- calculate the whole (100 %) when a multiple of 5 per cent (for example, 5 %, 10 %, 15 %, and so on) of the amount is known.

Starting point

Here is a game board for International Checkers. It is a little different from the traditional checkers board. Try to find out how and why.

There are 100 squares on this board. 50 % of the squares are light and 50 % are dark.

What fraction of the squares are light?

What is this fraction with denominator 100?

What is this fraction in its simplest form?

Players start with 20 pieces each on the board.

Each player's pieces cover $\frac{20}{100}$ of the board.

What is this as a percentage of the whole board?

What percentage of the whole board is covered with players' pieces in total?

Find out about the rules for playing International Checkers.

You could make your own game board and use counters for your pieces.

Introducing percentages

Key maths idea

Like a fraction, a **percentage** is a proportion of the whole.

'Per cent' comes from the Greek term 'centum', which means 'out of 100'.

All percentages can be expressed as a **fraction** with denominator 100.

We write percentages with the % sign, for example, 10 per cent is written as 10 %.

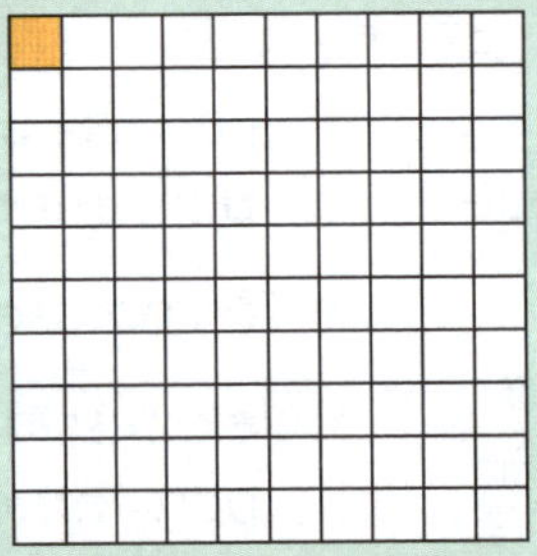

1 % of this grid is shaded.
We can write this as $\frac{1}{100}$ because 1 out of the whole 100 parts is shaded.

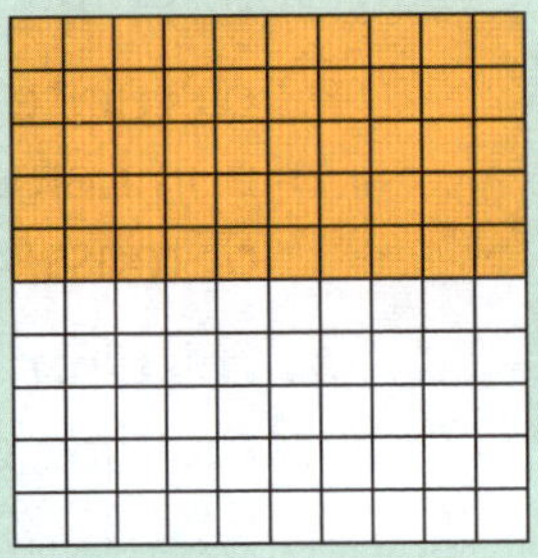

50 % of this grid is shaded.
We can write this as $\frac{50}{100}$ because 50 out of the whole 100 parts are shaded.
The **whole** is 100 %.

We use percentages in many everyday situations. You may have seen signs in a shop advertising a sale, for example, '25 % off all items'.

Talk to your partner about other examples of percentages you have seen in real-world situations.

Key words

percentage
per cent
fraction
whole

1 Write each of these using the percentage symbol (%).

a 6 per cent **b** 80 per cent **c** 45 per cent **d** $16\frac{1}{2}$ per cent

2 Write each percentage as a fraction. Your denominator should be 100.

a 6 % **b** 15 % **c** 35 % **d** 75 % **e** 100 %

3 Write each fraction as a percentage.

a $\frac{84}{100}$ **b** $\frac{17}{100}$ **c** $\frac{6}{100}$ **d** $\frac{20}{100}$

4 Leon says that 10 % is equivalent to the fraction $\frac{1}{10}$. What do you think? Explain your idea.

5 Order these proportions in ascending order:

a $\frac{15}{100}$ 9 % 80 % $\frac{8}{100}$ $\frac{1}{10}$

b $\frac{1}{100}$ 100 % $\frac{99}{100}$ 20 % $\frac{25}{100}$

c $\frac{100}{100}$ 10 % $55\frac{1}{2}$ % $\frac{1}{2}$ $\frac{11}{100}$

Equivalent percentages and fractions

Key maths idea

You already learned that any percentage can be written with the denominator 100 and vice versa, but sometimes we may need to convert a fraction with a different denominator to a percentage.

Example

A shop offers a discount of 25 %. Another shop offers the item at the same price with $\frac{1}{5}$ off. Which deal is better?

To convert a fraction to a percentage, we find the equivalent fraction with denominator 100.

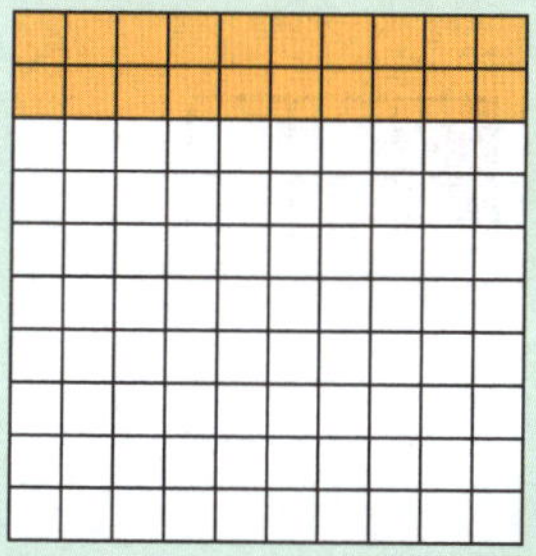

$$\frac{1}{5} = \frac{2}{10} = \frac{20}{100} = 20\,\%$$

We also know that 25% can be written as the fraction $\frac{25}{100}$ so we can also compare the fractions $\frac{25}{100}$ and $\frac{20}{100}$.

So 25 % off is a better discount.

Let's look at another example:

In a class of 20 children, 15 like football. What percentage of the class like football?

We know that $\frac{15}{20}$ like football.

Let's convert $\frac{15}{20}$ into an equivalent fraction with denominator 100.

$\frac{15}{20} = \frac{75}{100}$

So 75 % of the class like football.

Note how both the numerator and denominator are multiplied by 25, so the **equivalent** fraction with denominator 100 can be found.

Key word

equivalent

I've also noticed that $\frac{15}{20}$ can be simplified as $\frac{3}{4}$, and this is 75%, which is a percentage equivalent I know!

1 Copy and complete. Draw diagrams to help you.

a $1 = \frac{\square}{100} = \square\%$

b $\frac{3}{10} = \frac{\square}{100} = \square\%$

c $\frac{2}{5} = \frac{\square}{100} = \square\%$

d $\frac{3}{5} = \frac{\square}{100} = \square\%$

2 Write each fraction as a percentage.

a $\frac{22}{100}$ b $\frac{7}{10}$ c $\frac{6}{50}$ d $\frac{8}{20}$ e $\frac{9}{10}$ f $\frac{4}{5}$

3 What percentage of each diagram is shaded?

a
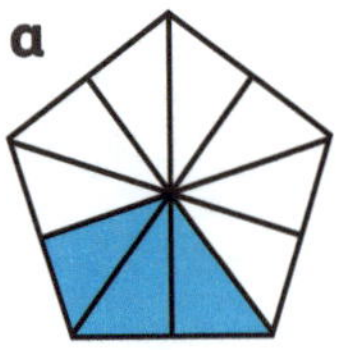
b
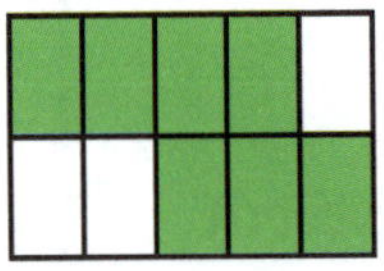
c
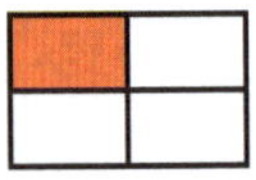
d
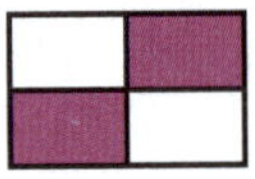
e

f
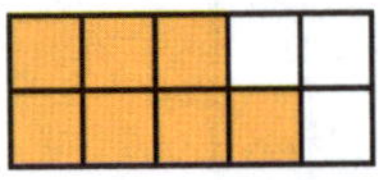
g

h
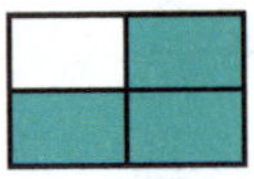
i
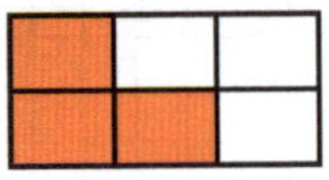

4 Leon got 80% of his spelling words correct in the test.
Candice got $\frac{3}{4}$ of the spelling words correct in the same test.
Who got the greater proportion of the spelling words correct? Why?

5 Rhianna and Malcolm are cycling around the perimeter of the park.
So far, Malcolm has cycled $\frac{3}{10}$ of the perimeter and Rhianna has cycled 35%.
Who has to cycle further still? Why?

Fractions, decimals and percentages

Key maths idea

Fractions, decimals and percentages are different ways of representing a proportion of a whole.

Look at the relationships between the decimals, fractions and percentages shown.
What do you notice? Explain this to your partner.

$0.25 = \frac{25}{100} = 25\%$

$0.4 = \frac{4}{10} = \frac{40}{100} = 40\%$

$17\frac{1}{2}\% = 17.5 \div 100 = 0.175$

$0.03 = \frac{3}{100} = 3\%$

This table shows some common equivalences that are useful to know.

Percentage	Decimal	Fraction
1	0.01	$\frac{1}{100}$
5%	0.05	$\frac{1}{20}$
10%	0.1	$\frac{1}{10}$
12.5%	0.125	$\frac{1}{8}$
20%	0.2	$\frac{1}{5}$
25%	0.25	$\frac{1}{4}$
$33\frac{1}{3}\%$	$0.\dot{3}$	$\frac{1}{3}$
50%	0.5	$\frac{1}{2}$
$66\frac{2}{3}\%$	$0.\dot{6}$	$\frac{2}{3}$
75%	0.75	$\frac{3}{4}$
90%	0.9	$\frac{9}{10}$
100%	1	$\frac{1}{1}$

Why do you think that 25% is written as a decimal to 2 decimal places but 20% is only written as a decimal to 1 decimal place?

When we convert $\frac{1}{3}$ and $\frac{2}{3}$ to 0.3 and 0.6, we get recurring decimals. $0.\dot{3}$ means 0.3333.... and $0.\dot{6}$ means 0.6666...

1 Write these percentages as decimals.

a 5% **b** 70% **c** 45% **d** 12% **e** 98%
f $12\frac{1}{2}\%$ **g** 45.75% **h** $40\frac{3}{4}\%$ **i** 83.3% **j** 7.6%

2 Write these decimals as percentages.

a 0.12 **b** 0.04 **c** 0.48 **d** 0.30 **e** 0.3
f 0.555 **g** 0.625 **h** 0.75 **i** 0.8 **j** 0.65

3 Write each percentage as a fraction in its simplest form.

a 80% **b** 14% **c** 25% **d** 40% **e** 64% **f** 70%

4 Write each fraction as a percentage.

a $\frac{1}{4}$ **b** $\frac{4}{5}$ **c** $\frac{15}{20}$ **d** $\frac{64}{200}$ **e** $\frac{3}{4}$ **f** $\frac{75}{150}$

5 Study this 10 × 10 grid. Write the fraction, decimal and percentage of the whole grid that each shape from A to C represents.

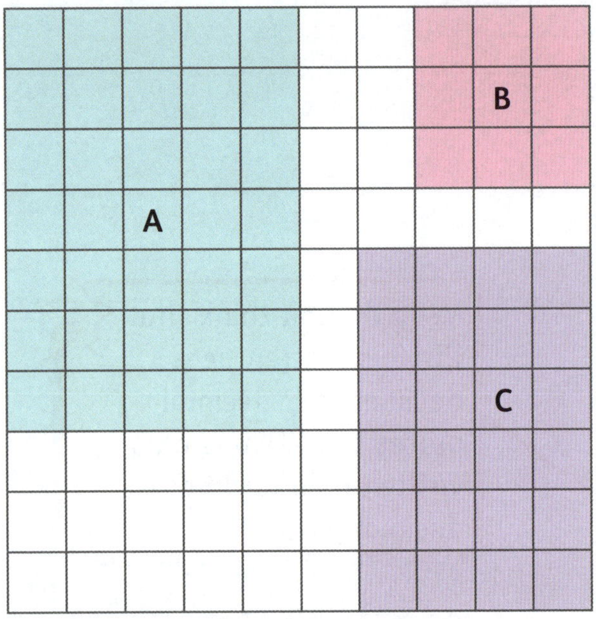

6 Lemar gave $\frac{1}{5}$ of his plantain chips to Joshua and $\frac{2}{5}$ to Mikayla.

a What percentage of his chips did Lemar give away?

b What percentage did he have left?

Percentages of quantities

Key maths idea

We can use fraction and percentage equivalence to help find a percentage of a **quantity**.

The bar model represents 10 % of a quantity. The whole quantity (100 %) is 40 and so $\frac{1}{10}$ (10 %) of 40 is 4.

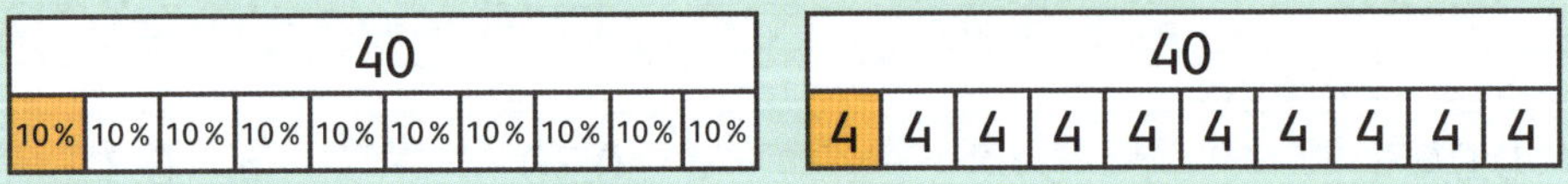

We can also use double number lines to help find percentages of amounts.

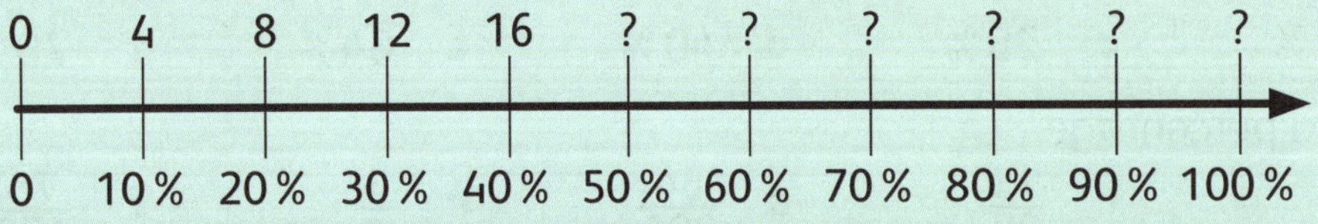

What is 20 % of 40? What is 30 % of 40?
What are the missing values on the line?
What percentage is 24 of 40?
What else can you say?

Key word
quantity

You can find 1 % of 40 by finding $\frac{1}{100}$ of 40 or 40 ÷ 100 = 0.4.
So 2 % of 40 = 0.8, and so on.

We can also use diagrams to find the whole when only a percentage is known.
If 10 % of an amount is $25, what is the whole amount?

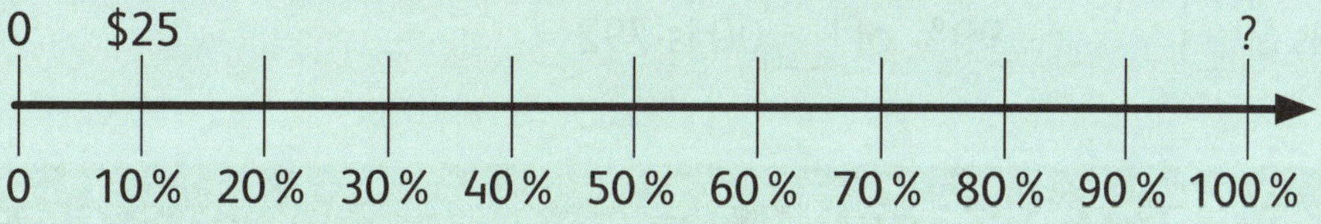

Talk to your partner about how you can use the number line to find the value of the whole 100 %.
What might you do differently if $25 represented 5 % of the whole amount?

1 Find:

a 10 % of 200 **b** 20 % of 200 **c** 40 % of 200 **d** 80 % of 200
e 8 % of 200 **f** 10 % of 60 **g** 5 % of 60 **h** 15 % of 60
i 1 % of 60 **j** 16 % of 60 **k** $12\frac{1}{2}$ % of 60 **l** 60 % of 60

2 What are the missing quantities?

a 5 % of ☐ is 10 **b** 10 % of ☐ is 45
c 15 % of $☐ is 30 **d** 25 % of ☐ metres is 100 metres
e 50 % of ☐ kilograms is 35 kg **f** 75 % of ☐ litres is 60 litres

3 In a school of 400 students, 70 % live within 3 km.

a How many students live within 3 km of the school?
b How many students live further away?

4 The children are cutting ribbons. What percentage of the ribbon is used each time?

a Anthony cuts 30 cm of a 60 cm length of ribbon.
b Leon cuts 1 m of a 10 m length of ribbon.
c Rhianna cuts 20 cm of a 50 cm length of ribbon.
d Candice cuts 50 cm of a 250 cm length of ribbon.

5 A box contains some oranges. 12 of them were not suitable for selling.

a If 20 % were not suitable for selling, how many oranges were in the whole box?
b What percentage of the whole box could be sold?
c How many oranges could be sold?

6 Use each digit in the box only once to make all the statements true:

2 3 4 7 8

☐0 is 50% of 60 ☐5% of 400 is 300 45 is ☐0% of 225

30% of 180 is 5☐ 99% of ☐00 is 792

Real-world maths: Import duty

1 The import duty to bring a car from one country to another is 20%. How much duty would a customer have to pay for a car that cost US $30 000?

2 Another customer pays US $5 000 import duty on a car. What was the cost of the car?

Using a calculator to find percentages of quantities

Key maths idea

It is important to know how to use the percentage key (%) on a calculator.

Try to find 50% of 200. Which keys do you have to press, and in which order?

Sometimes calculators work a little differently, so it is always important to check.

Not all calculators have a percentage key. So we need to know how to calculate a percentage on these devices.

Here are two methods. Which do you prefer? Explain why.

Method 1

Convert the percentage into a fraction, then divide the numerator by the denominator and multiply by the quantity.

Find 28% of 350

$\frac{28}{100} \times 350 = 98$

Method 2

Convert the percentage to a decimal and then multiply that by the quantity.

Find 28% of 350

$0.28 \times 350 = 98$

1 Use a calculator with a percentage key to find 32 % of each of these amounts.

a \$400 **b** \$320 **c** \$75 **d** \$48 **e** \$9.95

2 Now find 64 % of each amount without using the percentage key on the calculator.

a 400 kg **b** 320 kg **c** 75 kg **d** 48 kg **e** 9.95 kg

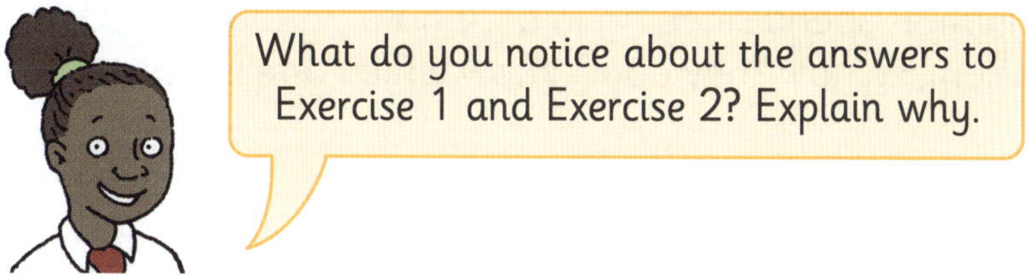

3 The area of Jamaica is approximately 10 990 km^2. Write down the values from a to c in the table in your notebook to show the proportion of land used for different types of farming. Round your answers to the nearest whole km^2.

Land use	Percentage (%)	Area in km^2
Crops	48	**a**
Pasture	14	**b**
Woodland	10	**c**
Other	28	3 077

4 Grandma uses 75 % of a 0.5 kg bag of flour to bake cakes.

a How many kilograms of flour does she use?

b How many kilograms of flour are left in the bag?

Real-world maths: Discounts

Jan and Tanice are shopping for a backpack. As they walk by ABC Supplies they see this sign:

Jan and Tanice both want to know how much they will save.

Jan says: '40 % of \$7 200 is \$2 880. That's a great saving!'

Tanice says: 'I think the cost for each bag is \$4 320.'

Who is correct?

1 Explain what each girl did to find their answer. What error did they make that led them to the incorrect answer?

2 How much would Janice and Tanice pay for both bags together?

Maths detective: Sleepy animals

Use your knowledge of percentage to solve these cases.

Case 1

Maddie and Jay were working on a project. They drew pie charts to show how much time different animals spend sleeping in a 24-hour period. The problem is that they forgot to label them, so now they don't know which one is which!

Here are the graphs.

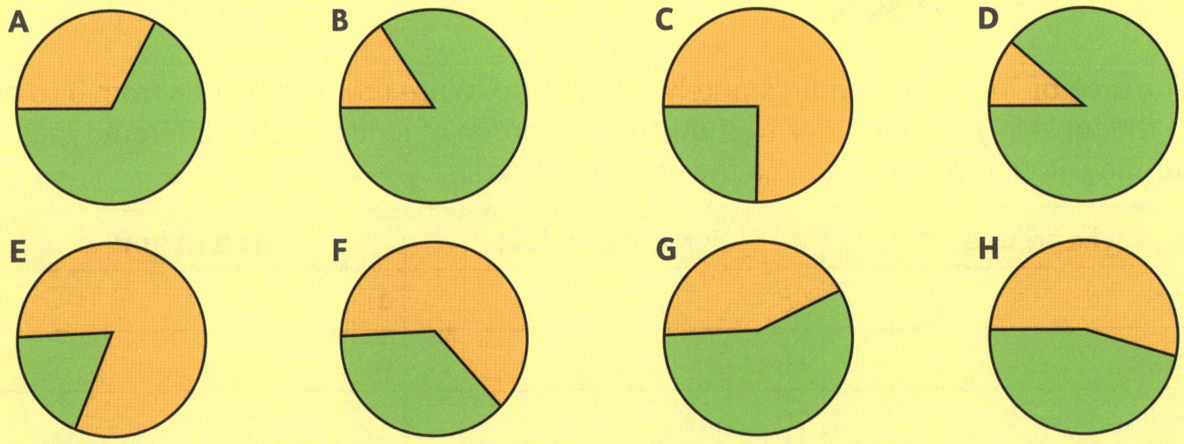

Key Hours awake Hours asleep

1 Use the data in this table to work out which graph matches each animal. Write down only the answers in your notebook.

Animal	% of day spent sleeping	Hours spent sleeping
Giraffe	10.5%	**a**
Elephant	14.5%	**b**
Human	33.3%	**c**
Python	75%	**d**
Cat	54%	**e**
Bat	83%	**f**
Dog	44%	**g**
Squirrel	62.5%	**h**

2 Calculate how many hours a day each animal spends sleeping.

Remember that one day is 24 hours.

Case 2

They do some more research and they find out that tigers sleep for 16 hours a day, male lions sleep for 20 hours a day and cows sleep for 4 hours a day.

1 What percentage of the day do each of these animals spend awake?

2 How can you show this data using bar diagrams? Draw your solutions and label them clearly.

Increasing and decreasing by percentages

Key maths idea

We also use percentages to describe an increase or a decrease of a quantity.

Let's look at an example of a **percentage increase**.

Sophia increases the prices of the items she sells in her café by 10 %.

Before the price increase, a cappuccino cost \$340. What is the cost of a cappuccino now?

We can use a diagram to represent the problem.

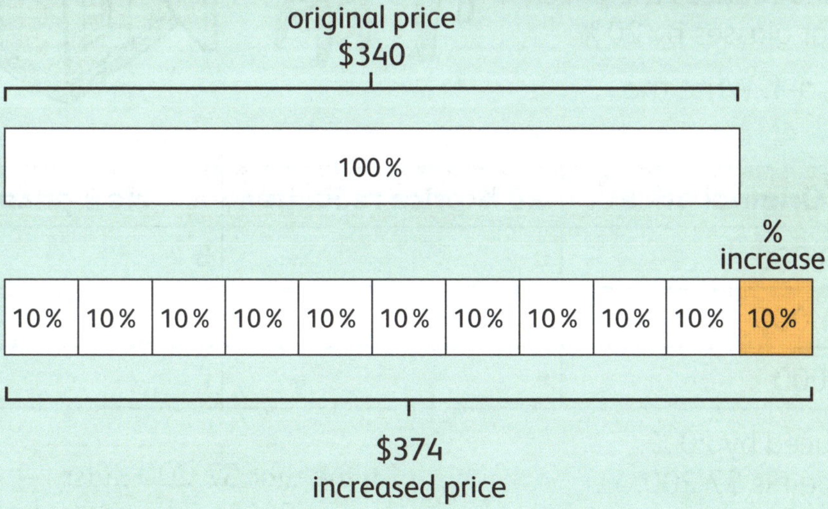

10 % of 340 = 34

100 % + 10 % is 340 + 34 = 374

A cappuccino now costs \$374.

Key words

percentage increase

percentage decrease

Notice how the percentage increase is added to the original quantity.

Now let's look at an example of a **percentage decrease**.

A skateboard is reduced by 25 % in price in a sale. The original price was \$8 000. Find the reduced price.

Work with your partner to sketch a diagram to represent this problem. What is different about it compared to the one used to represent the coffee shop problem?

25 % of \$8 000 = \$2 000

100 % – 25 % is \$8 000 – \$2 000 = \$6 000

The reduced price is \$6 000.

Note how the percentage decrease is subtracted from the original quantity.

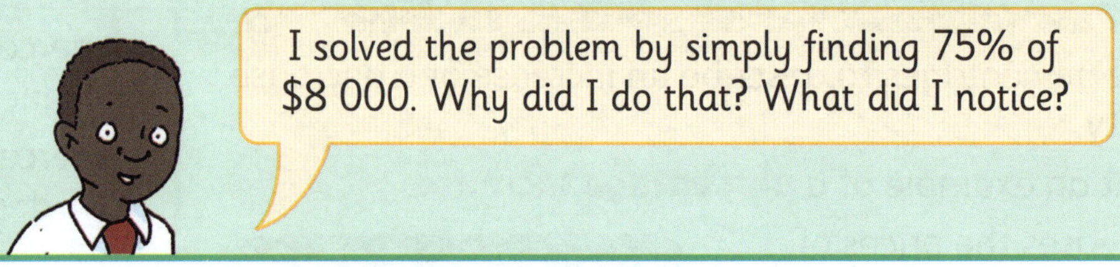

1 An armchair was priced at \$20 000. The price was increased by 25 %. What is the new price?

2 A furniture store has increased its prices by 12 % because the supplier's production costs have risen. Calculate the new price for each item shown here.

3 An optician has a sale and reduces the price of the frames she sells for glasses by 20 %.

Find the missing values, a–f. Write the answers in your book.

Item	Original price	20% price reduction	New price
Ladies' frames	\$19 000	a	b
Men's frames	\$21 000	c	d
Kids' frames	\$5 000	e	f

4 Sunglasses are also reduced by 20 %. The reduced price of a pair is \$7 200.

What was the original price?

5 A travel company is advertising a January sale. The cost of a flight before the sale was US \$240, but the discounted price is now US \$168. What was the percentage discount?

Real-world maths: Choose the better option

1 Andrew wants to buy a shirt with a price tag of $2 000. Which of the two options is better? Use calculations to explain your answer.

Option 1: Buying a shirt that was increased by 40 % and then reduced by 50 %.

Option 2: Buying a shirt that was increased by 50 % and then reduced by 40 %.

2 When you buy an item that will be increased by a percentage and reduced by another percentage, does the order matter? Use calculations to explain your answer.

Converting between ratios, fractions and percentages

Key maths idea

We can convert ratios and fractions into percentages.

Let's look at this problem:

100 air passengers chose an onboard meal for their flight. They could choose from jerk chicken or jerk fish.

The ratio of chicken to fish meals chosen was 3 : 2.

This means that two out of five passengers chose fish.

What percentage of the passengers chose fish and what percentage chose chicken?

$\frac{2}{5}$ chose fish and $\frac{3}{5}$ chose chicken.

We know that $\frac{2}{5}$ is 40 % and $\frac{3}{5}$ is 60 %.

So, 40 % chose fish and 60 % chose chicken.

Remember that if you are not sure about fraction and decimal equivalents, convert the fractions to those with denominator 100.

1 On Saturday, the ratio of people at the beach who hired an umbrella compared to those who did not was 1 : 3. There were 60 people on the beach.

- **a** How many people hired an umbrella?
- **b** What percentage of the people hired an umbrella?
- **c** The umbrella vendor charged $750 per umbrella. How much money did he make?

2 350 people watched a pop concert. The ratio of people who bought a programme compared to those who did not was 3 : 7.

a What percentage of people bought a programme?

b How many people bought a programme?

c How many people did not buy a programme?

d If a programme cost \$160, how much money did the programme seller make at the pop concert?

3 Look at this grid of letters.

L	O	T	E	K
A	F	P	W	I
Y	H	A	D	J
E	X	M	G	U

a What is the ratio of vowels to consonants?

b What percentage of the grid is filled with vowels?

c What percentage of the grid is filled with consonants?

d If one of the vowels is replaced with a consonant, what percentage of the letters are vowels now?

Maths detective: Tangram puzzles

Use your knowledge of percentage to solve these cases.

Case 1

1 Carlyn drew this tangram on a grid of squares. Each square is $\frac{1}{16}$ of the grid.

a Write the letters A to G in your book.

b Write the polygon name for each letter.

c Work out what percentage of the puzzle each polygon represents.

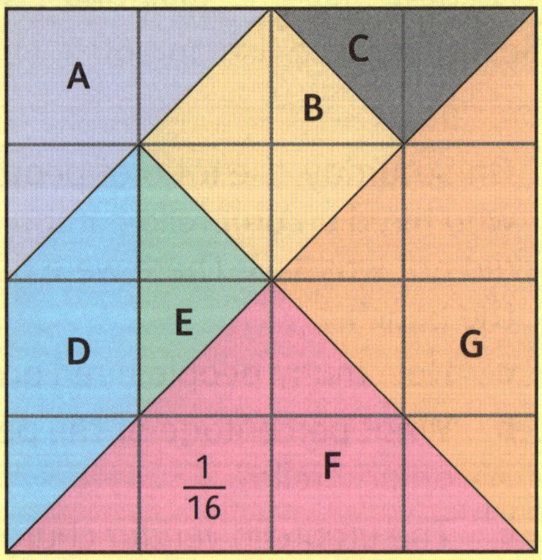

2 Explain how you worked out the percentages. Which ones were easiest? Why?

3 Copy the tangram template onto grid paper and cut out the puzzle pieces.

Case 2

Use your pieces to build the following shapes. Take photographs to keep a record of the shapes you build.

1 A square using 50 % of the pieces.

2 A square using 12.5 % of the pieces.

3 A rectangle using 100 % of the pieces.

4 A rectangle using 25 % of the pieces.

5 A parallelogram using 25 % of the pieces.

6 A parallelogram using 100 % of the pieces.

Case 3

Tangram pieces can be arranged to form many designs. Try arranging your tangram pieces into flowers or animals. Sketch or photograph each design. Write what percentage of the original shapes you used for each design.

What I have learned

1 Answer the questions to show what you have learned about fractions, percentages, ratio and decimals.

- a Explain the concept of percentage and give examples.
- b Describe how you would order percentages. Make up your own list of percentages and order it.
- c Can percentages be expressed as decimals? If your answer is yes, give an example.
- d Can percentages be expressed as fractions? If your answer is yes, give an example.
- e Can percentages be expressed as ratios? If your answer is yes, give an example.

2 a Explain to a partner how to simplify ratios and fractions.

- b Can you simplify percentages? Why or why not?

3 How would you order a set of numbers that contains fractions, decimals and percentages. Draw up a list that contains all three and order the list correctly.

Practice questions

1 Complete the table below by writing the answers for a to k in your notebook.

Simple fraction	Decimal	Fraction with denominator 100	Percentage
$\frac{1}{2}$	0.50	$\frac{50}{100}$	50 %
$\frac{3}{4}$	**a**	$\frac{75}{100}$	**b**
c	0.35	**d**	35 %
e	0.625	$\frac{62.5}{100}$	**f**
$\frac{2}{3}$	**g**	**h**	$66\frac{2}{3}$ %
i	0.14	**j**	**k**

2 Convert each decimal to a percentage.

a 0.55 **b** 0.7 **c** 0.45 **d** 0.08

3 Convert each fraction to a percentage.

a $\frac{4}{10}$ **b** $\frac{8}{10}$ **c** $\frac{17}{100}$ **d** $\frac{6}{100}$

4 Write each percentage as an equivalent fraction and decimal.

a 79% **b** 42% **c** 7% **d** 80%

5 What percentage of each shape is shaded red?

a

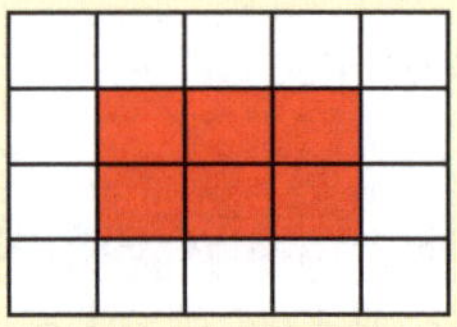

b

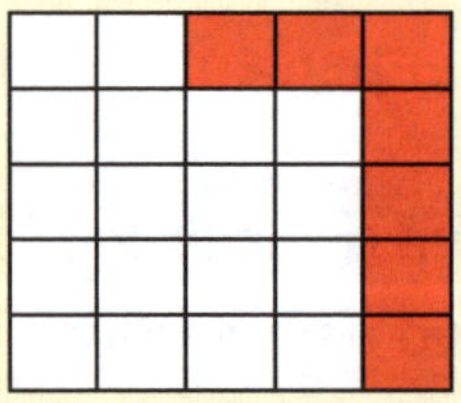

c

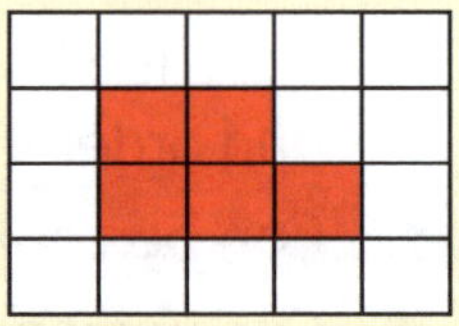

6 **a** Jayden spent 15% of his money on sweets and 31% on clothes. What percentage of his money does he have left?

b Gabrielle spent 20% of her homework time on geography and 24% of her homework time on maths. What percentage of her time does she have left for other subjects?

7 Leon scored $\frac{28}{50}$ in his English test and $\frac{17}{25}$ in his history test. In which subject did he get a higher percentage?

8 Iris is playing an online game. Here are her scores at three different levels:

Level	Points
A	440 out of 550
B	210 out of 300
C	45 out of 90

At which level did Iris have the most success?

9 Savion is shopping for a pair of sneakers. He sees this sign:

a By how much was the pair of sneakers reduced?

b What was the original price of the pair of sneakers?

c If the price of the pair of sneakers had been reduced by 15 %, how much more would Savion have to pay for the pair of sneakers?

12 Self check

Check if you can do this:	If you have difficulty, try this:
1 Show how you can apply your understanding of percentages to real-world contexts.	Think about some real-world contexts where you have seen percentages.
2 Apply your understanding of ratio to percentages and use the symbol % correctly.	Go back to page 190. Remind yourself that we use the % symbol to show percentages.
3 Write a percentage as a fraction with denominator 100 or in its simplest form.	Remember that 'per cent' means 'out of 100' so any percentage can be written as a fraction with denominator 100, and vice versa.
4 Recognise that 100 % is a whole.	Use 10 x 10 grids to practise representing different percentages. Remind yourself that there are 100 equal parts, each worth 1 %, so the whole is 100 %.
5 Convert between fractions and percentages to help solve problems.	Go back to pages 192–193. Remind yourself that it useful to find an equivalent fraction with denominator 100 so you can compare percentages and fractions easily.

6	Write a percentage as a decimal.	Remember that a percentage can be written as a fraction with denominator 100, for example, 15 % as $\frac{15}{100}$. We can then use what we know about dividing by 100 to find the decimal equivalent, for example, 15 ÷ 100 = 0.15.
7	Identify a given percentage of a set or of an object.	Go back to pages 194–195. Remind yourself of the percentage and fraction equivalents you know, e.g. 10 % is $\frac{1}{10}$, 50 % is $\frac{1}{2}$, to help you make decisions.
8	Express a given number as a percentage of another given number (multiple of 10).	Remember that you can use what you know about expressing a number as a fraction of another number to help you, for example, 5 is $\frac{1}{10}$ of 50, so 5 is 10 % of 50.
9	Calculate a given percentage of a number, amount of money, measure of mass, capacity, etc.	Remember that a quantity can be a number or a measure, so you can use the same strategies to help you. Remind yourself how you can use a calculator to help calculate percentages of quantities.
10	Calculate the whole (100 %) when a multiple of 5 per cent (for example, 5 %, 10 %, 15 %, and so on) of the amount is known.	Go back to pages 194–195. Remember that you can use double number lines or diagrams to help calculate the whole.

Unit 2: Measurement

Chapter 13 Area and tessellations

In this chapter you will explore this focus question:

How do I derive and use formulae in measurement situations?

To answer this question, you will:

- understand the concept of area
- estimate and measure to solve related everyday problems
- explore the tiling of a plane using different shapes
- identify shapes that will cover a surface exactly and those that will not
- differentiate between the following units: square centimetre, square metre, hectare and square kilometre
- name, measure and calculate the area of rectangles and right-angled triangles
- break compound shapes into rectangles and triangles to calculate the area
- calculate the surface area of 3-D objects with rectangular or triangular faces
- solve problems involving area measurements.

Starting point

1 Look at this panel on the wall of a new hotel.

a What polygons have been used to cover the rectangular area?

b Choose one of the polygon shapes. Is it possible to cover a rectangular area with that shape? Show how.

2 Work in pairs. Draw the perimeter of your partner's hand on square grid paper. Then draw your own hand on square grid paper.

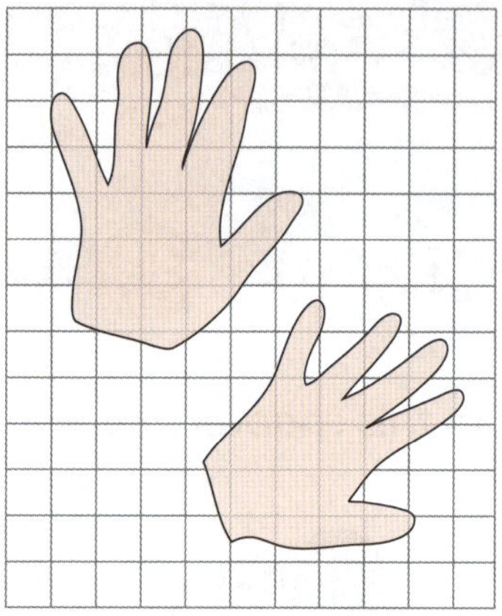

- **a** How can you use the grid to estimate the area and perimeter of each hand?
- **b** Decide with your partner:
  - Whose hand has a bigger area?
  - Whose hand has a longer perimeter?
  - Does a larger area always mean a larger perimeter?
- **c** How did you estimate the area? How did you **estimate** the perimeter?
- **d** What did you do with areas that took up a fraction of a square?

Tessellation

Key maths idea

Tessellation (or **tiling**) is when a surface is covered with a pattern of **two-dimensional** (**2-D**) shapes so that there are no overlaps or gaps. If shapes cover a surface without gaps or overlaps, we say that they tessellate.

We use tiling to cover floors, walls and other surfaces. Tiling is also a way of producing patterns.

A regular tessellation is a pattern in which one regular polygon is repeated.

Only three regular shapes can be used to form a regular tessellation: equilateral triangles, **squares** and regular hexagons.

Key words

estimate
tessellation
tiling
two-dimensional (2-D)
square

Why do you think that only these three shapes will form a regular tessellation? Look at where the shapes meet at a point. Talk to your partner about the angles you see that meet at a point. What is their total?

You will see other patterns made by tessellating more than one shape.

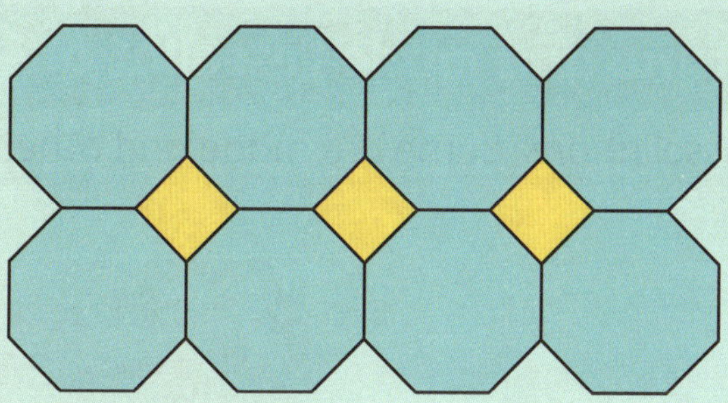

1 What polygons have been used in the following tessellations? Name them.

a

b

c

d

2 Trace and cut out copies of these shapes to make tiles. Use your tiles to see if you can tessellate them on a piece of paper.

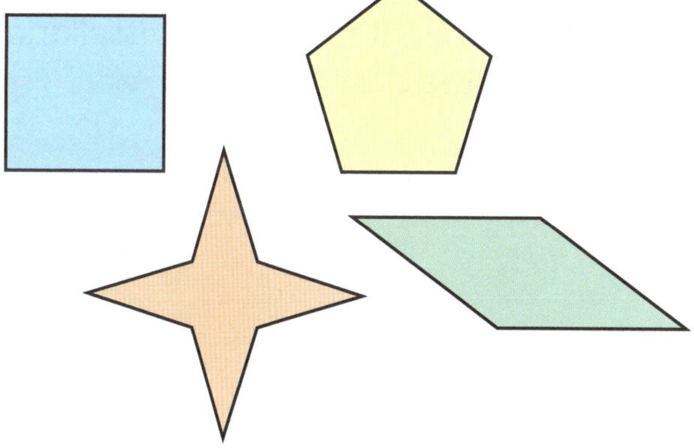

3 Name the two polygons shown in this tessellation.

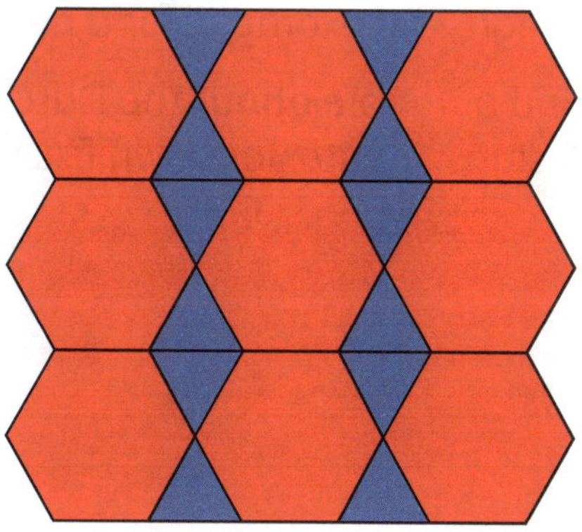

Real-world maths: Natural and made tessellations

Look at these examples of tessellations. Some are made and others are found in the natural world.

a

b

c

d

1 What shapes are tessellated each time?

2 Which examples have regular tessellations?

3 Use the internet or magazines to find other examples of tessellations in the world around you.

Real-world maths: Design and art

1 Talk with a partner about square numbers and how they link to tessellations. For example, 25 square units cover a 5 × 5 square. Try this for different square numbers up to 12 × 12. If you have access to a simple drawing software programme, use it to draw some square number tessellations. You can use different colours to create designs, for example, for a new range of tiles for a bathroom design company.

2 Find out more about the Dutch artist M.C. Escher, who created intricate tessellation designs. Create your own Escher design to present to the class.

Area of rectangles

Key maths idea

You can use a **formula** to calculate the area of many polygons.

To calculate the **area** of a rectangle, we multiply length by width.

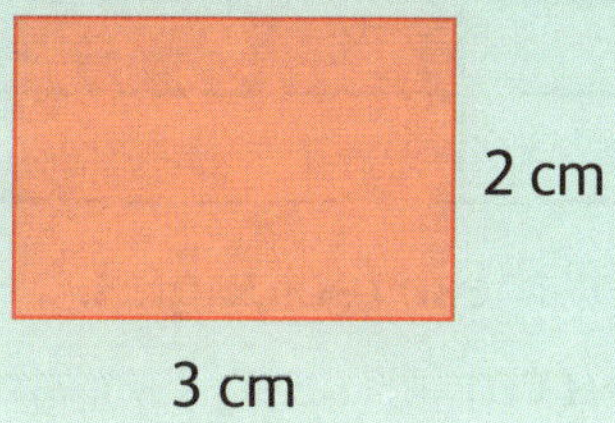

Example

$3 \text{ cm} \times 2 \text{ cm} = 6 \text{ cm}^2$

The area (A) equals the length (l) times the width (w).

The formula to calculate area of a rectangle is therefore:

$A = l \times w$

You can also use this knowledge to solve other area problems.

Example

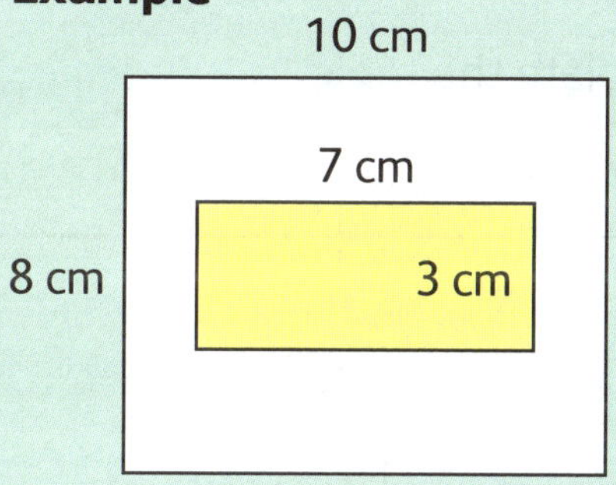

Look at this diagram.
What is the area of the unshaded part?

Area of large rectangle = $10 \text{ cm} \times 8 \text{ cm} = 80 \text{ cm}^2$

Area of shaded rectangle = 7 cm × 3 cm = 21 cm^2

Area of unshaded part = 80 cm2 − 21 cm^2 = 59 cm^2

Key words

formula

area

1 Measure the length and width of these rectangles.

a

b

c

Write down the information to calculate their areas.

Figure	Length	Width	Length × width	Area
a				
b				
c				

2 Find the area of each of these rectangles.

- **a** length 7 cm, width 4 cm
- **b** length 6.5 cm, width 3 cm
- **c** length 9.4 cm, width 2.2 cm

3 Calculate the missing side for each of these rectangles.

a Area = 15 cm^2
Length = 5 cm
Width = _____

b Area = 36 cm^2
Length = 12 cm
Width = _____

c Area = 30 cm^2
Length = 7.5 cm
Width = _____

d Area = 20 m^2
Length = _____
Width = 2.5 m

4 Look at the figures a to c and calculate the values to complete the table.

i

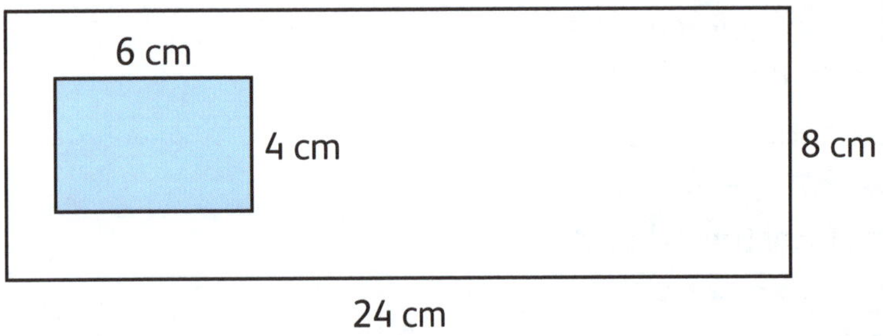

ii

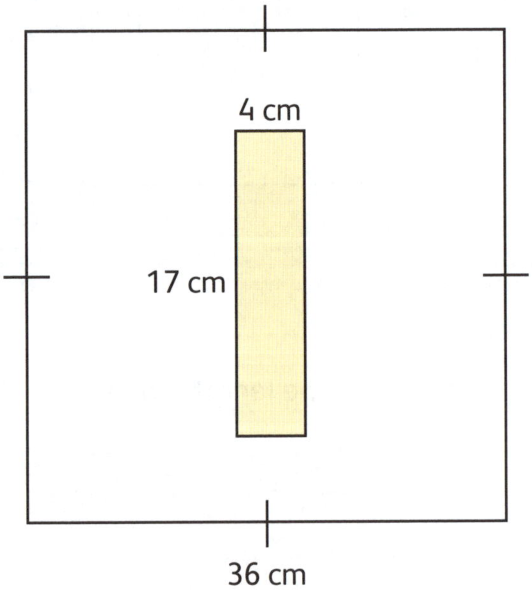

iii

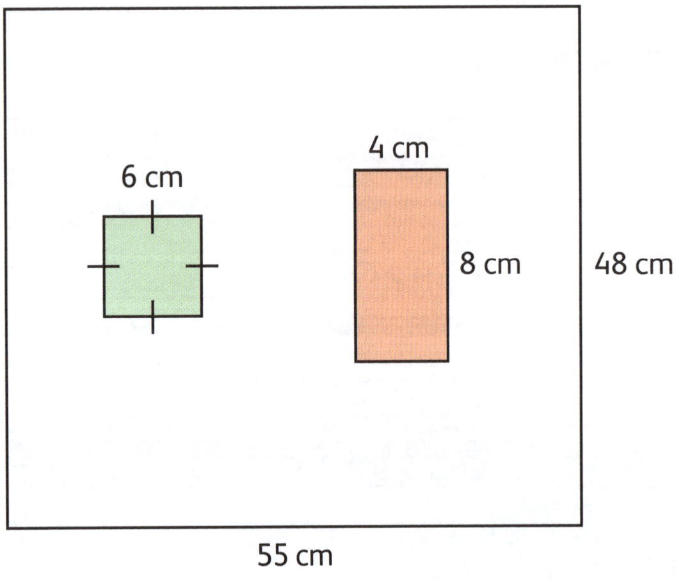

	Area of larger shape	Area of smaller shape	Area of unshaded shape
i	a	b	c
ii	d	e	f
iii	g	h	i

Area of squares

Key maths idea

This square has 4 rows of 4 square centimetres.

The total area is 16 cm^2.

The area (A) of a square can be calculated in a similar way to a rectangle, by multiplying the length and width.

The length and width of a square are the same.

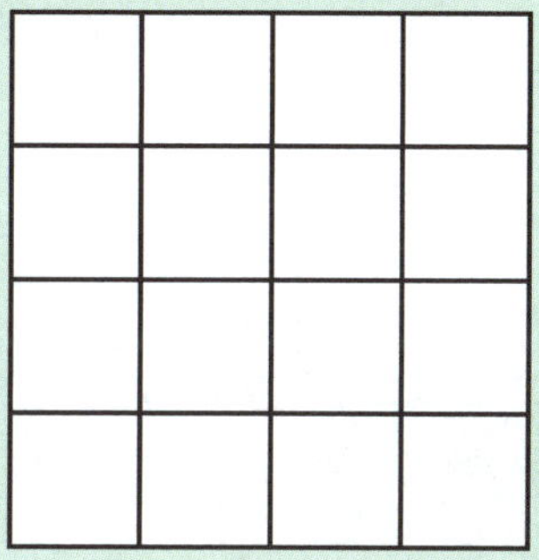

So we can express the formula as:

$A = l \times l$ or $A = l^2$

Remember that the **unit of measure** we choose depends on the size of the object or area.

Key word

unit of measure

1 Use the formula for a square to calculate the area of each of these squares.

a
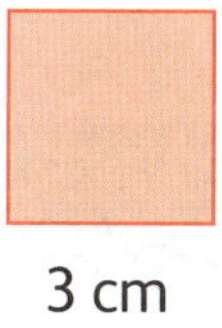

b

c

d
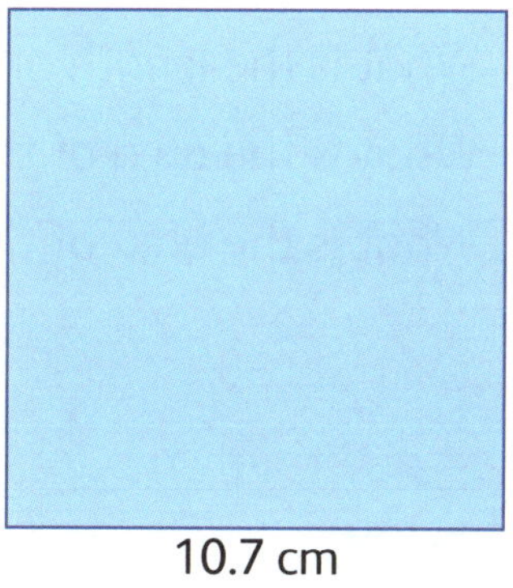

2 The length of one side of a square is 8 cm. What is the area of the square?

3 The area of a square-shaped yard is $144 m^2$. What is the length of the yard?

4 Malcolm records the area of some different square regions for his maths project.

$25 m^2$ 1 ha $4 km^2$ $121 cm^2$

a Order the areas from smallest to largest.

b Write the length of one side for each of the square regions.

5 A rug covers a square floor in a room exactly. The length of the room is 7.7 metres. Calculate the area the rug covers.

6 The Martin family have 600 metres of wire fencing. They used the wire to fence this square garden.

a Calculate the length of one side of the garden.

b Use your answer in part a to calculate the area of the garden.

Real-world maths: Designing a sports complex

A developer is planning to build an outdoor sports complex on a square piece of land. The perimeter of the land is 800 metres.

1 What is the length of one side of the land?

2 What is the area of the land in metres?

3 What is the area of the land in hectares?

Remember: a **hectare** measures 100 m × 100 m, or $10\,000 m^2$.

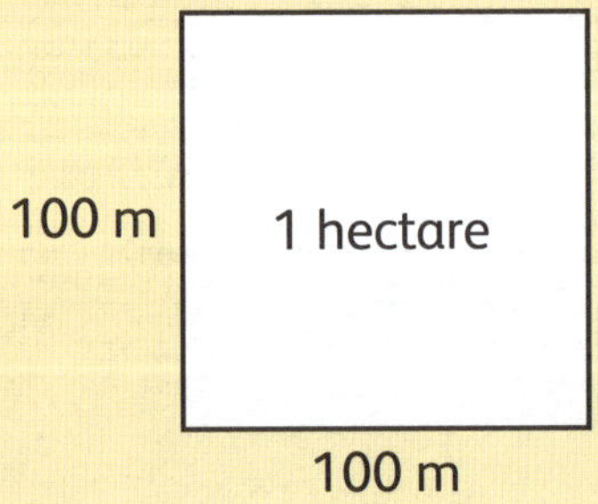

Key word
hectare

Real-world maths: Newspaper layout

How do you think newspaper and magazine designers decide on the layout of the articles and advertisements they want to include?

1 With a partner, cut out a selection of articles and advertisements from newspapers or magazines.

2 Imagine you are working for a newspaper and want to cover as many topics as possible on a page. Decide how you would best arrange your articles and advertisements to achieve this.

3 Find out how newspapers make money. How do your findings change your ideas about your page layout? What changes would you make to your layout based on what you know now?

Area of triangles

Key maths idea

A triangle has three straight sides and thee interior angles.

Look at the shapes on the centimetre grid.

The rectangle WXYZ has:

Length = 4 cm, width = 3 cm

Area = 4 cm × 3 cm = 12 cm²

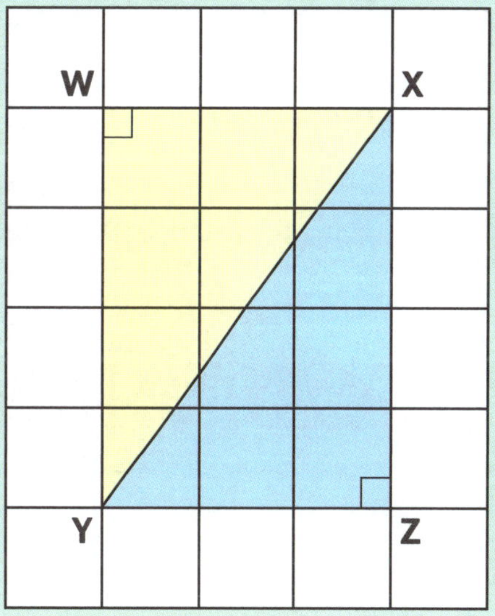

The triangle XYZ covers half the area of the rectangle.

The formula to calculate the area of a triangle is:

The area (A) equals half the area of the length of the triangle base (b) times the height (h)

Area (A) = $\frac{1}{2} \times (b \times h)$

The area of triangle XYZ is:

$\frac{1}{2} \times (3\text{ cm} \times 4\text{ cm}) = \frac{1}{2} \times 12\text{ cm}^2$

$= 6\text{ cm}^2$

1 Calculate the area of each triangle.

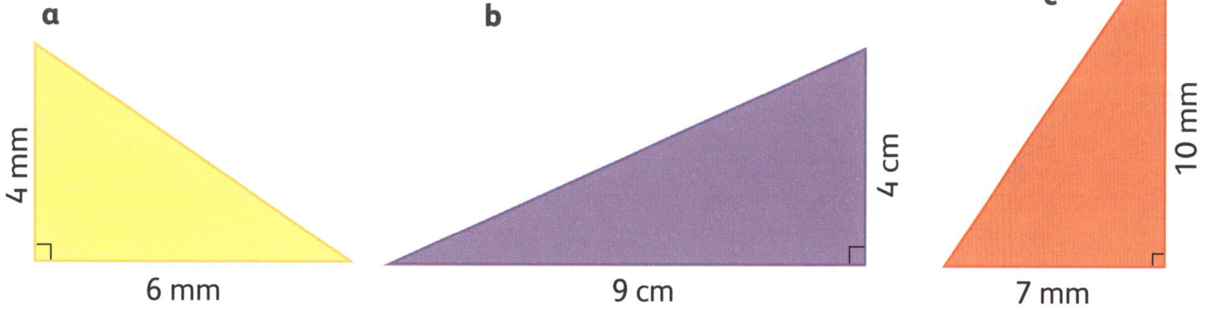

d

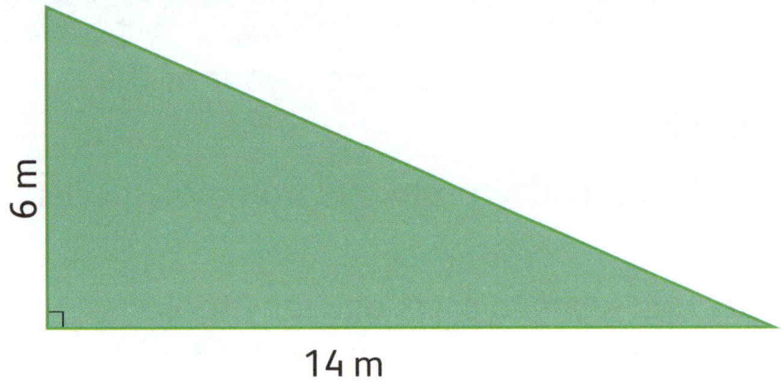

e

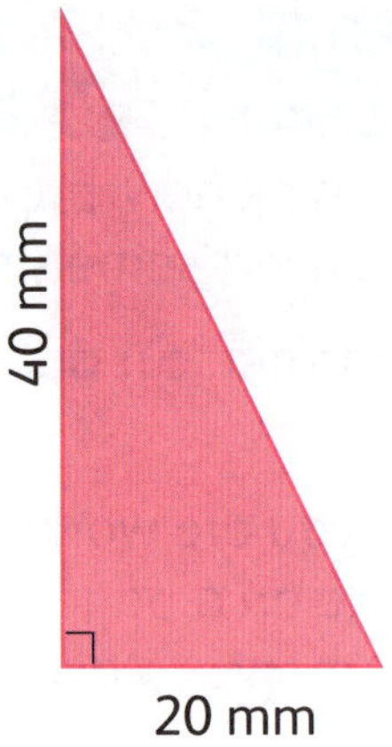

f

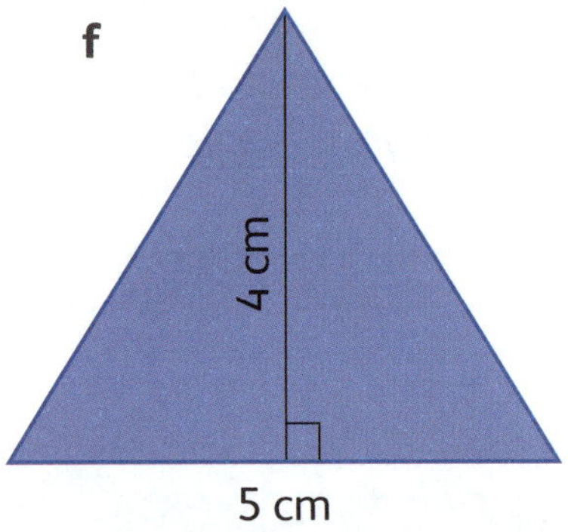

g

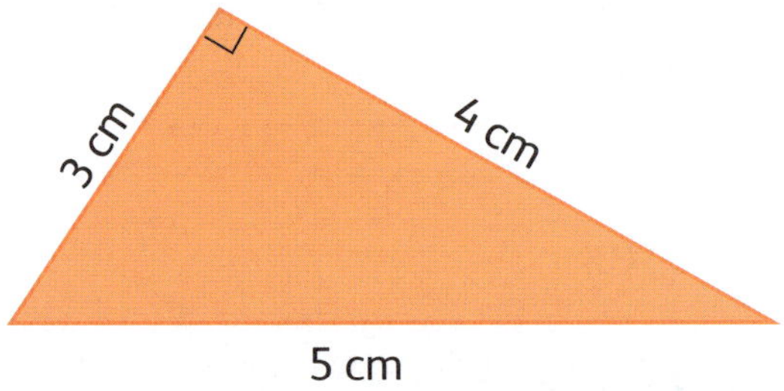

2 Calculate the area of a triangle with a base of 5 cm and perpendicular height of 7 cm.

3 Calculate the area of these triangles.

a

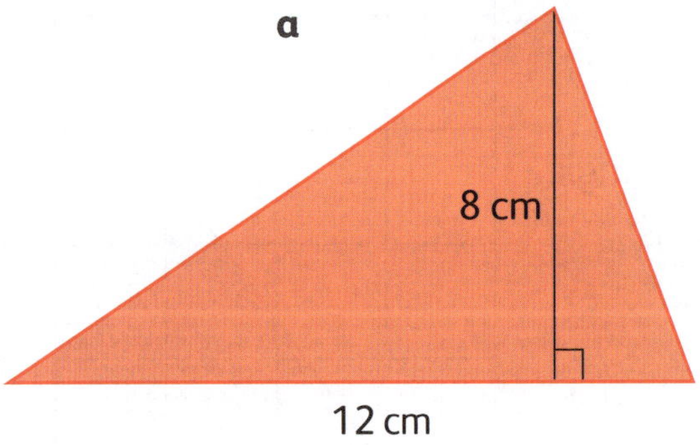

b

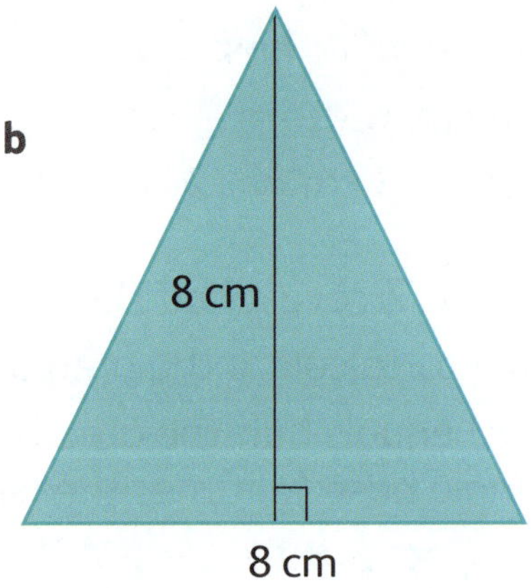

c

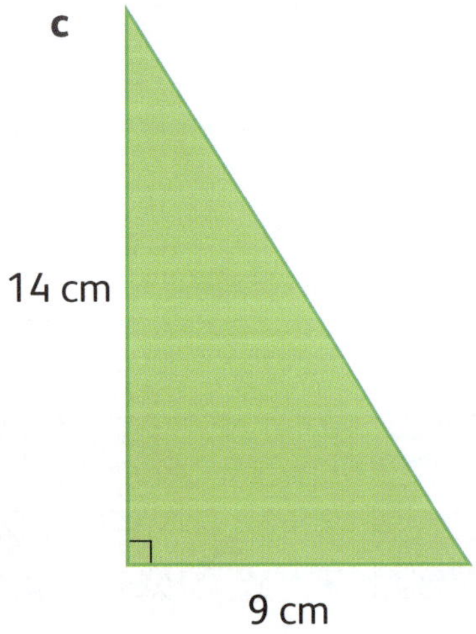

d

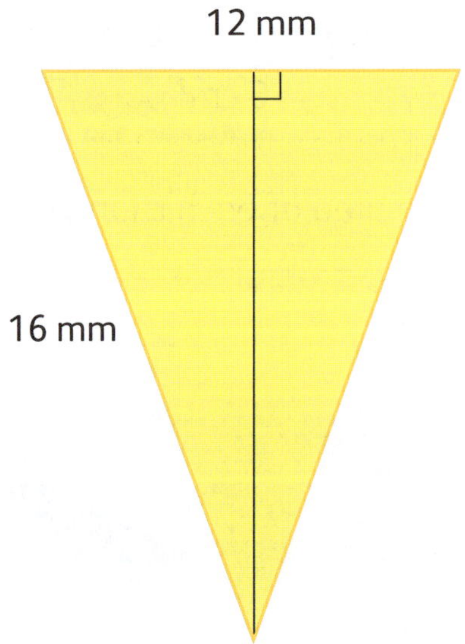

Real-world maths: Designing the layout of a playground

Your school wants to turn a section of the school ground into a play area. Here are the dimensions and shapes of each feature of the play area they want to install.

Play area	Shape of feature	Area of feature	Perimeter of feature
Sandbox	Square	25 m^2	20 m
Jungle gym	Rectangular	8 m^2	12 m
Seesaw	Rectangular	5 m^2	12 m
Slide	Triangular (right-angled)	6 m^2	12 m
Swing	Square	9 m^2	12 m
Netball hoop	Rectangular	24 m^2	20 m
Table tennis station	Rectangular	12 m^2	16 m

Use squared paper or a computer software package to develop and design the play area for your school. Remember to include your scale as shown. Scale: 1 m × 1 m square = 1 m^2

Area of compound shapes

Key maths idea

A compound (or composite) shape is made by combining two or more geometric shapes. This shape is made by combining rectangles.

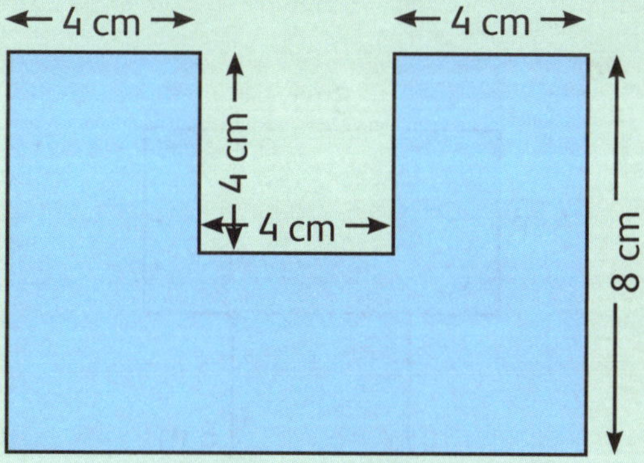

There are different methods for finding the area of a compound shape.

Method 1

Sometimes the compound shape can be seen as two or more shapes joined together.

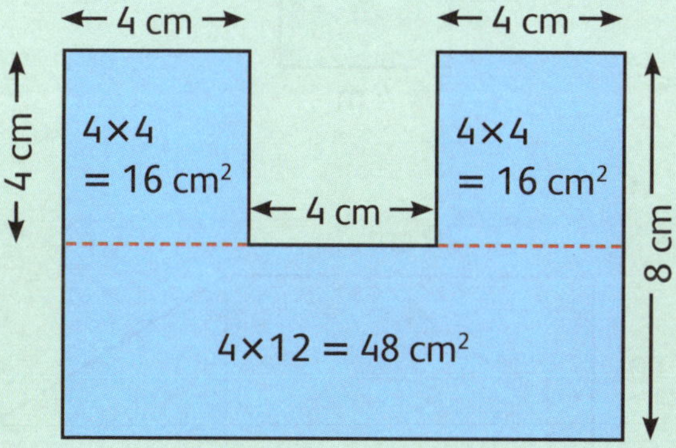

Step 1 Divide the shape into squares and rectangles.

Step 2 Find their individual areas and then add them together.

The length of the larger rectangle is $4 + 4 + 4 = 12$ cm.

Area of compound shape = $(16 + 16 + 48)$ cm^2 = 80 cm^2.

Method 2

Imagine this shape as a large rectangle with a section cut out.

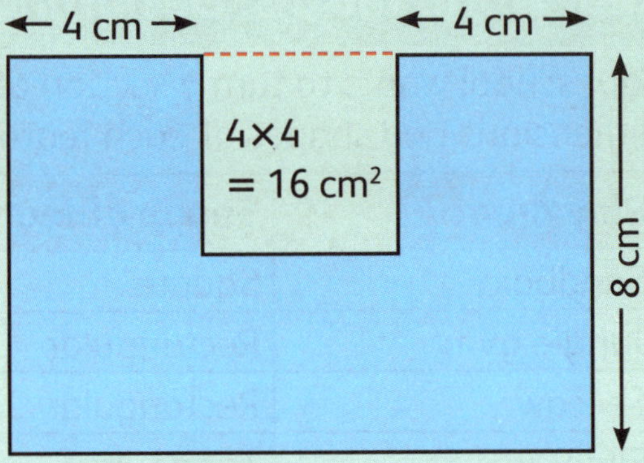

The length of the outer rectangle is $(4 + 4 + 4)$ cm = 12 cm.

Find the area of the large rectangle (12×8) and then subtract the part that has been cut out (4×4).

Area of compound shape = $(12 \times 8)\text{ cm}^2 - (4 \times 4)\text{ cm}^2 = 96\text{ cm}^2 - 16\text{ cm}^2 = 80\text{ cm}^2$.

Here is another compound shape. It is made up of a square and a triangle.

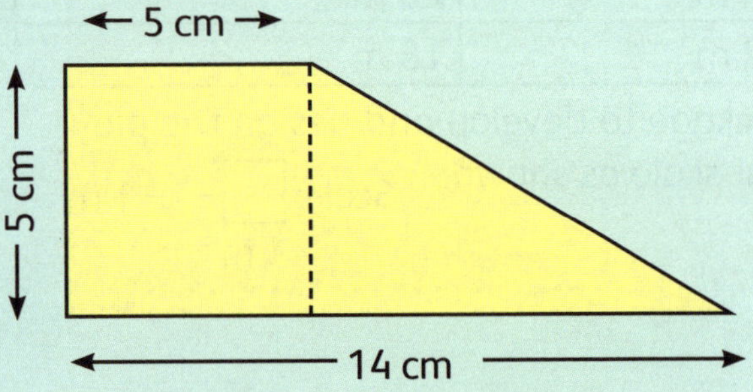

Discuss with your partner how you can use method 1 or method 2 to calculate the area of this compound shape.

1 Calculate the area of these compound shapes.

a

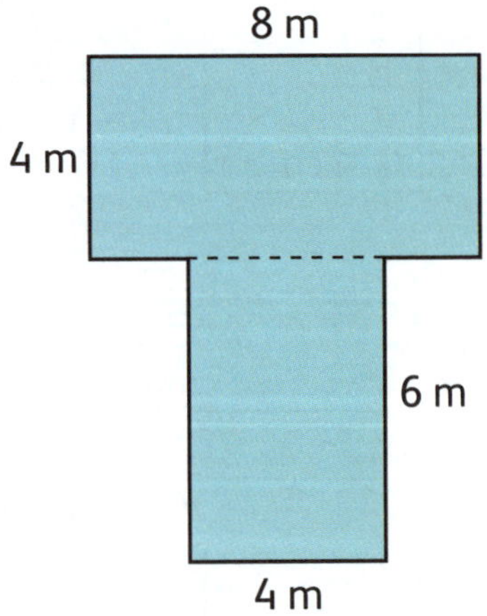

b

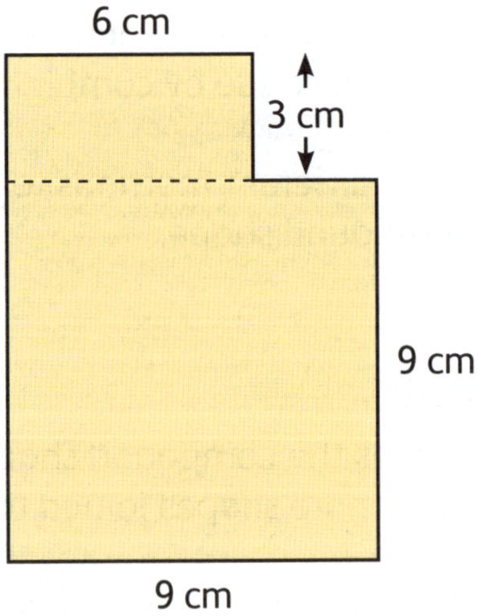

c

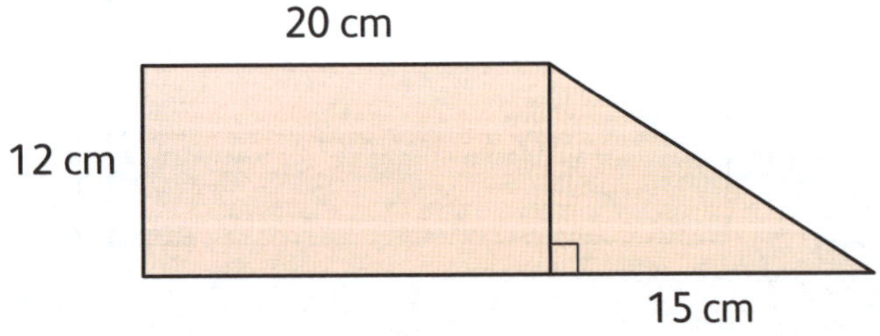

d

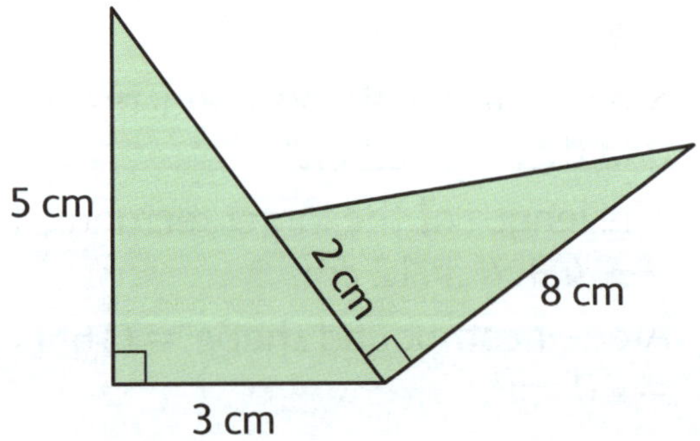

2 Calculate the area of each shape. Think about which method to use for each.

a

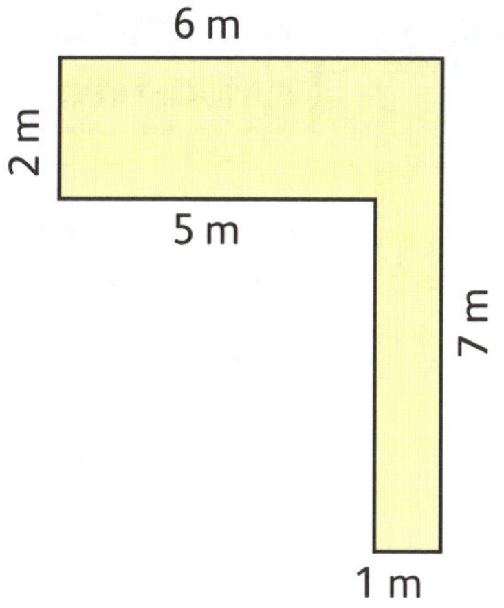

b

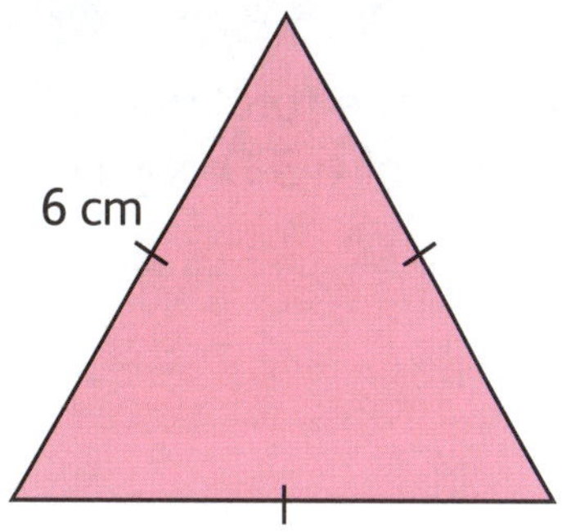

c

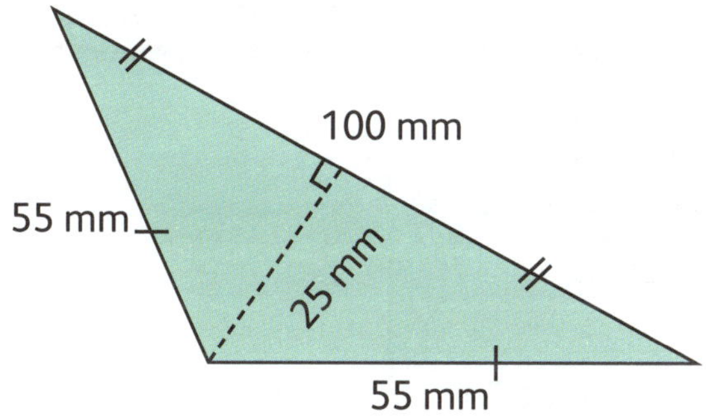

d

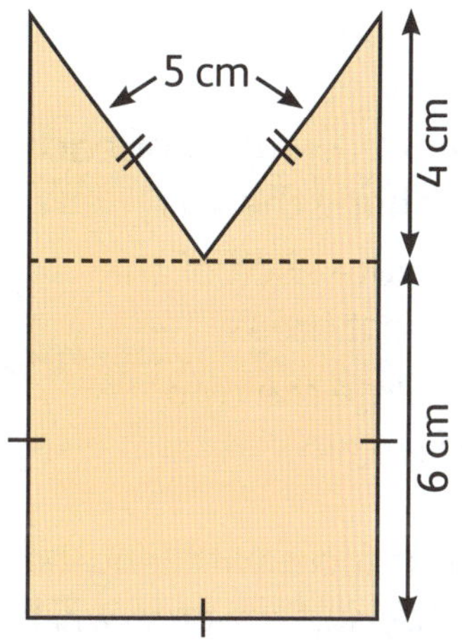

e

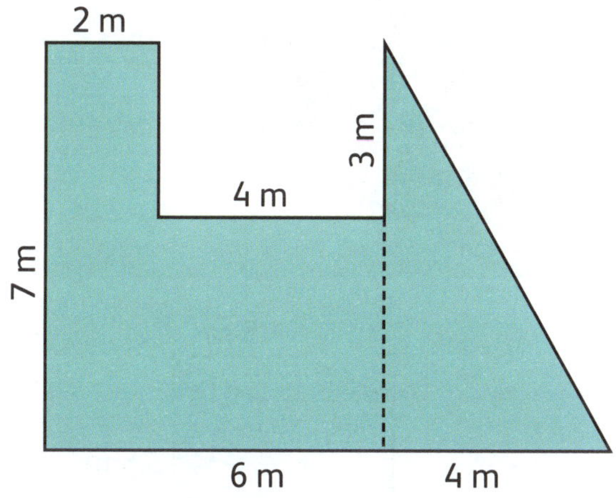

f

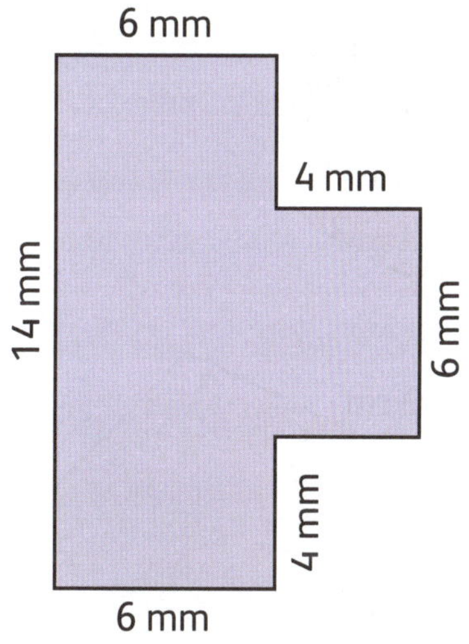

Surface area

Key maths idea

The **surface area** is the total area of the outside surfaces of a 3-D object. Look at this cube.

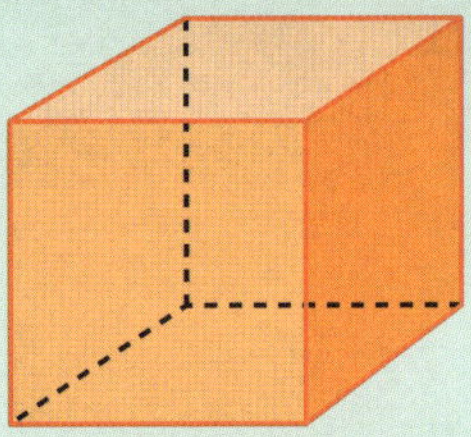

If all six faces are painted, the total painted area is called its surface area.

To calculate the surface area:

Step 1 Decide which shape each face is.

Step 2 Count how many of each size shape there is.

Step 3 Calculate each of the different areas.

Step 4 Calculate the total surface area.

If this cube has faces with lengths of 2 cm, explain to your partner why the surface area will be 24 cm².

Key word

surface area

Remember:

area of rectangle = length × width

area of square = side × side

area of triangle = $\frac{1}{2}$ (base × height)

1 Use the area formulae to calculate the surface area of each 3-D object.

a

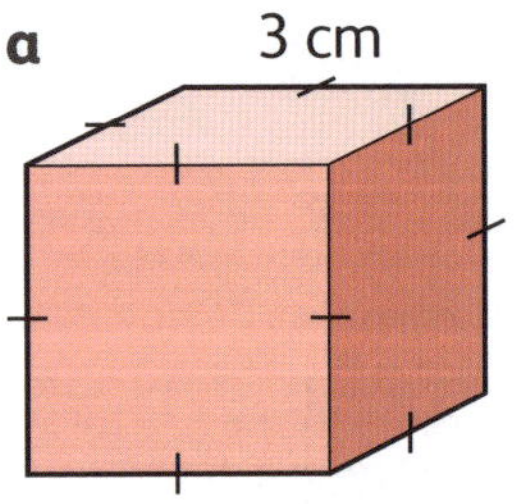

b

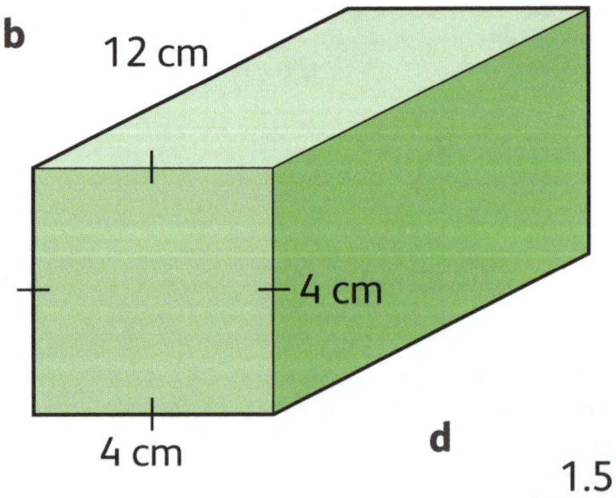

c

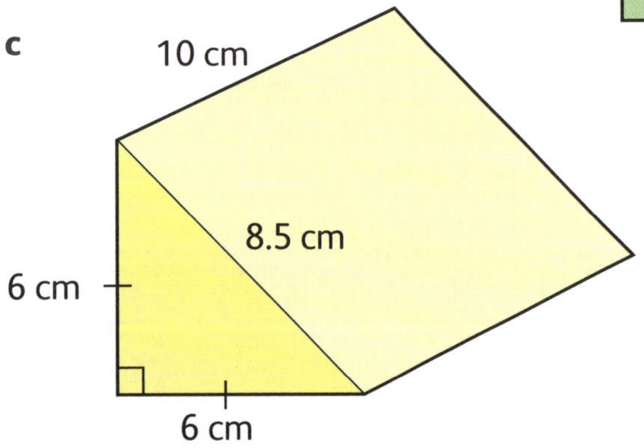

d

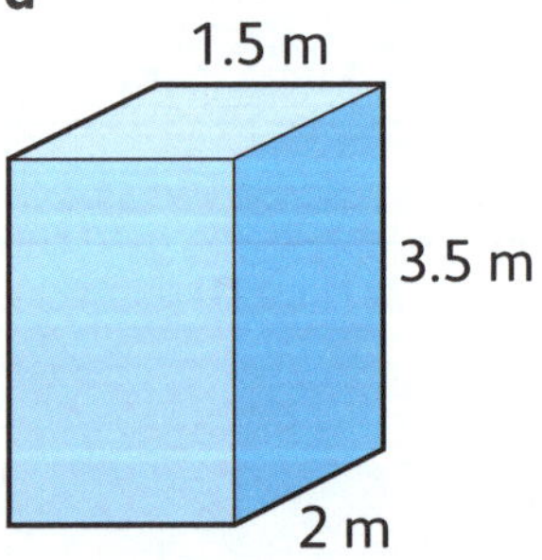

e

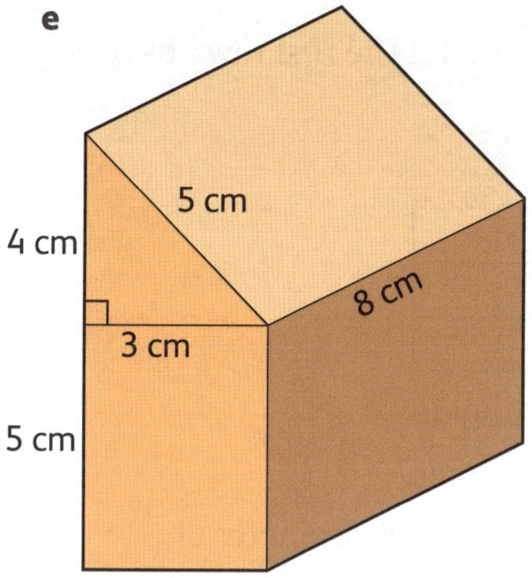

f

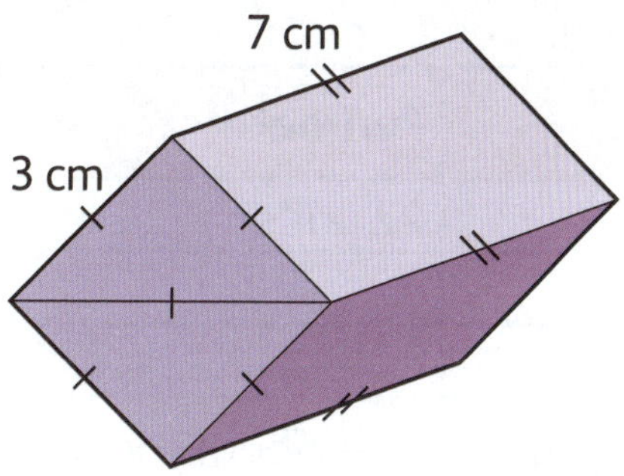

2 All the surfaces of large cube-shaped and cuboid-shaped blocks must be painted for the set on a film production. There are 3 cubes and 4 cuboids.

- The cubes have faces with dimensions of 3 m × 3 m.
- The cuboids have two square faces with dimensions of 2 m × 2 m, and 4 rectangular faces with dimensions of 2 m × 5 m.
- A 5 litre tin of paint will cover a surface area of 70 m².

a What is the surface area of a cube-shaped block?

b What is the surface area of a cuboid-shaped block?

c How many square metres must be painted in total?

d How many tins of paint are needed in total?

Maths detective: Investigating area

Use what you know about perimeter, area and polygons to solve these problems.

Case 1

1 This diagram shows three figures made up of three 2 × 3 rectangles consisting of unit squares.

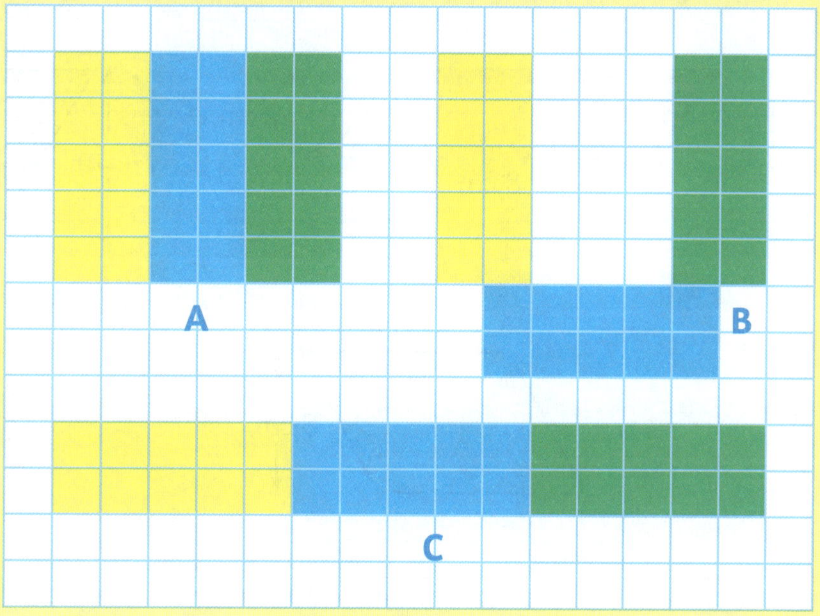

2 Write only the answers in your book to complete the table. The first row has been done for you.

	Perimeter	Area
Figure A	22 units	30 square units
Figure B	a	b
Figure C	c	d

3 What can you conclude about the area of each rectangle?

4 What do you notice about the perimeter of each rectangle?

5 What general observation can you make about the area and perimeter of rectangles A, B and C?

Case 2

Explore finding the area of a parallogram by splitting it into triangles and rectangles.

Use square paper to draw a parallelogram like the one shown here. Cut off triangle C and move it to the other end of the shape to make a rectangle.

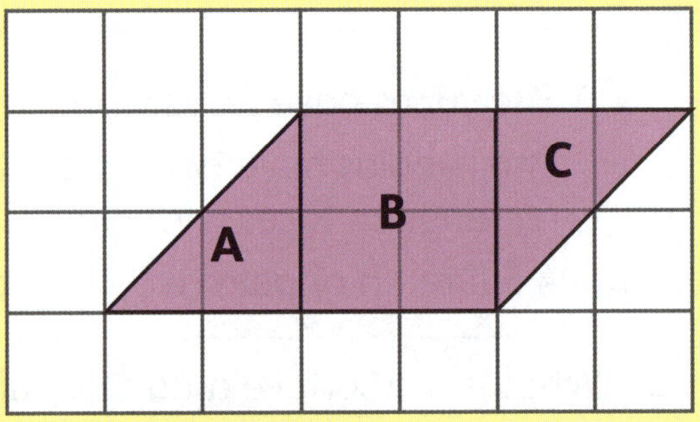

1 What is the area of the rectangle you have made?

2 Explain why the original parallelogram must have the same area as the rectangle.

3 How can you use your findings to come up with a formula for finding the area of any parallelogram?

4 Try out your formula to find the area of these parallelograms.

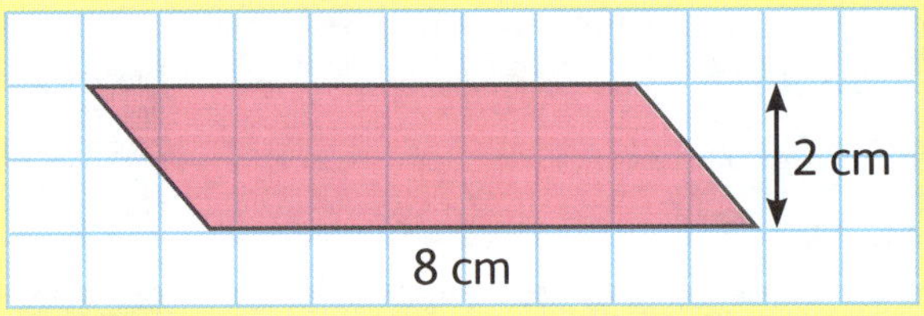

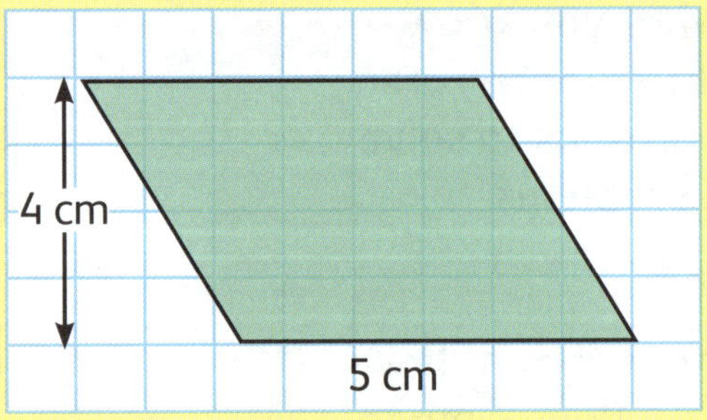

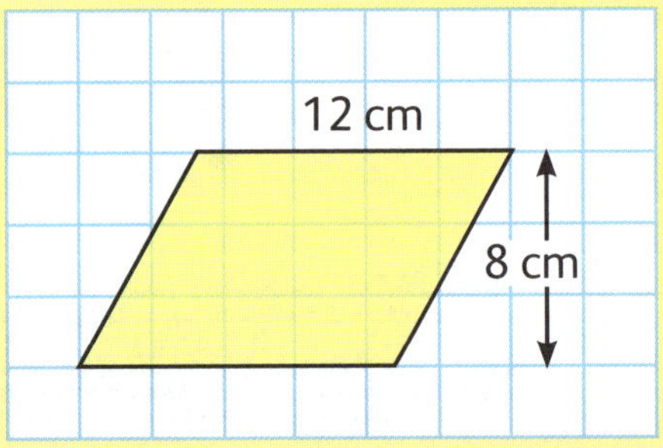

Case 3

This is a plan of a parking lot at a shopping centre. Each square on the plan has sides 10 metres long.

1 Calculate the area of the parking lot.

2 Draw a flow diagram to show the steps you followed to work out the area.

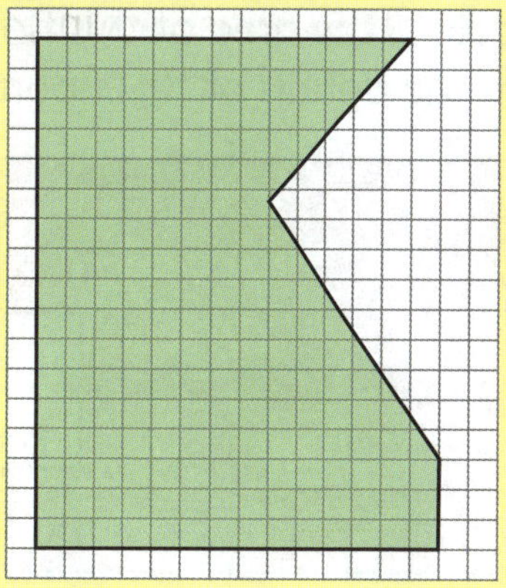

What I have learned

Read the statements. Then fill in the missing words to summarise this topic.

1 The area of a shape is given in _____ units.

2 The area of a shape is the amount of _____ it covers.

3 A unit square has dimensions of _____ _____ in length and _____ _____ in breadth.

4 The surface area of a solid is the sum of the area of all its ______________ .

5 _____ and _____ are two polygons used in a regular tessellation.

6 To find the area of a compound shape, split the shape into _____ shapes, work out the _____ of each shape, then add them together.

Practice questions

1 Which one of the following regular polygons would *not* form a regular tessellation?

a squares b hexagons c triangles d pentagons

2 What is the area of a rectangular cane field measuring 98 m by 385 m?

3 Use a formula to work out the areas of these squares and rectangles.

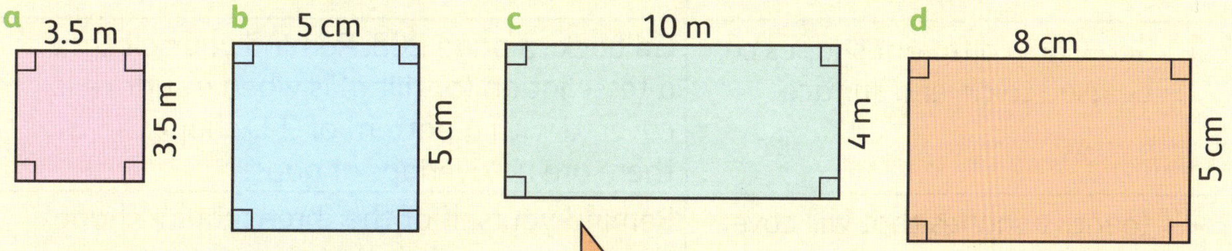

4 Calculate the area of this triangle.

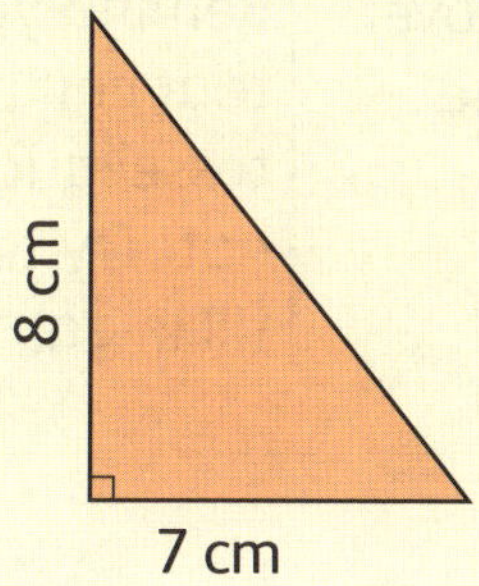

5 Calculate the area of this floor from the details of this floorplan.

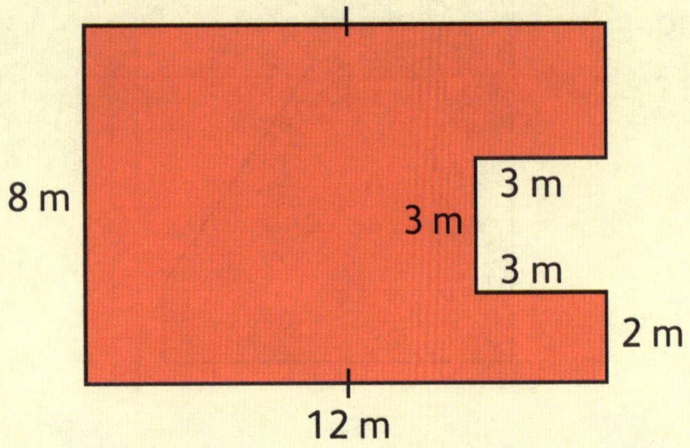

6 What is the surface area of this cuboid?

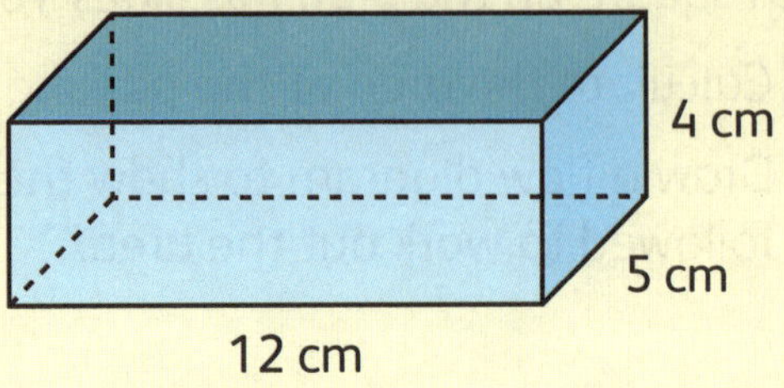

7 Kira wants to decorate the exterior faces of this gift box in the shape of a cube. The figure shows the net of the gift box.

a What is the surface area, in square centimetres, of the gift box?

b Decorations for the gift box cost $143.75 per square centimetre. How much will Kira pay for decoration if she decorates all the faces?

Net of Kira's gift box

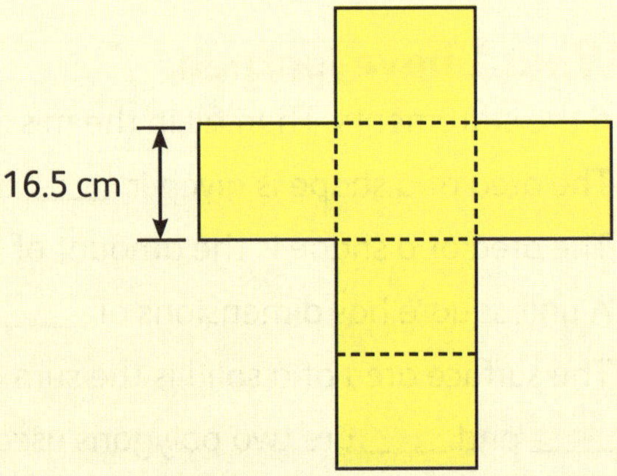

13 Self check

Check you can do this: 	If you have difficulty, try this: 
1 Show how different shapes can be used to cover a surface.	Go back to page 208. Remind yourself that a tessellation (or tiling) is when a surface is covered with a pattern of 2-D shapes so that there are no overlaps or gaps.
2 Identify shapes that will cover a surface exactly and those that will not.	Remind yourself of the three regular shapes that can be used on their own to make regular tessellations. Now think about those that will not. Try to identify shapes that can be used to fill in gaps.

3 Understand the concept of area; estimate and measure to solve related everyday problems.	Go back to page 211. Remind yourself that the area of a shape is the amount of surface it covers and is measured in square units, such as square centimetres (cm^2), square metres (m^2), etc. Talk to a partner about some examples of estimating and measuring that you did.
4 Differentiate between the following units: square centimetre, square metre, hectare and square kilometre.	Remember that an area can be very small or very large, or somewhere in between. You already know that centimetres are used to measure shorter lengths and kilometres are used to measure longer lengths and distances. Think about how you relate this knowledge to units of measure for area.
5 Name, measure and calculate the area of rectangles and right-angled triangles individually.	Remember that we can use a formula to calculate the area of a rectangle. You may find it useful to think about the arrays you have seen in your multiplication work and how you calculated the total number of squares or dots. Go back to pages 213–215 to remind yourself of the formula to calculate the area of each of these shapes.
6 Break compound shapes into rectangles and triangles to help calculate the area.	Go back to pages 217–218. You can find the areas of compound shapes by splitting them into rectangles and triangles so you can use the formulae you know.
7 Calculate the surface area of 3-D objects with rectangular or triangular faces.	The surface area of a 3-D object is the total area of its faces. You can use the formulae you know for calculating the area of rectangles, squares and triangles to help you. Remember that you can use what you know about the properties of 3-D objects to identify the shape of each face.
8 Solve problems involving area measurements.	Look back at some of the different problems that you have seen in this chapter. Remember that it is important to use diagrams to help make decisions.

Chapter 14 Area and volume

In this chapter you will explore this focus question:

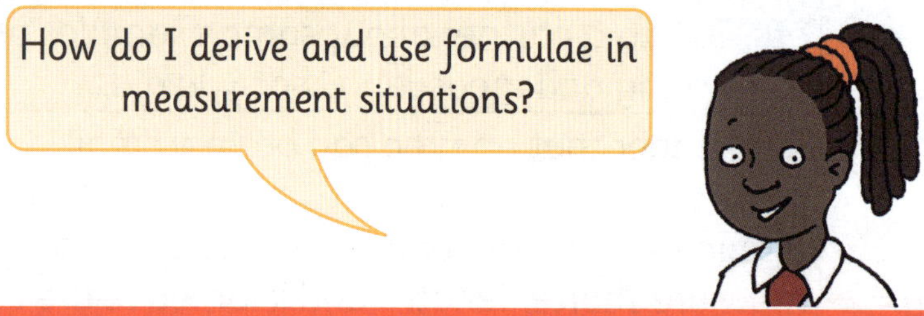

To answer this question, you will:

- investigate, estimate and calculate the volume of cubes and cuboids
- explain or use examples to show what is meant by a unit solid
- build unit solids of volume 1 cm^3, 1 dm^3 and 1m^3
- calculate the volume of a cuboid when given the number of unit solids in one layer and the number of layers
- investigate and use the formula for the volume of a cube or cuboid to solve problems.

Starting point

Look at these gift boxes. Which do you think holds the most? Which takes up the most space? How could you find out?

Unit solids

Key maths idea

Solids have three dimensions: length, width and height. They are **three-dimensional (3-D)** objects.
The **volume** of a solid is the amount of space it takes up. We express volume in **cubic units**.
We can use cubic centimetres (cm^3), cubic decimetres (dm^3) or cubic metres (m^3) as cubic units of volume.

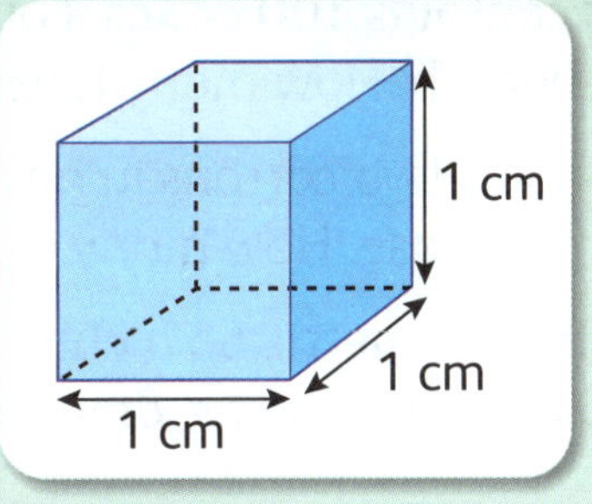

This cube has length, width and height equal to 1 cm.

We say this is a **unit solid** because each dimension is 1.
A cube with length, width and height equal to 1 m is also a unit solid. Explain why to your partner.

Key words

three dimensional (3-D)
volume
cubic unit
unit solid

We can use the word 'depth' or 'height' to describe the vertical dimension of a solid.

1 Use centimetre cubes to make each of these solids. Then calculate the volume of each in cubic centimetres.

a

b

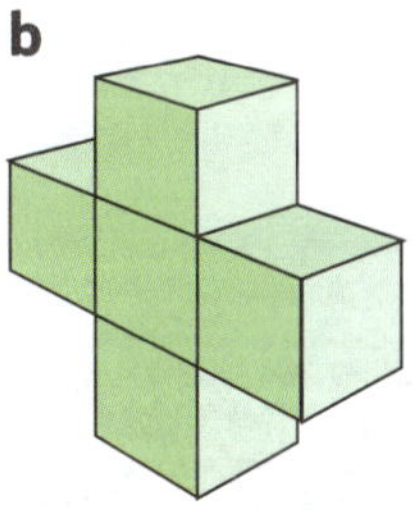

c

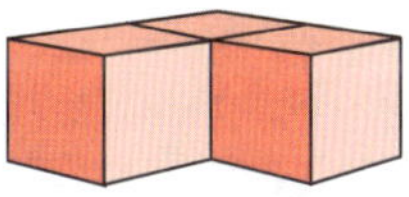

d

e

f

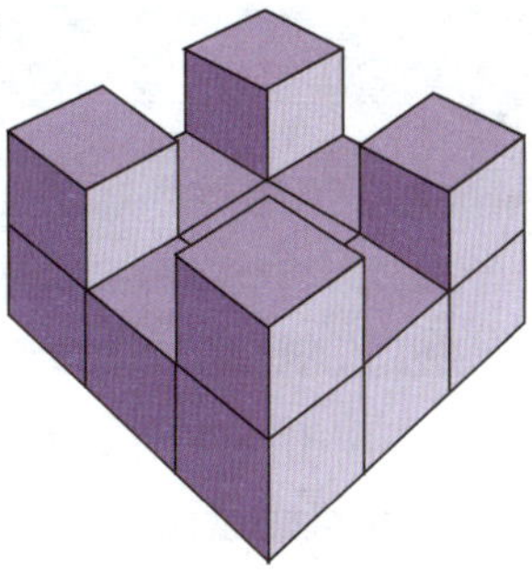

g

For each of the problems in Exercises 2 to 6 you could draw diagrams or use centimetre cubes to help you.

2 Now calculate the surface area of each solid in Exercise 1.

3 a In a box, six square containers can fit in one row. The box can hold four rows. How many containers can fit in one layer of the box?

b If the box can hold three layers of containers, how many containers can the box hold altogether?

4 A crate of cereal boxes contains 100 boxes. Four boxes fit the width of the crate and five boxes fit along the length. How many layers of boxes are in the crate?

5 A case of orange juice holds two cartons in height and four cartons along its width. There are 48 cartons in the case. How many cartons fit along the length of the case?

6 A decimetre is one tenth of a metre. So a cubic decimetre (dm^3) is the volume of a cube with sides of 10 cm each. How many cubic centimetres are equivalent to a cubic decimetre?

7 Choose a box or another cuboid-shaped container in the classroom.

a Estimate how many centimetre cubes it will hold.

b Measure the base of the box and then decide how many centimetre cubes are needed for one layer.

c Adjust your estimate if needed.

d Use centimetre cubes to fill the box. How close was your estimate?

Real-world maths: Solids around you

Work with a partner

1 Find or take photographs of buildings, furniture, artworks, children's play areas and other structures.

2 Make a poster to show how solids are used in real -life situations. Which solids are most common in buildings? Why?

3 Try to use the vocabulary of 3-D shapes to talk about the number of **faces**, **vertices** and **edges** each solid has.

Key words

face

vertex

edge

Volume of cuboids

Key maths idea

To calculate the volume of **cubes** and **cuboids** we can use this formula:

volume = length × width × height

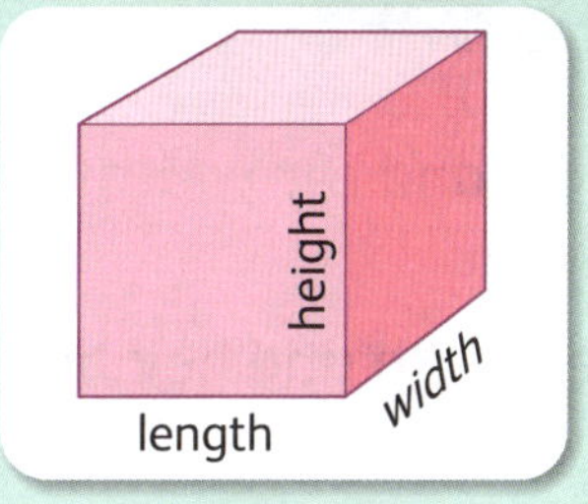

Remember: for a cube, the length, width and height will all be the same! We also call a cuboid a **rectangular prism**.

Key words

cubes
cuboids
volume
rectangular prism

1 Copy and complete the sentences for each cuboid.

a The length is _____ .

b The width is _____ .

c The height is _____ .

d The volume is _____ × _____ × _____ = _____ .

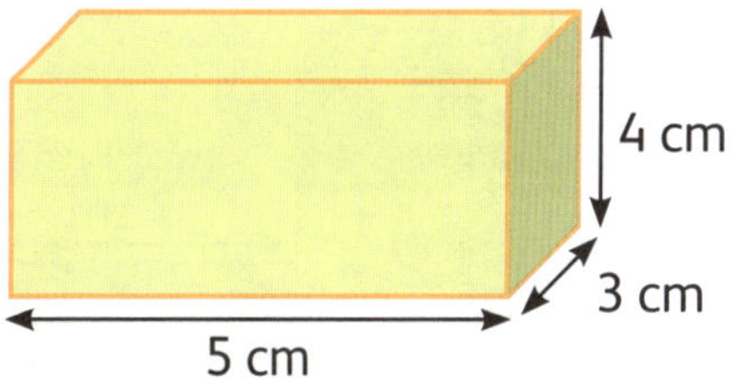

2 Find the volume of each cube.

a

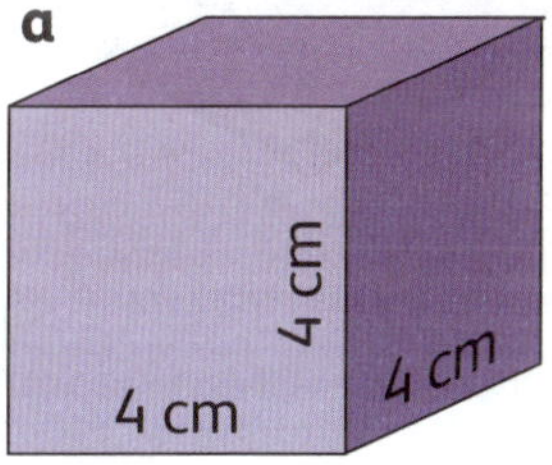

b

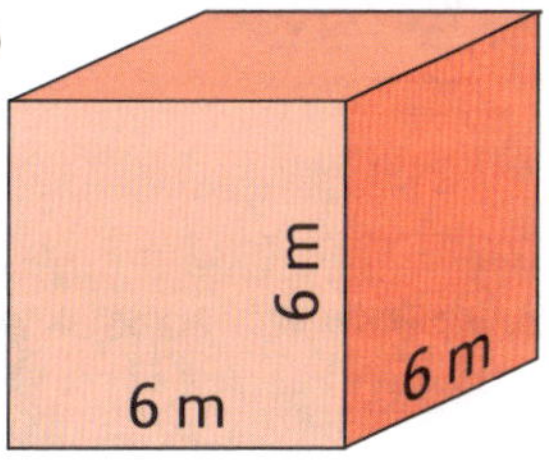

3 Find the volume of these cuboids.

a

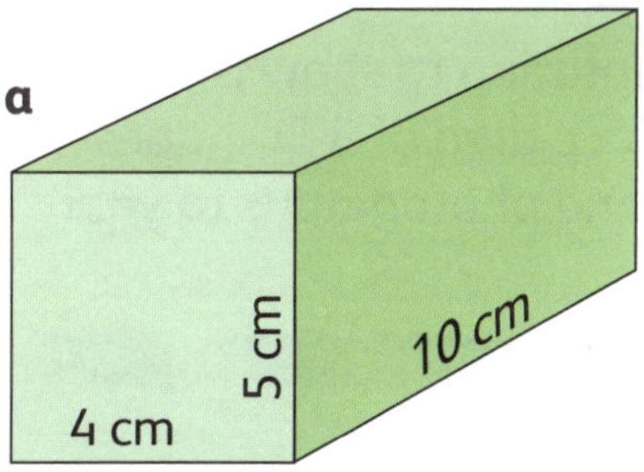

b

4 The volume of the cuboid is 32 cm^3.
Calculate the height.

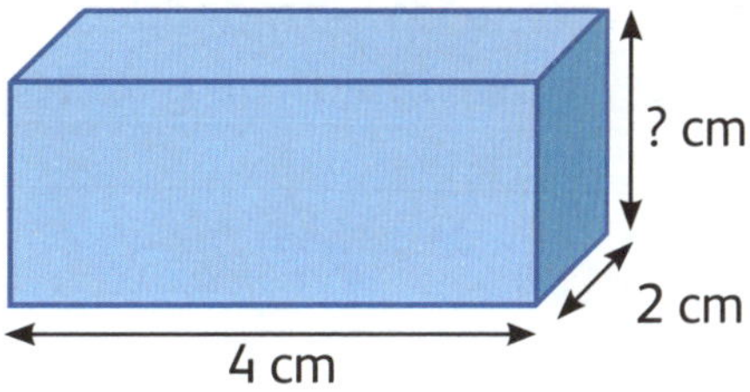

5 Iris is using decimetre cubes to estimate the volume of a toy box. She thinks the volume will be 20 dm^3.

a Do you agree with Iris?

b What do you think the approximate volume of the toy box will be? Explain your answer.

6 The volume of a rectangular prism is 462 cm^3. The length is 11 cm and the width is 6 cm. What is the height of the rectangular prism?

7 The volume of a fish tank is 735 cm^3. The length is 15 cm. If the length and height of the fish tank are equal, what is the width?

8 Calculate the volume of this compound solid.

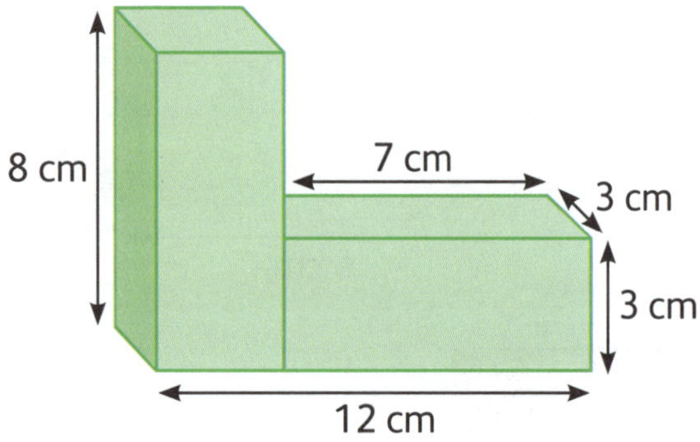

9 Calculate the volume and the surface area of this wooden box. Think carefully about the units of measure you need to use for each calculation.

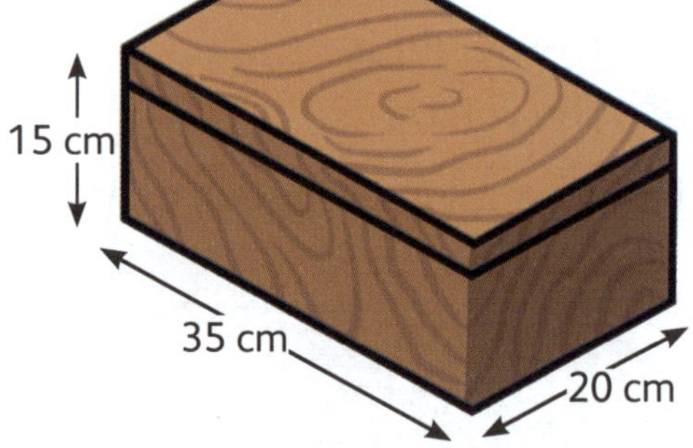

Maths detective: 2-D shapes and 3-D solids

Use your knowledge of 2-D shapes and 3-D solids in this section.

Case 1

Work in groups. Make a poster or prepare a computer presentation to show the difference between finding the area of 2-D shapes, the surface area of 3-D shapes and the volume of 3-D shapes. Add information, diagrams and photographs to your explanation.

Case 2

Think about the size of different cubes. Decide how many cubic centimetres each one will hold. Think about what materials you could use to build unit solids of 1 cm^3 and 1 dm^3. Use these to come up with an idea to construct a unit solid of 1 m^3. What materials could you use? How would you glue together the faces to make sure your cube is stable?

Real-world maths: Water volumes

1 An Olympic swimming pool is 50 metres long, 25 metres wide and 2 metres deep. What is its volume in cubic metres?

2 We can also say that the volume of the pool is 2 500 000 litres.

a What conclusions can you draw about 1 000 litres and 1 cubic metre?

b What do you think a kilolitre is? Explain your thinking.

3 The quote for filling the pool is $1 950 per cubic metre. How much will it cost to fill the pool? Show your working.

4 Carry out an investigation to find out how long it takes to fill a bottle or a bucket with water. Use this information to estimate how many hours it will take to fill the Olympic pool. Show your working.

5 Did you know that a thunderstorm can release 450 000 kilolitres of water? If all this water could be collected and used to fill Olympic-sized pools, how many pools could be filled? Show your working.

Maths detective: Eureka!

Use your scientific investigation skills and what you have learned about volume to solve these problems.

Case 1

You already know that the volume of a solid is how much space it occupies and that you can work out the volume of a cube or cuboid using its dimensions.

Volume = length × width × height

1 Marcha has a metal cube with sides 2 cm long. She works out that is has a volume of 8 cm^3. Marcha then does this experiment:

Step 1: Marcha puts 50 millilitres of water into a measuring cylinder. She marks the level with a small elastic band like this:

Step 2: She carefully places the cube in the water and marks the new level with another elastic band like this:

Step 3: Marcha writes this conclusion:

Conclusion

The water moves because the cube takes up some of the space in the cylinder.
There is still 50 ml of water in the cylinder, but it now shows 58 ml. 50 ml of water and 8 ml of cube.
I already know that the cube has a volume of 8 cm^3, so 1 millilitre must be equivalent to 1 centimetre cubed.

So: 1 ml = 1 cm^3.

2 In science we say that the cube displaced 8 ml of water. Write your own definition of displacement.

3 How can this method help you find the volume of irregular shaped objects such as a small stone, a button or a penlight battery?

Case 2

Use the information in the diagrams to work out the volume of each item in cubic centimetres.

1

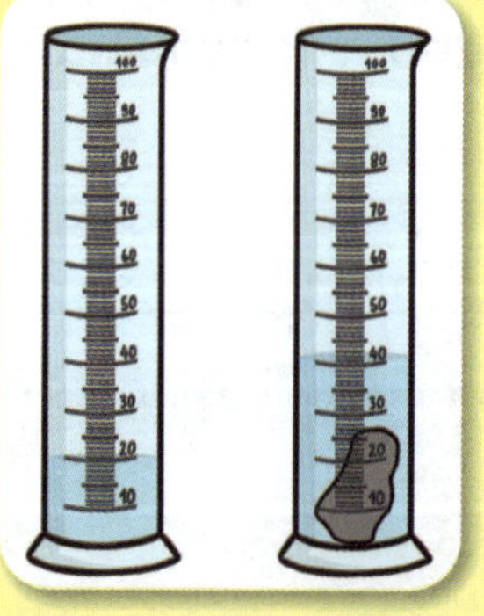

2

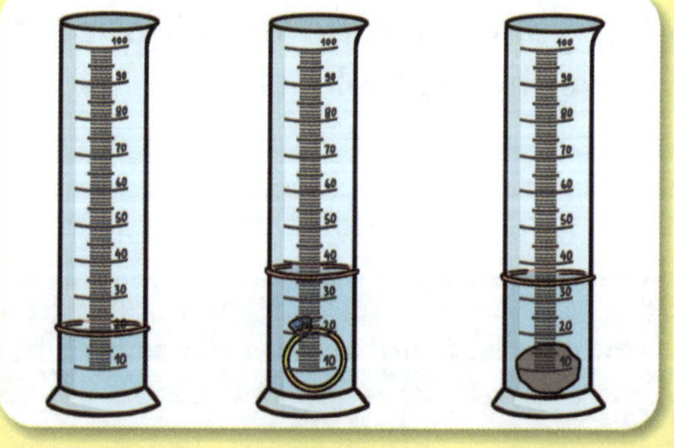

Case 3

Some objects are too large to fit into a measuring cylinder. Scientists use a piece of equipment called a 'Eureka jar' to find the volume of larger, irregular objects.

Look at this experiment that Keeno and Leah set up using a Eureka jar and a stone.

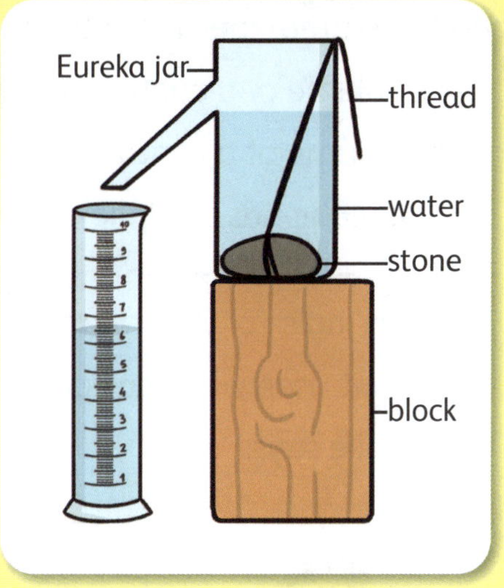

1 Discuss what they did with your partner.
2 What is the measuring cylinder for?
3 What is the volume of the stone?
4 What will happen to the water level in the Eureka jar when they remove the stone? Why?

Case 4

The word 'Eureka' is often associated with an ancient Greek scientist called Archimedes.

1 Find out about the story of Archimedes realizing that solids displaced water when he was in the bath.

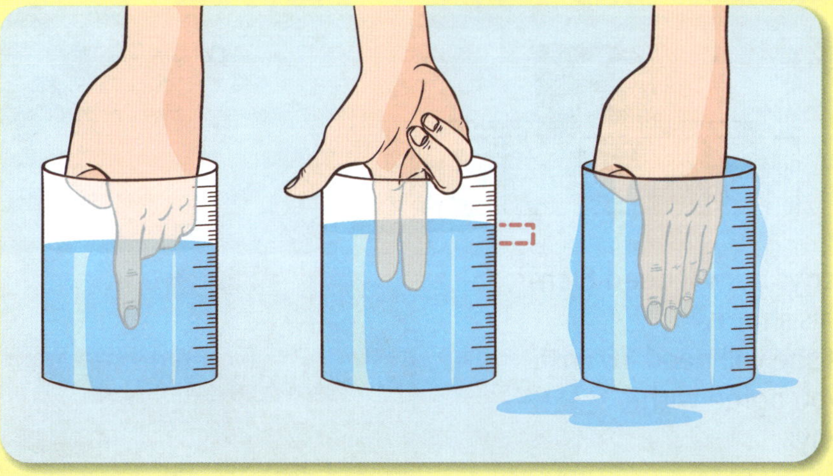

2 Use the diagram and do your own research to find out about Archimedes' principle. Write short notes to explain what you understand this to mean.

What I have learned

Read the statements. Fill in the missing words to complete the summary.

1 The volume of a container is given in _____ units.

2 The volume of a container is the amount of _____ that is occupied.

3 A unit cube has the following dimensions: length = _____ , width = _____ and height _____ = _____ .

4 The formula for the volume of a cuboid is _____ × _____ × _____ .

5 The unit used when calculating the volume of a shoe box is _____ .

6 Gabrielle says:

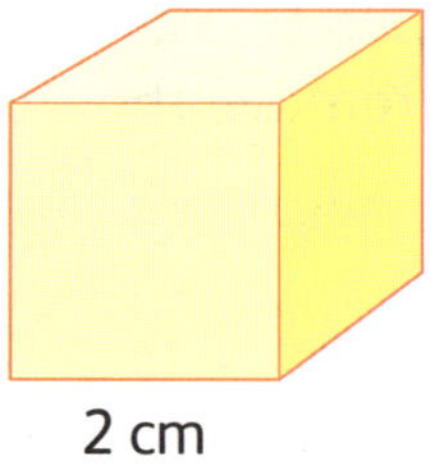

Do you agree? Explain your answer.

Practice questions

1 Look at the four solids A to D. Sort them into ascending order according to their volume.

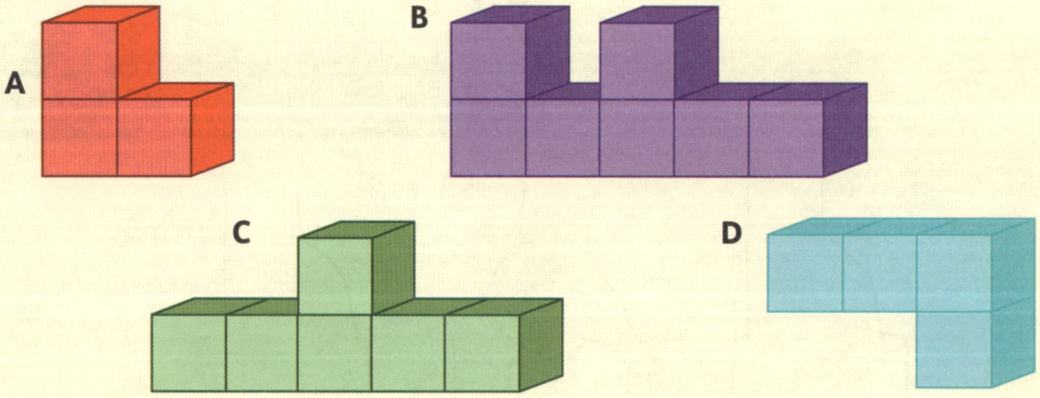

2 Anthony says he will need 8 cm^3 to build this shape.
Jada says she will need 10 cm^3.
Who do you agree with?
Explain why.

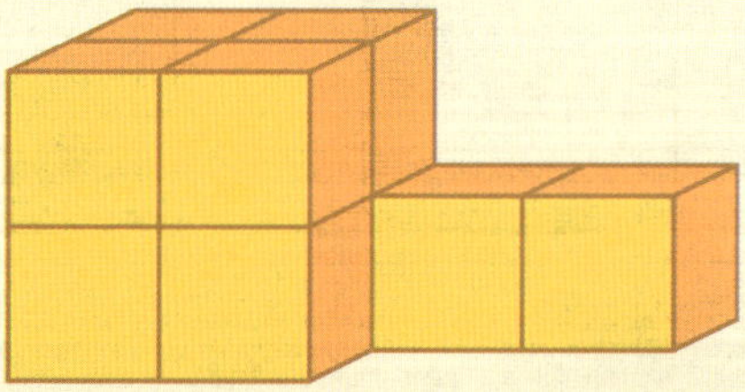

3 a The diagram shows a rectangular prism and the number of unit cubes in each layer. Use the diagram to complete the table below. Write only the letter and your answer in your notebook.

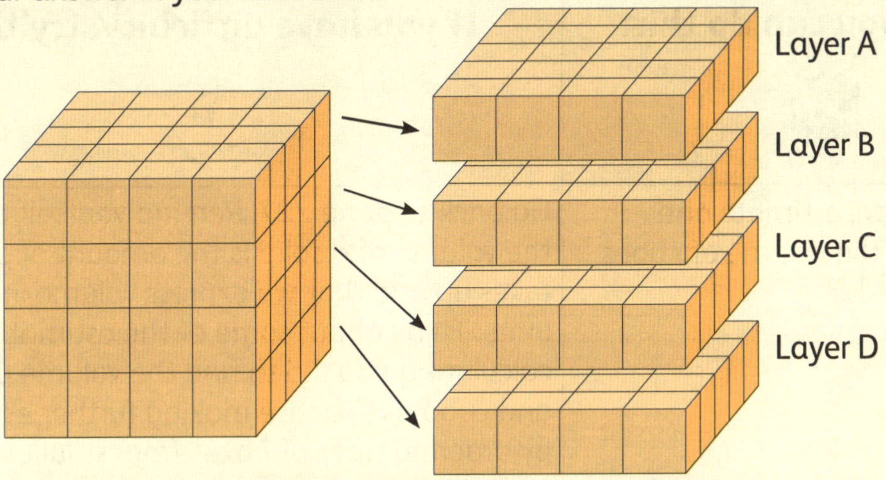

Layer	Number of unit cubes
A	
B	
C	
D	
Total cubes	

b In your own words, say what the total number of unit cubes represents.

4 Copy and complete these sentences.

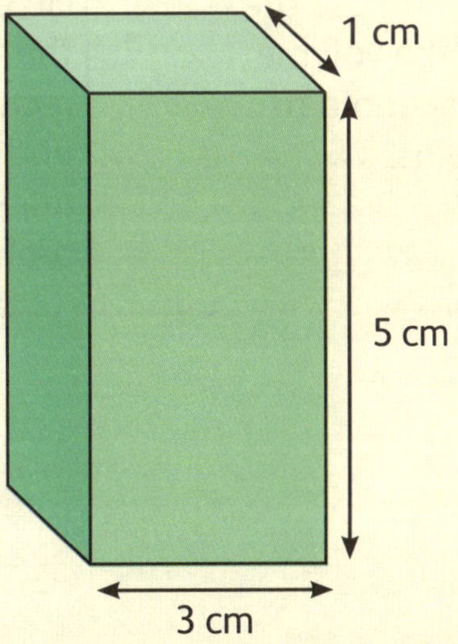

a This 3-D solid is called a _____ .

b The length is _____ .

c The width is _____ .

d The height is _____ .

e The volume is
_____ × _____ × _____ = _____.

5 A storage container is 9 m long, 5 m wide and 2 m high. What is its volume?

6 In how many different ways can you make a cuboid with a volume of 63 cm^3?

14 Self check

	Check you can do this:	If you have difficulty, try this:
1	Investigate, estimate and calculate the volume of cubes and cuboids.	Go back to page 227. Remind yourself that the volume of a solid is the amount of space it takes up and that we express volume in cubic units. Think about some of the estimating and calculating you did to find the volume of cubes and cuboids. Practise making further estimates by ordering a set of boxes from smallest to greatest volume. Use unit solids to check.
2	Explain or use examples to show what is meant by a unit solid.	Go back to page 227. Revise the different unit solids you can use to express volume. For example, the unit solid 1 cm^3 has a length, width and height of 1 cm.
3	Build unit solids of volume 1 cm^3, 1 dm^3 and 1 m^3.	Use nets of shapes and cubes to build unit solids.
4	Calculate the volume of a cuboid when given the number of unit solids in one layer and the number of layers.	Remember that you can use unit cubes to build models and find the volume of rectangular prisms. Start with the bottom layer and work up to the top layer.
5	Investigate and use the formula for the volume of a cube or cuboid to solve problems.	Go back to page 229. Remind yourself that the formula for calculating the volume of a rectangular prism is length x width x height, and that the measure will be expressed in cubic units. You can draw diagrams to help you solve problems about volume.

Unit 3: Geometry

Chapter 15 Symmetry and congruence

In this chapter you will explore this focus question:

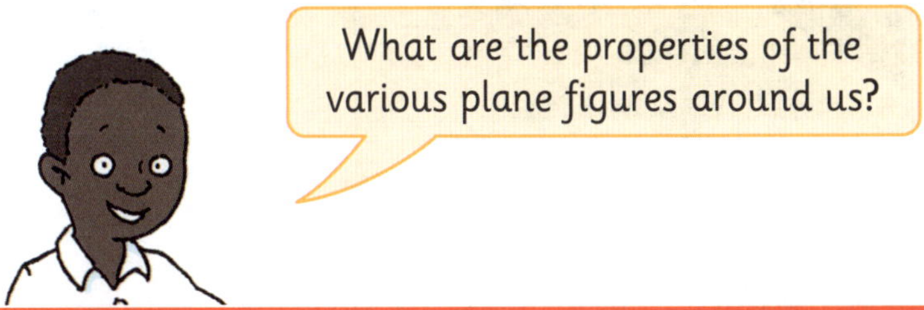

To answer this question, you will:

- explore the ideas of symmetry in geometric shapes found in the environment
- identify and count the number of lines of symmetry in compound plane figures
- describe congruence in plane and solid shapes
- distinguish between similar and congruent figures (triangles and quadrilaterals).

Starting point

1 Do you remember what a **line of symmetry** is? Explain it in your own words and draw a diagram to show what it means.

2 Look at this picture of a man's face and answer the questions.

  a The dotted lines are not lines of symmetry. Why not?

  b The shape made by the dotted lines could be symmetrical. Where would you draw the mirror line to check?

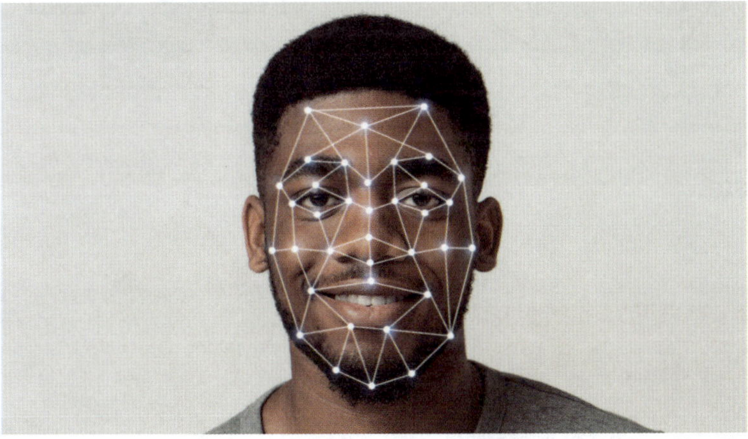

3 Human faces are not really **identical** on both sides. How could you check?

Key words
line of symmetry
identical

Symmetry in common shapes and objects

Key maths idea

A line of **symmetry** is an imaginary straight line that divides a **plane shape** into two congruent shapes. We say the parts are **symmetrical** because they appear to mirror each other. You could fold the image so that both parts match perfectly.

A line of symmetry is also called a **mirror line**.

A shape can have more than one line of symmetry.

The Jamaican national flag has two lines of symmetry. You will learn more about congruent shapes later in this chapter.

Key words

symmetry
plane shape
symmetrical
mirror line

1 Look at the shapes in the grid. Use a ruler or small mirror to help you work out whether they are symmetrical. How many lines of symmetry can you find?

2 **a** Name each shape shown here.

A

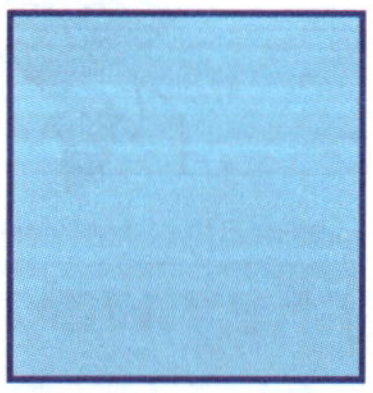

B

C

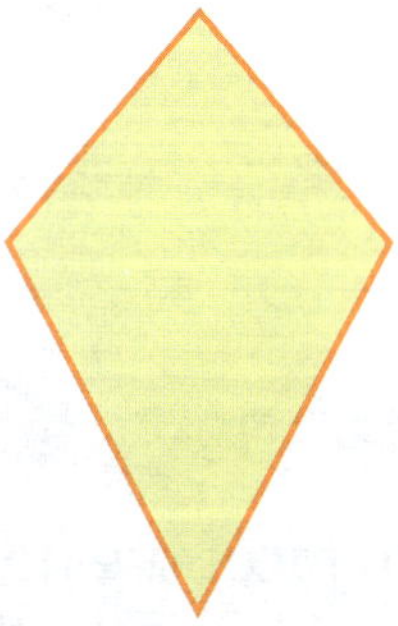

D

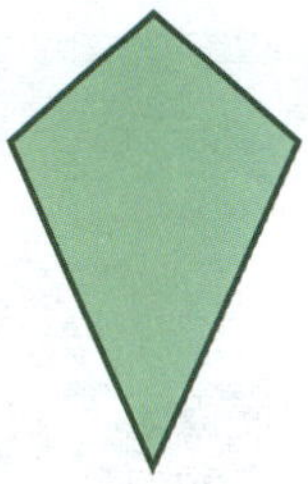

E

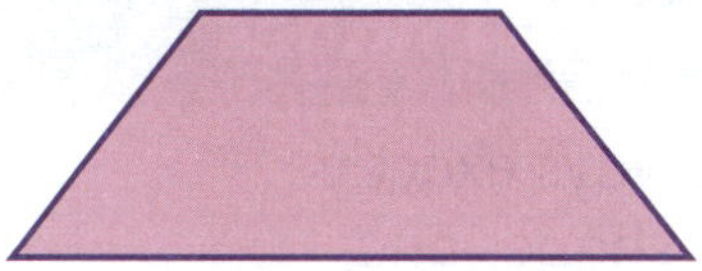

F

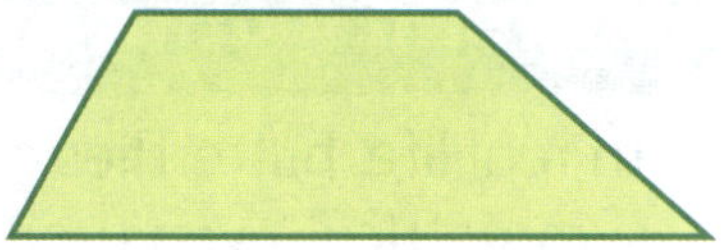

G

H

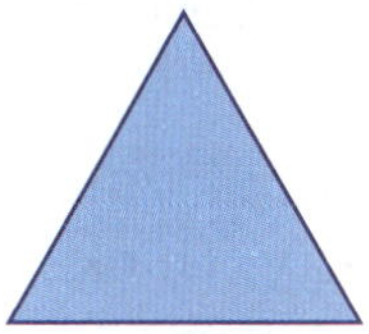

I

b Which shapes do not have a line of symmetry?

c Do any shapes have more than one line of symmetry?

d How do you know if you have found a line of symmetry?

3 **a** Which of these capital letters have a line of symmetry?

A C F H M O S X Y

b Which of the letters have more than one line of symmetry? Copy those letters and draw the lines of symmetry.

4 Which of these shapes have no lines of symmetry?

a

b

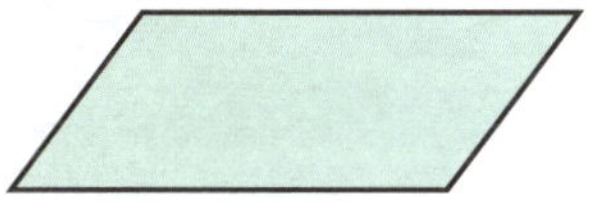

c

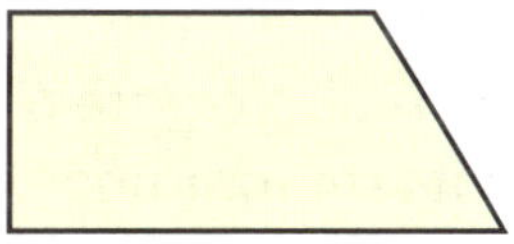

d

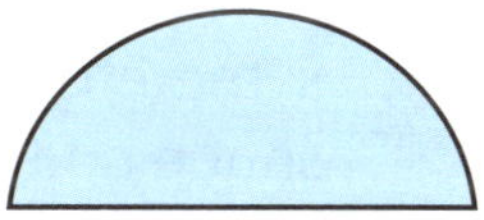

5 How many lines of symmetry does each of these shapes have?

a

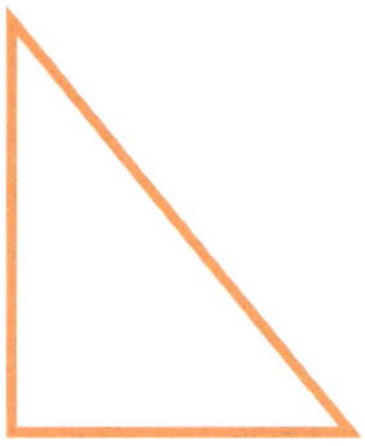

b

c

d

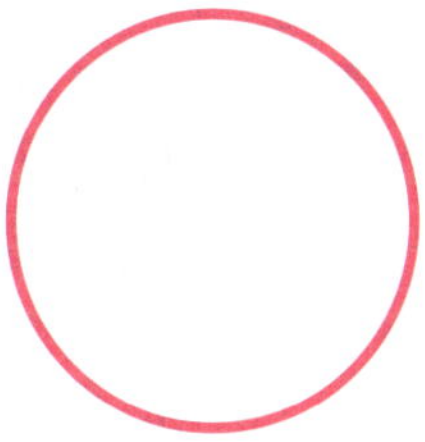

6 Use a ruler and a small mirror to find the line (or lines) of symmetry in each object in these four pictures.

a

b

c

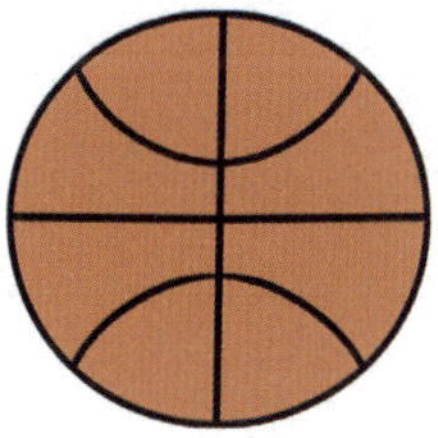

d

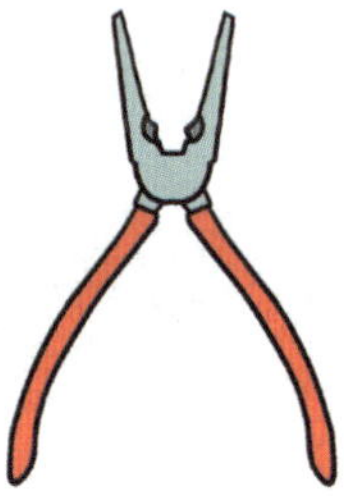

Real-world maths: Symmetrical images

1 In real life, butterflies are not always exactly symmetrical, even though their shapes and patterns do look symmetrical to us.

a Discuss with a partner where the lines of symmetry on these pictures of butterflies are. Write down your findings.

b Draw two butterflies in your book and draw in their lines of symmetry. Think of other examples of symmetry in nature and draw two more examples of symmetry in nature in your book.

2 In a photographic project called 'Both sides of', the photographer explored how people's faces look when they are digitally altered to look symmetrical. If you have internet access, you can look up this project. If you have photo editing software, you can try altering photographs of yourselves and people you know to try make their faces symmetrical. Write about what you notice.

3 A geoboard is a type of pegboard. You can use elastic bands to create different shapes on the geoboard. Find or make a simple geoboard. Use elastic bands to create simple polygons and show their lines of symmetry. You can also create a shape and then its mirror image.

4 You can use paint or ink to make blob paintings: start by making paint drizzles and then fold the sheet of paper to create a mirror image. You can use the blob paintings to make greeting cards or to decorate wrapping paper. Talk about the symmetry in your blob design with a partner.

Similar and congruent shapes

Key maths idea

Congruent shapes have their **corresponding** sides equal in length and their corresponding angles equal in size. In other words, congruent shapes can fit onto each other exactly. A mirror line divides a shape into congruent shapes that are mirror images of each other.

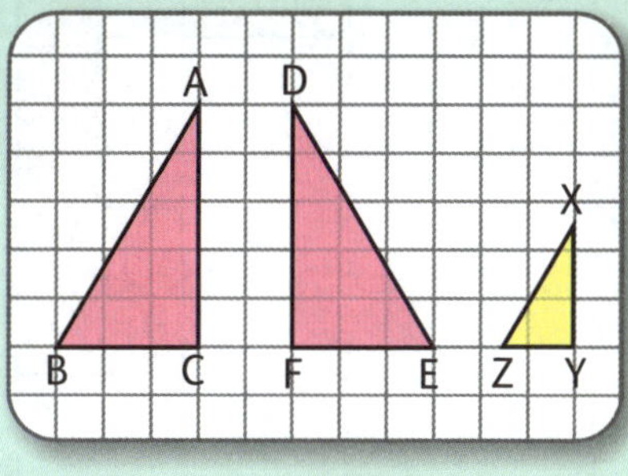

Look at the triangles in this diagram. Triangle ABC is congruent to triangle DEF. This is because each corresponding side of these two triangles is equal in length, and each corresponding interior angle is equal:

Side lengths: AB = DE, AC = DF, BC = EF

Angles: $\angle A = \angle D$, $\angle B = \angle E$, $\angle C = \angle F$

Now look at triangle XYZ. It is not congruent to the other two triangles, but it does have its corresponding angles equal.

Angles: $\angle A = \angle D = \angle X$, $\angle B = \angle E = \angle Z$, $\angle C = \angle F = \angle Y$

If you look at the sides, each of the sides of triangle XYZ is half as long as the corresponding sides in triangles ABC and DEF.

We say these triangles are **similar**. Similar triangles have their corresponding interior angles equal.

Key words
congruent
corresponding
similar

Geometry

1 Look at each pair of shapes. Explain to a partner whether they are similar, congruent or neither.

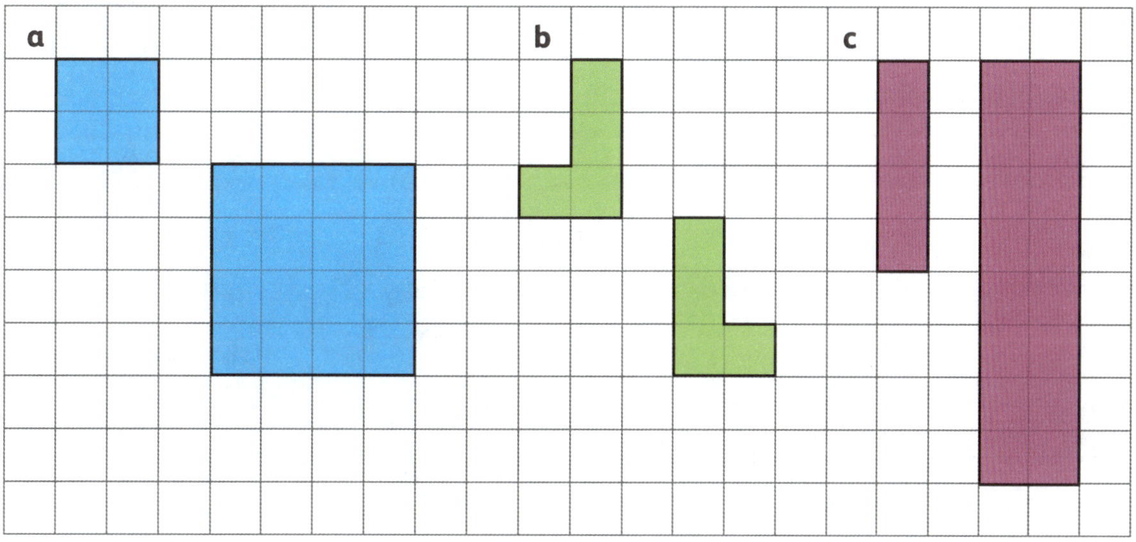

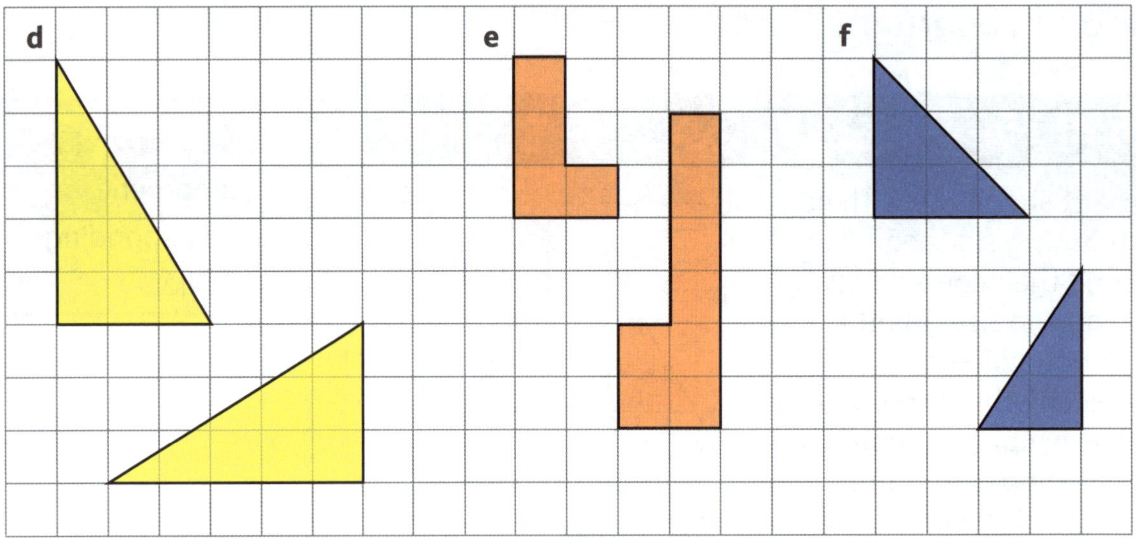

2 Which two shapes in each set are congruent?

a

b

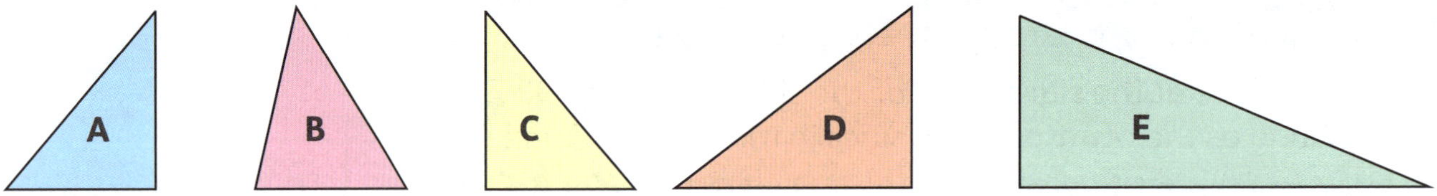

c

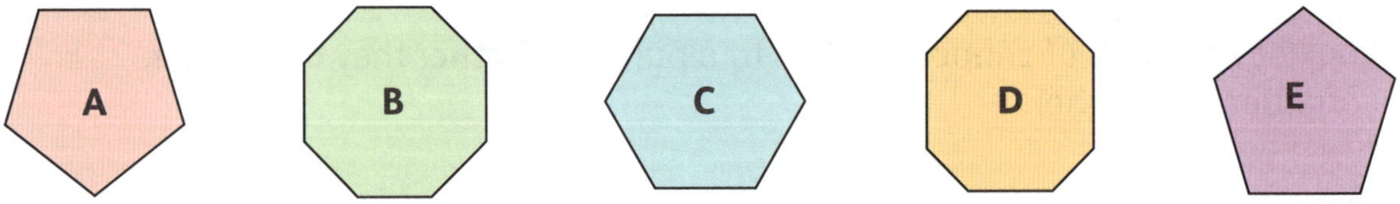

3 Identify which five shapes are congruent.

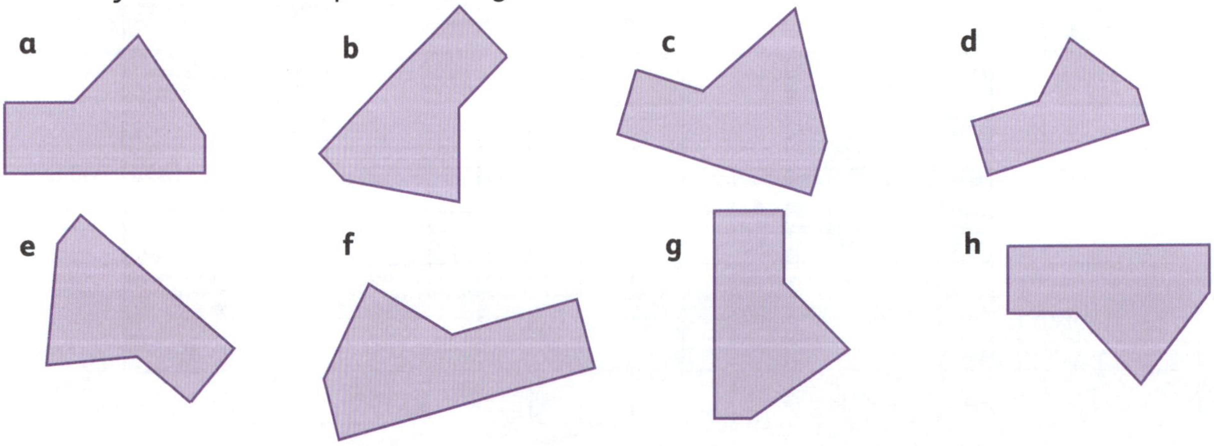

4 Solids can also be congruent if all their corresponding faces are congruent. Which two solids in each set are congruent?

a

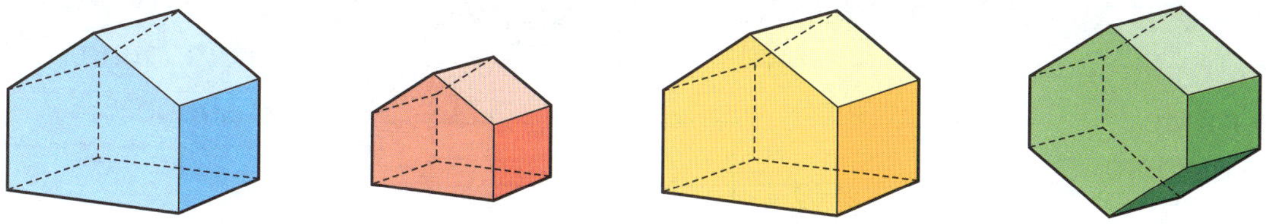

b

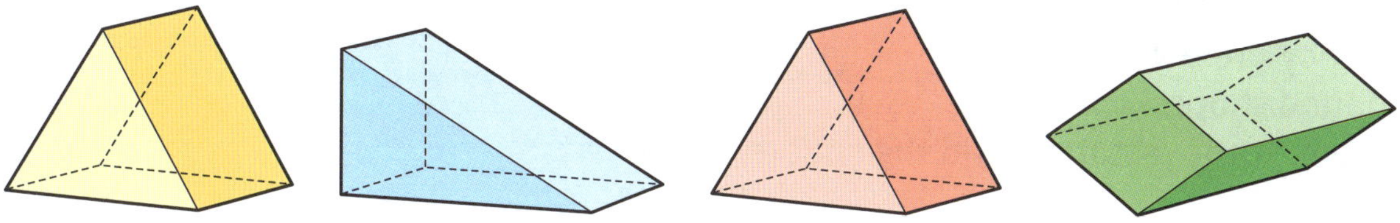

c

d

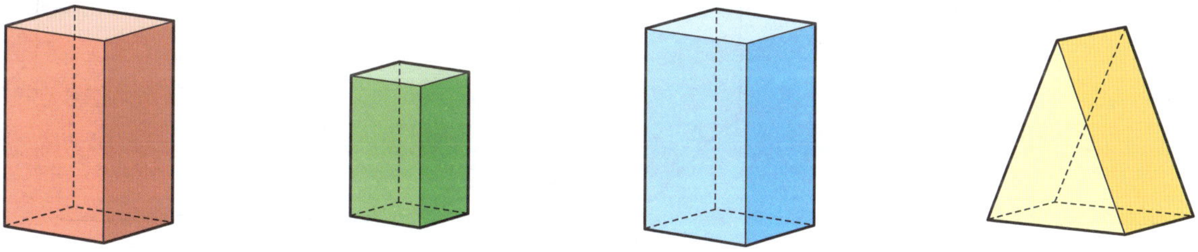

5 Discuss each statement with a partner. Do you agree or disagree? Give reasons.

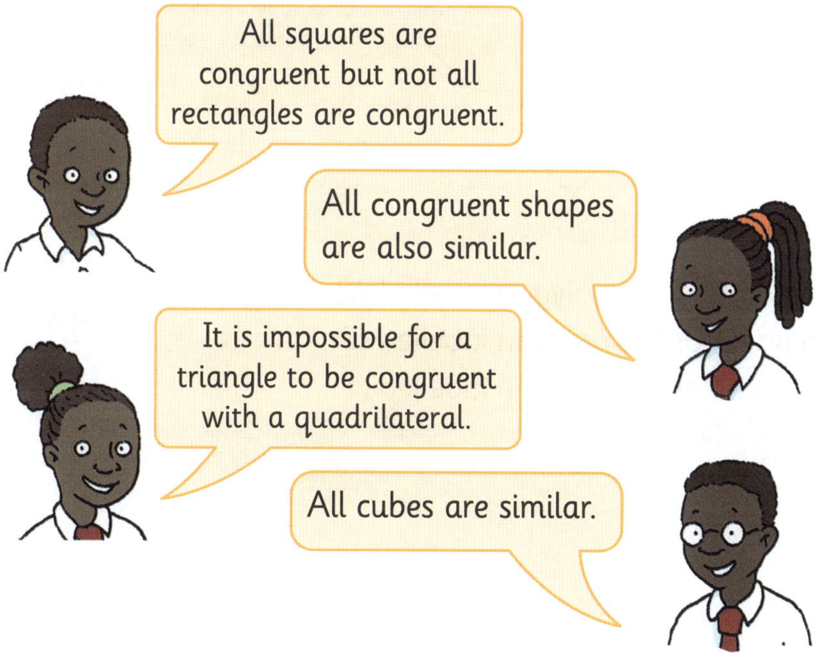

Remember that when a figure is enlarged or reduced in proportion, the resulting figure is similar to the original. If you have access to online drawing programs, you can explore enlarging and reducing diagrams of 2-D shapes and 3-D solids.

Symmetry in compound shapes

Key maths idea

A **compound shape** is a shape made out of two or more other shapes. For example, look at the two shapes on the right. They are both compound shapes made out of the rectangle and the square on the left.

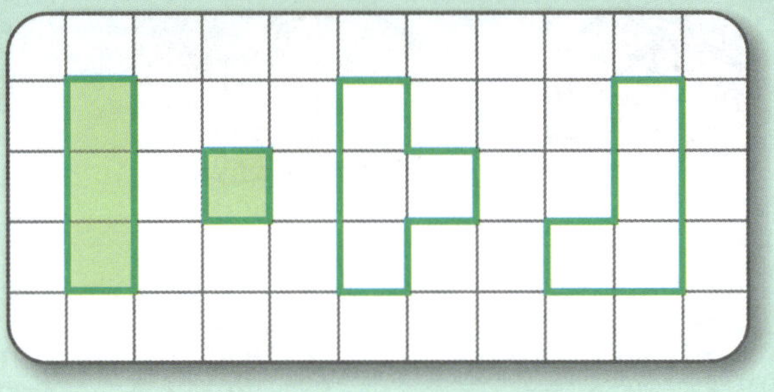

Some compound shapes also have lines of symmetry. Which of the two compound shapes on the right has a line of symmetry?

Key word
compound shape

1 Here are some compound shapes. Discuss with a partner which other shapes you could use to build each shape. Then answer the questions that follow.

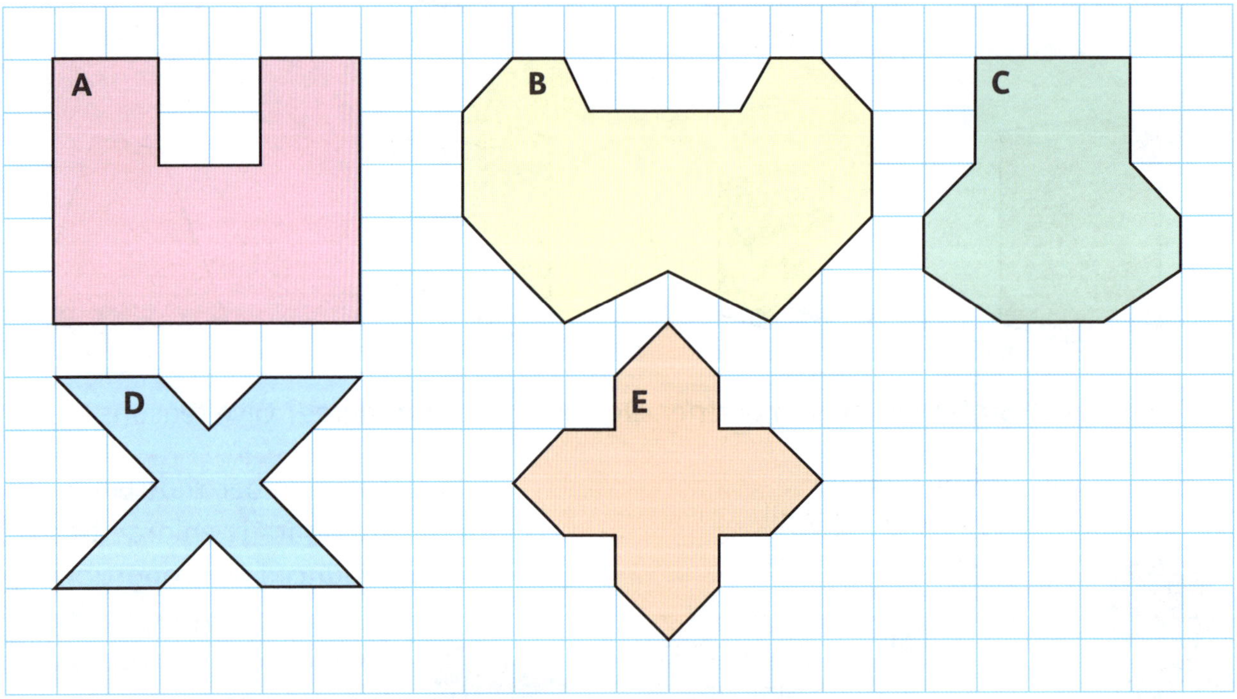

a What is the mathematical name of the polygon labelled A?
b What is the mathematical name of the polygon labelled C?
c What is the mathematical name of the polygon labelled D?

2 **a** Write the answers in your notebook to complete the table about the polygons above.

Polygon	Number of lines of symmetry
A	a
B	b
C	c
D	d
E	e

b Draw another compound shape with the same number of lines of symmetry as polygon E. Use a square 10 × 10 coordinate grid for your drawing.

3 This box shows all the letters of the alphabet.

A B C D E F G H I J K L M
O P Q R S T U V W X Y Z

Copy the Venn diagram. Then write each letter in the correct region of the diagram.
U = {letters of the alphabet with at least one line of symmetry}
P = {letters with two lines of symmetry}
Q = {vowels}

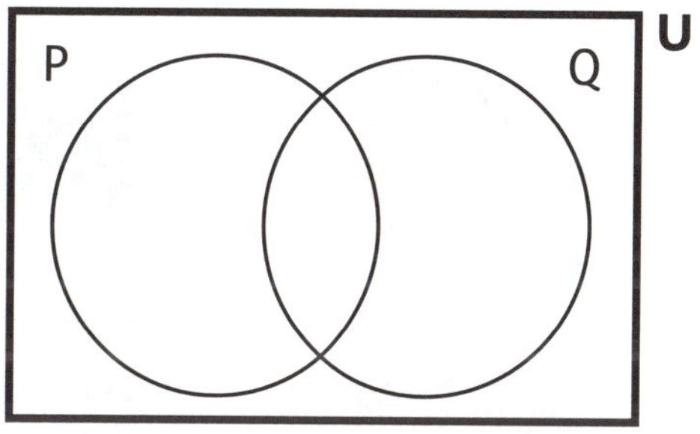

4 **a** Quadrilateral A is made up of two equilateral triangles. What is the mathematical name for quadrilateral A?

b Quadrilateral B is made by joining two quadrilaterals A along one side. What term describes the two quadrilaterals that were joined together to form quadrilateral B?

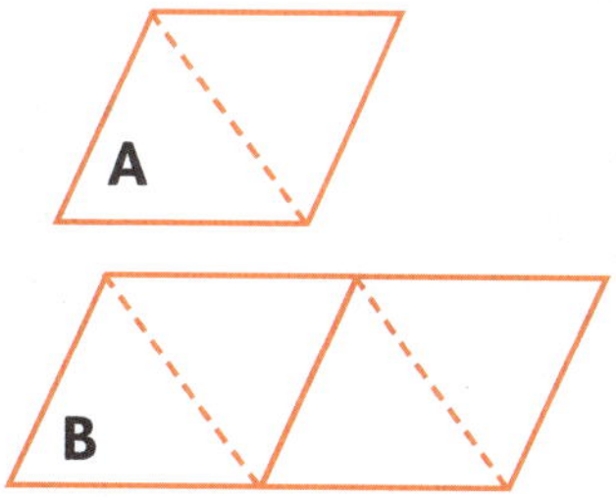

c What is the mathematical name for quadrilateral B?

d Write each statement from i to vi in your notebook. Then write A, B or Both next to each statement to show whether it is true for quadrilaterals A, B or both.

Statement:	
i	It has just one pair of parallel sides.
ii	It has two pairs of parallel sides.
iii	It has four equal sides.
iv	It has two pairs of equal sides.
v	It has one line of symmetry.
vi	It has two lines of symmetry.

Real-world maths: Design a symmetrical logo

You are a designer for an advertising agency. Your client is a small business called Farm Fresh. They source and deliver vegetables directly from farmers to households. They have asked you to create a symmetrical logo that they can use on their website and print onto the oval stickers they add to the delivery boxes.

1. Make a small portfolio containing suitable images for your client that have lines of symmetry. On each picture identify the line(s) of symmetry.
2. Create a logo design for your client that makes use of symmetry. Show how the logo will look on the banner of the website and on the oval stickers.

Maths detective: Flags of the Caribbean

Use what you know about symmetry to solve these problems involving flags.

Case 1

Here are the flags of sixteen Caribbean nations.

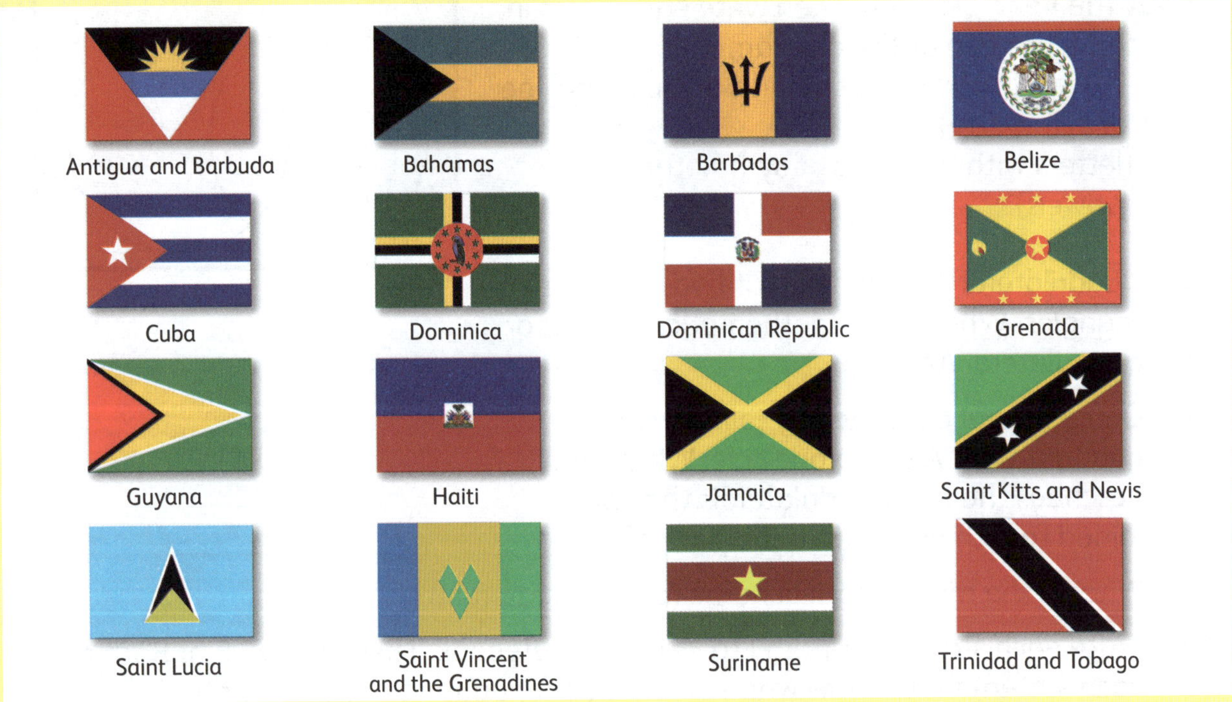

1. Eight of these flags have no line of symmetry. Find them and list them.
2. Explain why the flag of Bahamas has a line of symmetry, while Cuba with a similar design does not.
3. Which five countries have flags with only one vertical line of symmetry?
4. Which two flags have only one horizontal line of symmetry?
5. Adjani says the Jamaican flag has four lines of symmetry: one horizontal, one vertical and two diagonal lines. Is he correct? Give a reason for your answer.

Case 2

Three countries of the world have flags that are not rectangular.

1 Switzerland has a square flag with a white cross on it. Sketch the flag and draw all the lines of symmetry.

2 The two other countries are Nepal and the Vatican City. Find the flags of these countries and explain why they are not symmetrical.

Case 3

Work with a partner. Complete a symmetry scavenger hunt in the classroom to find shapes or objects that match each of these descriptions.

Write the letters and the objects you find in your notebook.

1 I am rectangular and I have two lines of symmetry.

2 I am a solid and I have only one line of symmetry.

3 I am a solid with at least two lines of symmetry.

4 I am a square shape but I have no lines of symmetry.

5 I have so many lines of symmetry it's hard to count them all.

6 I am a small object with two identical halves.

What I have learned

Write a definition for each of these concepts. Add a drawing to help you explain each concept.

1	2	3	4
symmetrical	line of symmetry	congruent shapes	similar shapes

Practice questions

1 Copy and complete each shape. The dotted lines are lines of symmetry.

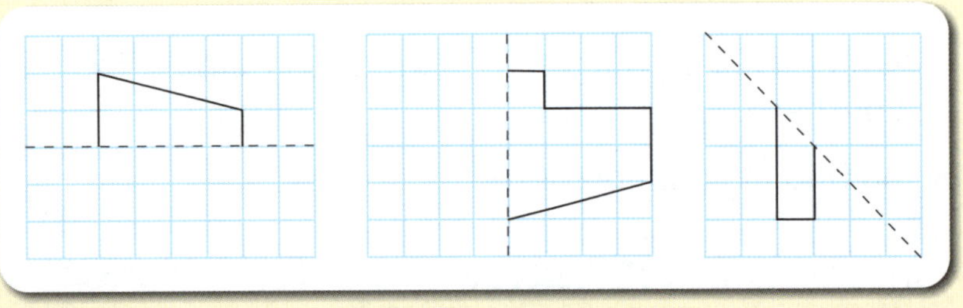

2 Shape A has more than 1 line of symmetry. Now sort shapes B to H by answering the questions.

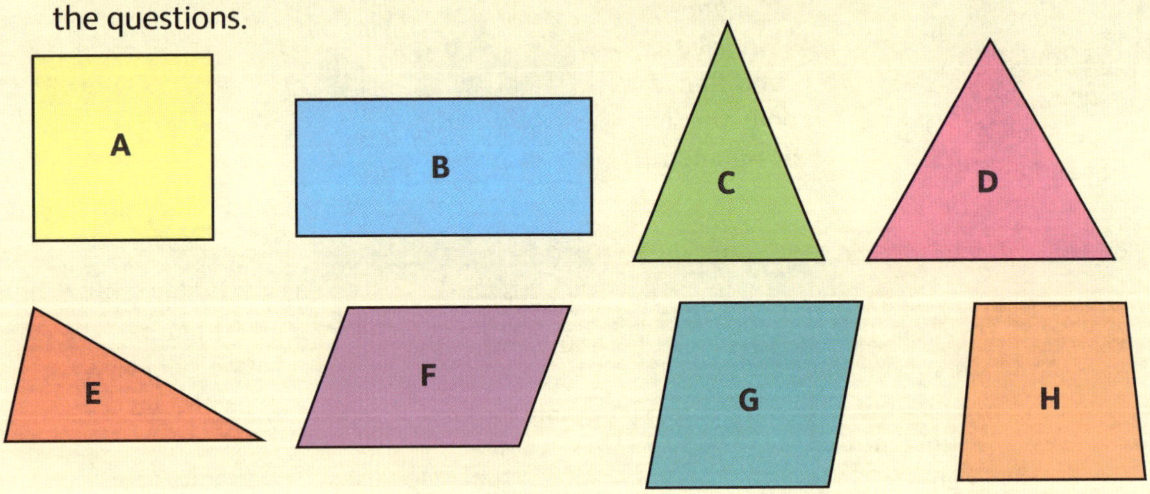

a Which shapes have 1 line of symmetry?

b Which shapes have more than 1 line of symmetry?

c Which shapes have no lines of symmetry?

3 Which shape is not congruent to the other three shapes?

4 These two rectangles are similar. Use equivalent ratios to calculate the missing length.

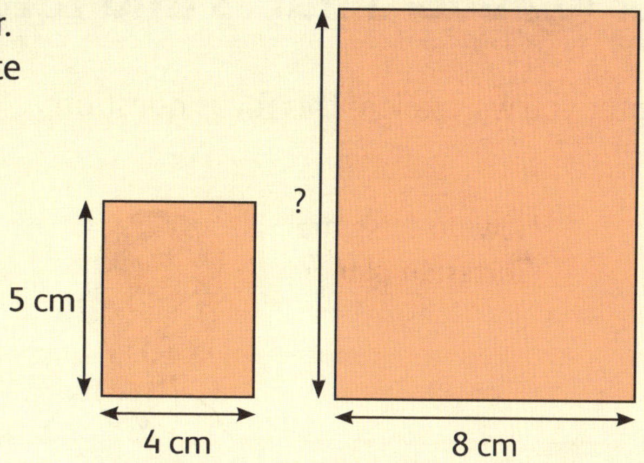

15 Self check

Check you can do this:	If you have difficulty, try this:
1 Describe symmetry in geometric shapes in the environment.	Go back to page 238. Make sure you can identify shapes that are symmetrical. You should be able to identify the mirror line (line of symmetry) in symmetrical 2-D shapes. Also look at page 240 to remind yourself of real-life examples that appear to have line symmetry.
2 Identify and count the number of lines of symmetry in compound plane figures.	Go back to page 244. Make sure you know how to identify whether a more complex plane figure has a mirror line.
3 Describe congruence in plane and solid shapes.	Go back to page 241. Make sure you understand what the words similar and congruent mean. You should be able to compare sets of shapes and identify them as similar, congruent or neither.
4 Distinguish between similar and congruent triangles.	Go back to page 241. Remind yourself of the definitions of similar and congruent, and apply them to pairs of triangles.
5 Distinguish between similar and congruent quadrilaterals.	Go back to page 241. Make sure you can apply the definitions of similar and congruent to pairs of quadrilaterals.

Chapter 16 Coordinates and reflections

In this chapter you will explore this focus question:

To answer this question, you will:

- understand and use the concept of reflection within the Cartesian plane
- explore how a coordinate system identifies location
- use the first quadrant of the Cartesian plane to plot points
- reflect a shape or design in a horizontal or vertical line in the first quadrant of the Cartesian plane.

Starting point

In the game Battleships, two players each position their ships on separate grids. Each player guesses coordinates on the grid to try and hit their opponent's ships. You can play with a special set if you have one, but you can also play using a simple grid.

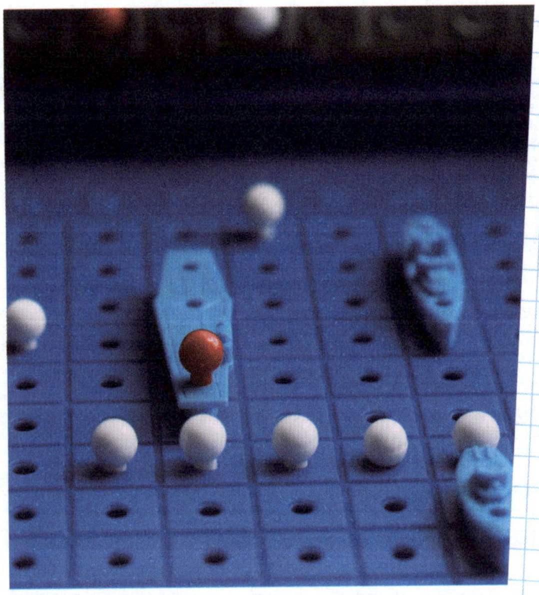

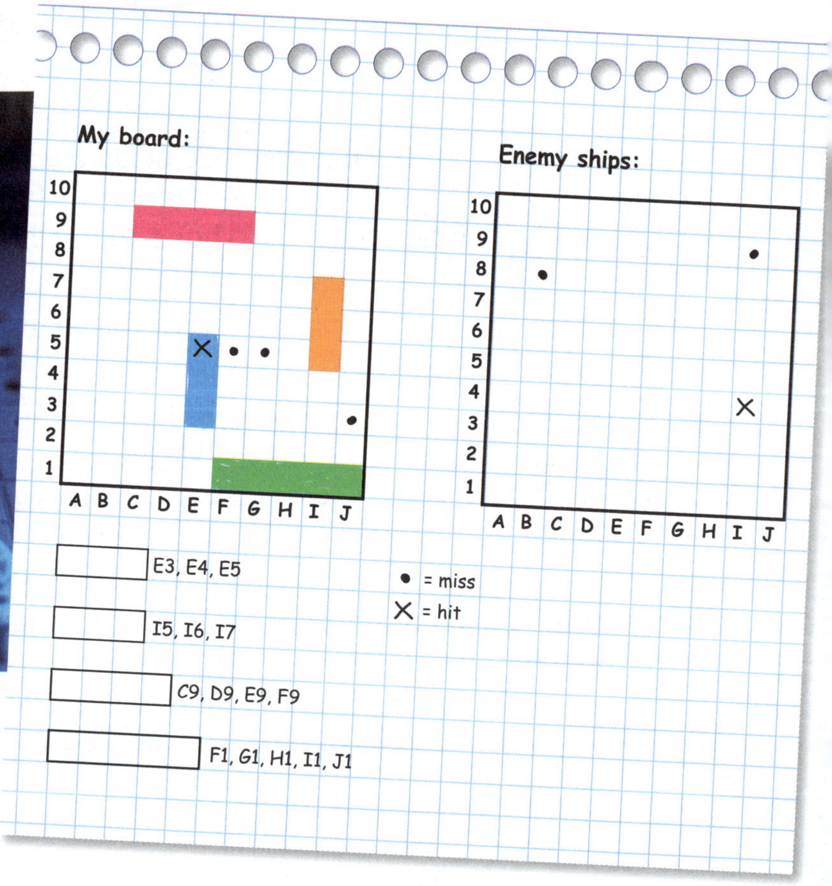

Follow these rules:

- Each player draws two 10 × 10 grids. On each grid, label the x-axis A to J, and the y-axis 1 to 10.
- Do not let your opponent see your grid. Draw your ships on your grids. Your opponent will try to guess their position by eliminating coordinates.
- When you name a coordinate block, you name the x-coordinate then the y-coordinate, for example, the first block of the green ship is on F1.
- Take turns naming coordinates on each other's grids. Keep a record of hits and misses. The first one to hit all the opponent's ships is the winner.

Using coordinates

Key maths idea

On this **grid**, the position of Point A is at (4, 6).
The **x-coordinate** (or **abscissa**) is 4. It tells us how far along the x-axis to go in order to locate point A.
The **y-coordinate** (or **ordinate**) is 6. It tells us how far up or down to go.
An **ordered pair** is a pair of **coordinates** written in this order (**horizontal** distance first, then **vertical** distance). To **plot** a point means to draw it from the coordinates.

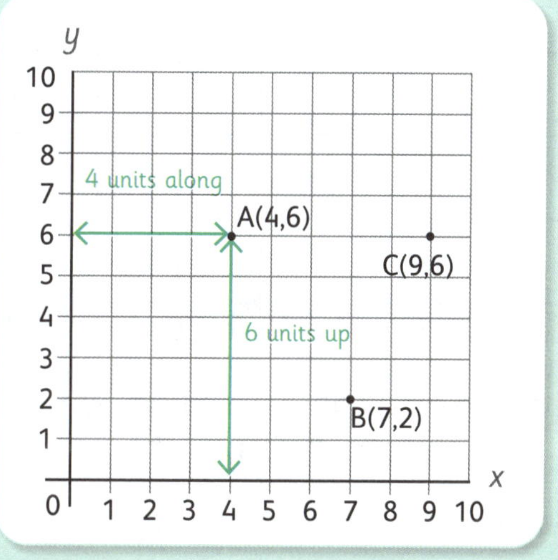

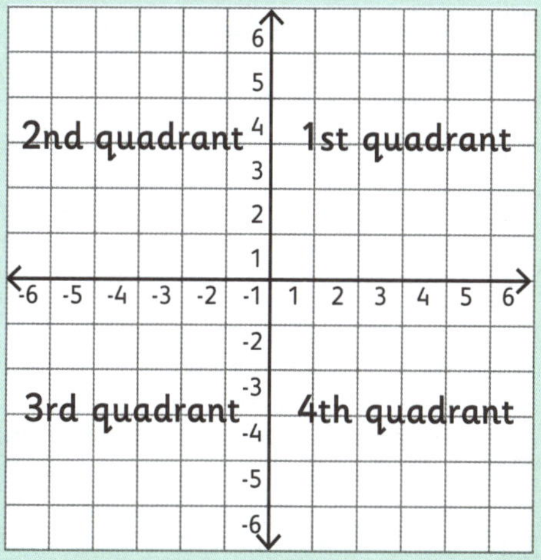

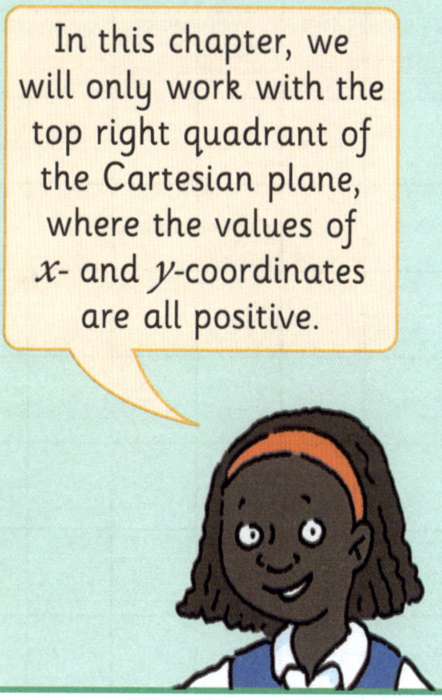

Key words

grid
x-coordinate
abscissa
y-coordinate
ordinate
ordered pair
coordinates
horizontal
vertical
plot
Cartesian plane
quadrant

The **Cartesian plane** is an imaginary surface. We draw it using a pair of axes that divides an area into four **quadrants**.

1 Sophia plots three points. Write the coordinates for A, B and C.

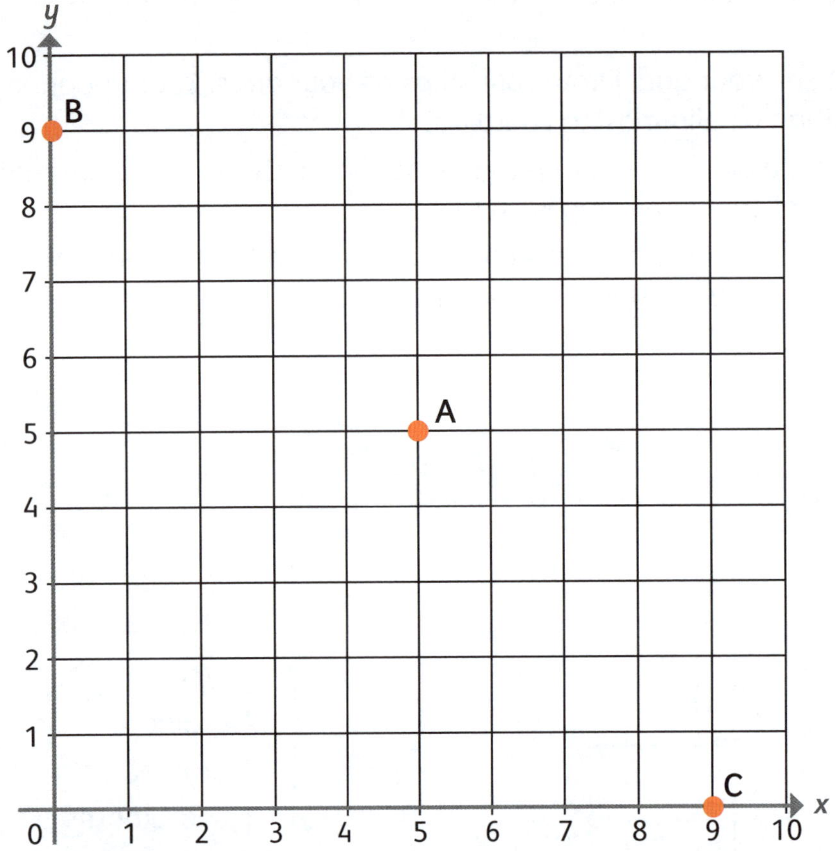

2 Leon is drawing a rectangle on a coordinate grid. Where should the final vertex of the rectangle be placed? Write the coordinate of the final vertex.

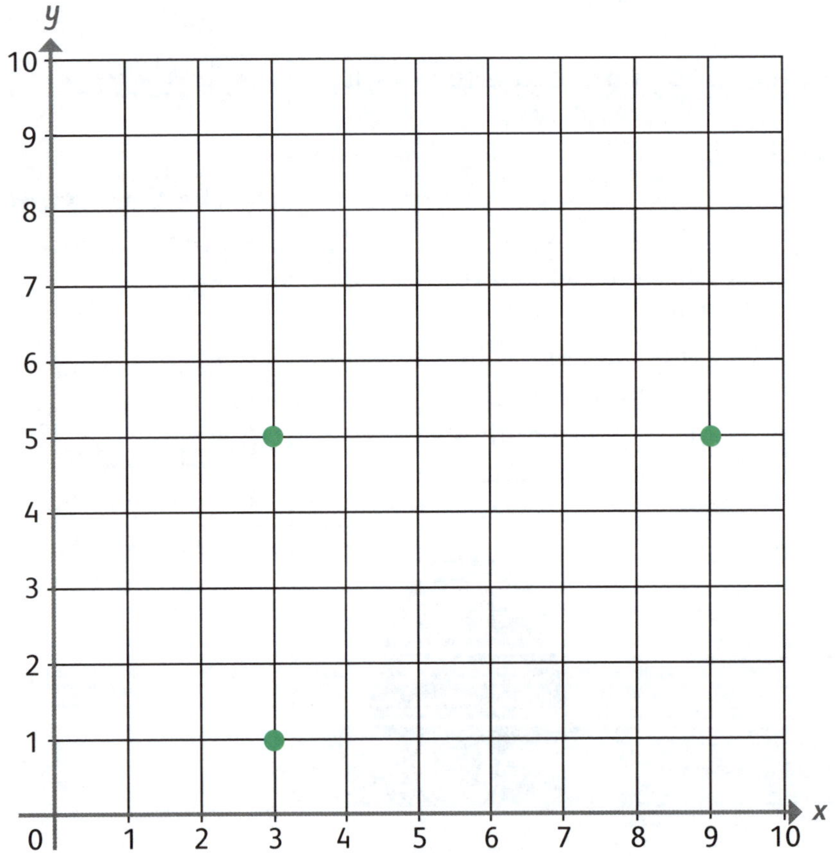

3 **a** Read this statement. Explain to a partner why it is true.

> As the value of x increases, the point moves to the right on the grid.
> When x decreases in value, the point moves to the left.

b What happens to the point when the value of y increases? What happens when it decreases?

4 Use squared paper. Draw an x-axis labelled from 0 to 10, and a y-axis labelled from 0 to 10, as shown above.
On your grid, plot these points.

A(2, 5)	B(2, 3)	C(1, 4)	D(10, 6)
E(7, 9)	F(8, 2)	G(5, 8)	H(10, 10)

5 Give the coordinates of each point on the graph below.

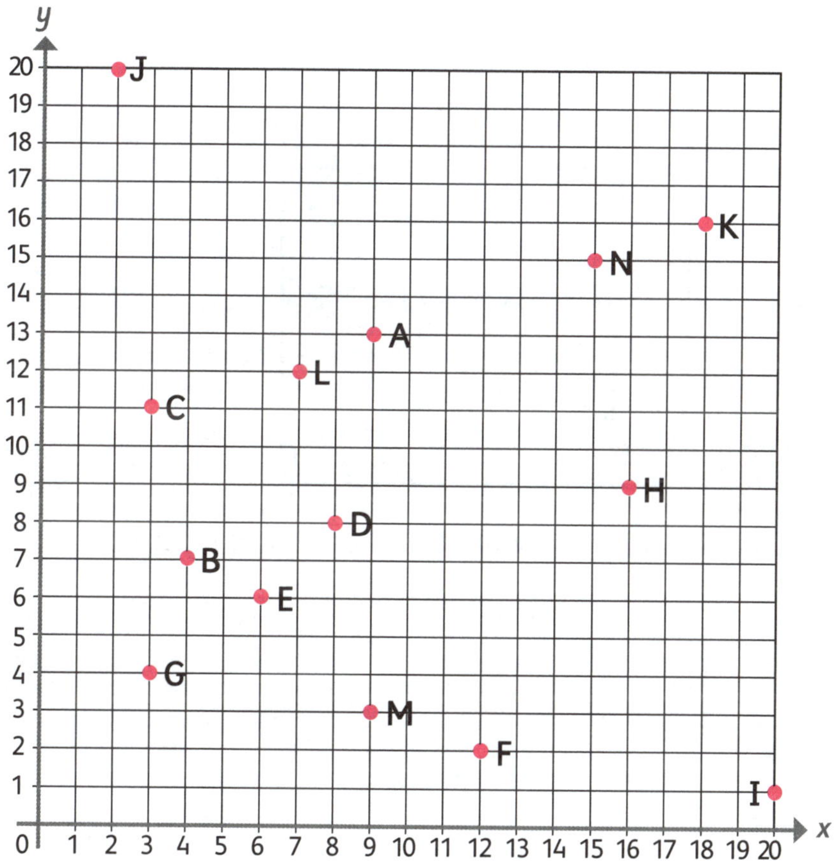

6 Use a 10 × 10 coordinate grid. Now draw the vertices of the polygon with the coordinates (7, 1), (7, 4) and (10, 1). What type of polygon is this shape?

7 Use a 10 × 10 coordinate grid. Dominic is finding coordinates where the x-coordinate and the y-coordinate add up to 8. For example, (3, 5), because $3 + 5 = 8$.

a Find all of Dominic's coordinates and plot them on the grid.

b What do you notice?

c Try the same but with a different total.

Reflections

Key maths idea

A line of symmetry or mirror line divides a shape into two congruent shapes. Sometimes we start with one shape (called the **object**) and **reflect** (or **flip**) it across a mirror line. The reflected **image** is the same size and shape as the object, but facing the other way round. We call this type of **transformation** a **reflection** or a flipped image.

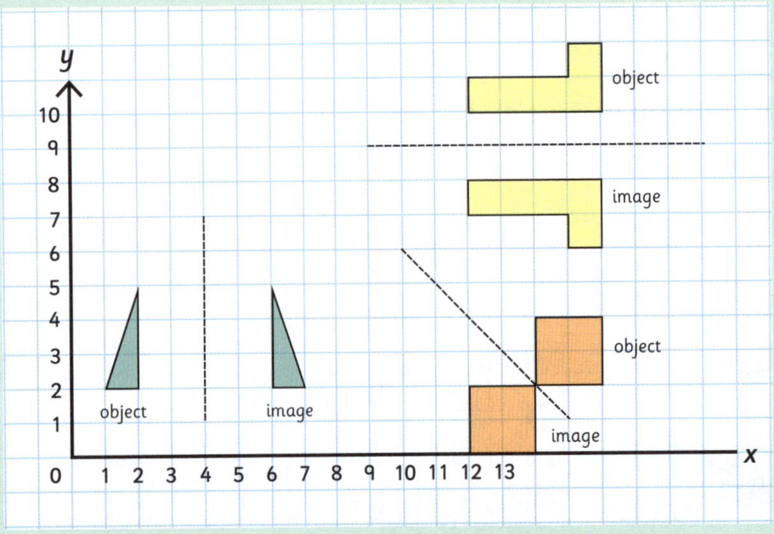

What do you notice about the distance between each image and the mirror line?

Key words
- object
- reflect
- flip
- image
- transformation
- reflection

1 Write the coordinates of each point. Then work out the coordinates of its reflection in the mirror line.

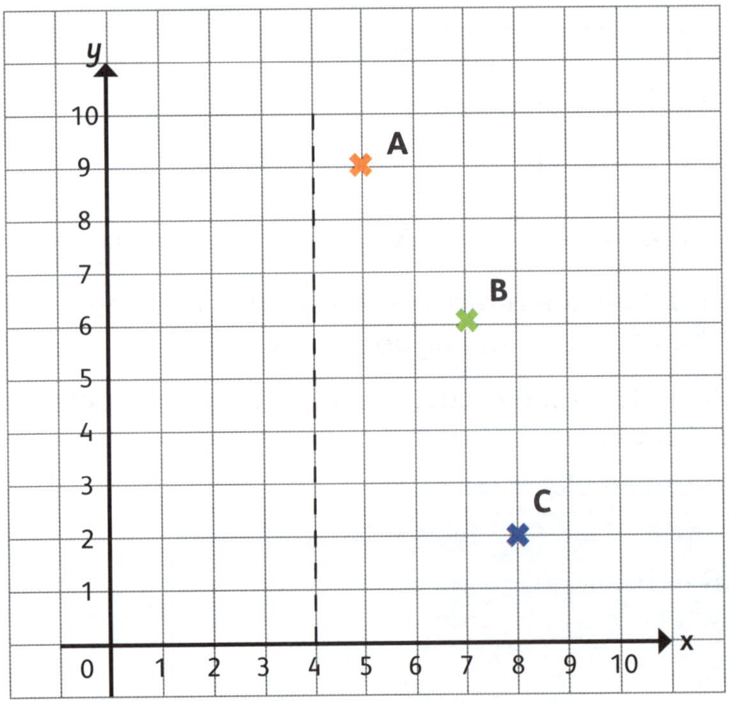

2 Explain why each pair of shapes does not show a reflection.

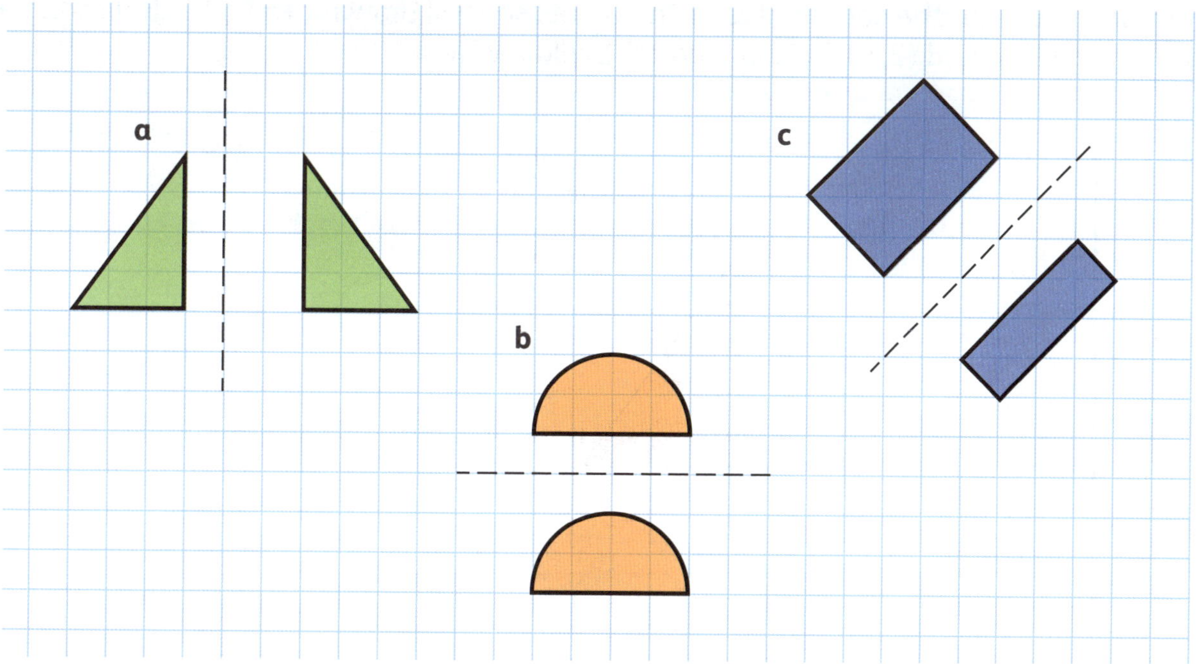

3 For each of these shapes, write the ordered pairs that describe each vertex (corner). Then write the coordinates of their reflections, reflected in the dashed mirror line.

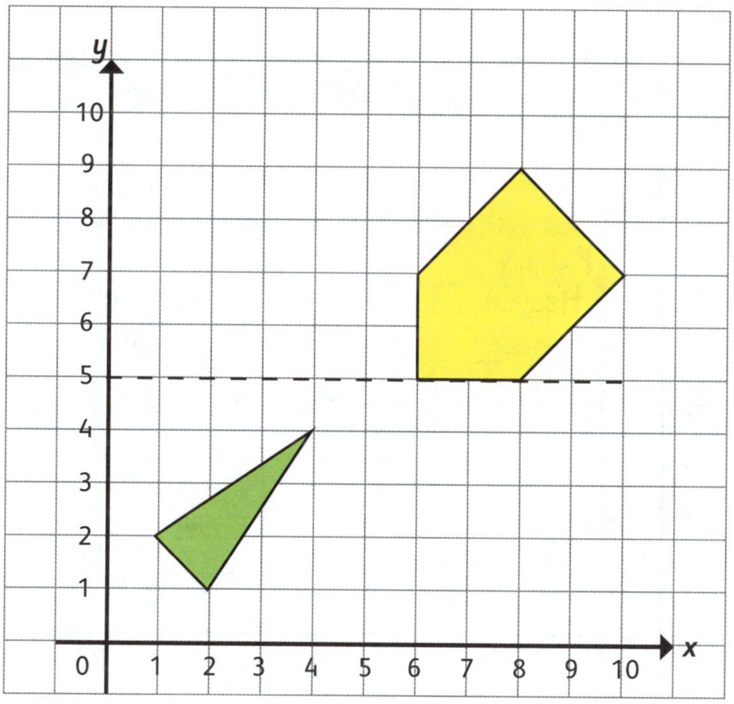

4 Lola wants to draw the reflection of this orange triangle. The dashed line is the mirror line. She says that the coordinates of the points she will draw for the reflected triangle are (5, 5), (2, 5) and (2, 9). Is Lola correct? Explain why.

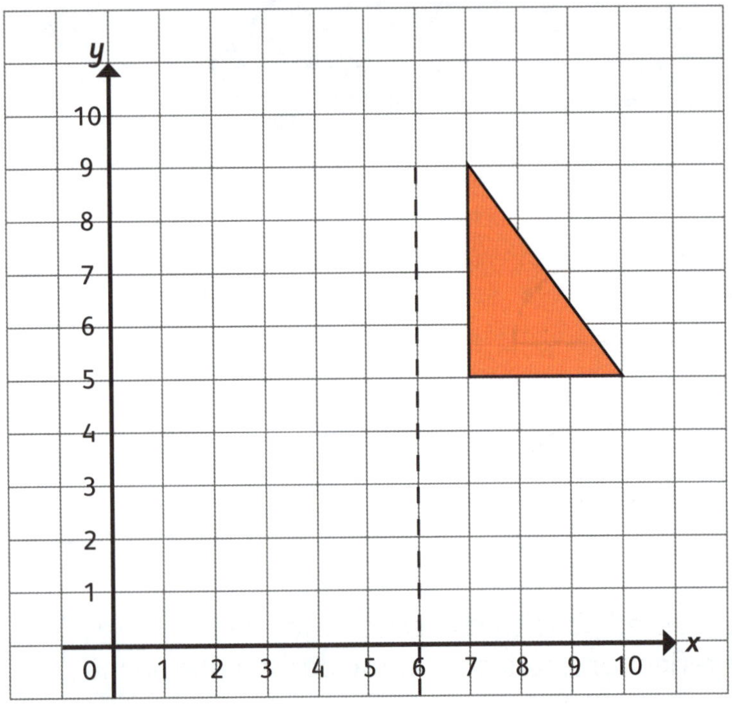

Real-world maths: Coordinates on a map

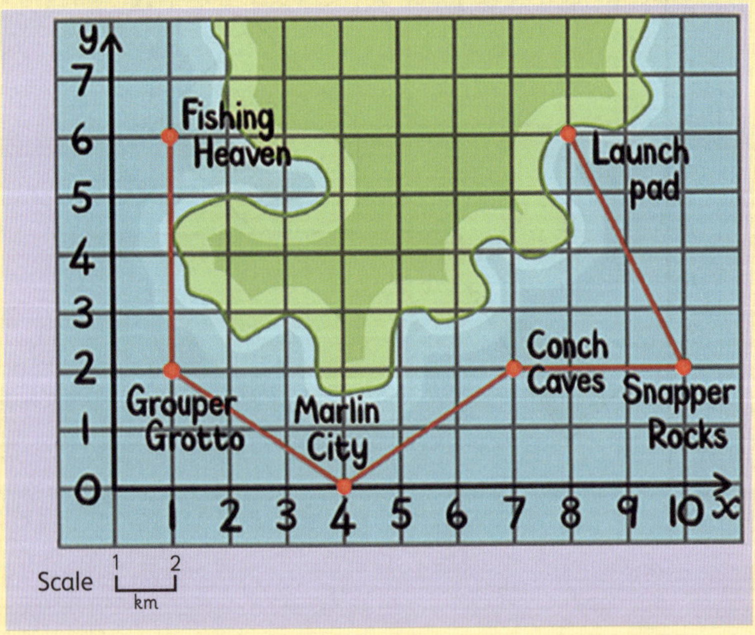

This map shows the route that a small boat followed from the launch pad to Fishing Heaven.

1 In what direction does the boat travel from the launch pad to Snapper Rocks?

2 How far and in what direction does the boat travel from Snapper Rocks to Conch Caves?

3 Where is the boat when its coordinates are (4, 0)?

4 What are the coordinates when the boat is halfway between Grouper Grotto and Fishing Heaven?

5 Use the scale on the map to work out the distance the boat travels from (7, 2) to (1, 2).

6 When it leaves Fishing Heaven, the boat travels 4 km north and then 4 km east. What are its coordinates after that?

Maths detective: Reflections and coordinates

Use your knowledge of coordinates and what you know about reflections to solve these cases.

Case 1

Tisha painted these shapes on two pieces of squared paper.

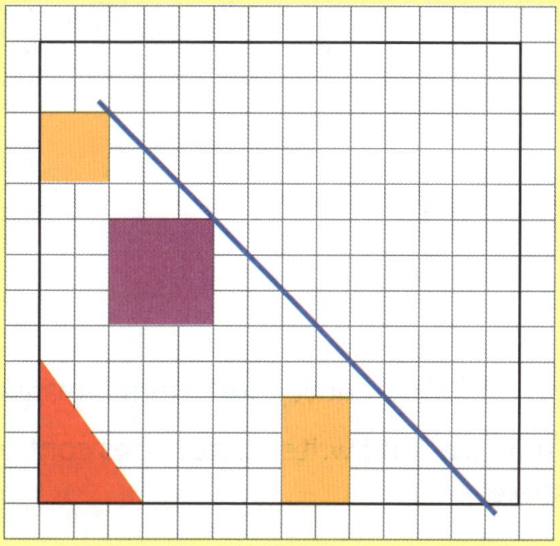

While the paint was still wet, Tisha folded each piece of paper along the blue line.

1 Predict what each piece of paper will look like when she opens it up. Draw a sketch to show your prediction.

2 How can you test whether your prediction is correct or not? Write down what you could do.

Case 2

Mario put six counters on a coordinate grid like this:

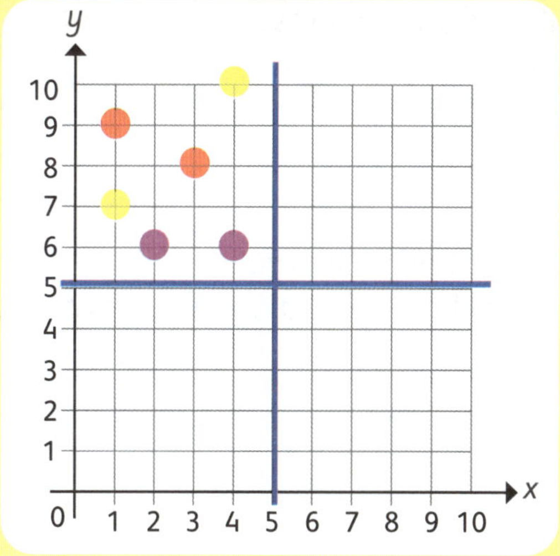

He challenged Tisha and Junior to reflect the counters in both the blue lines.

This is what they did.

Tisha

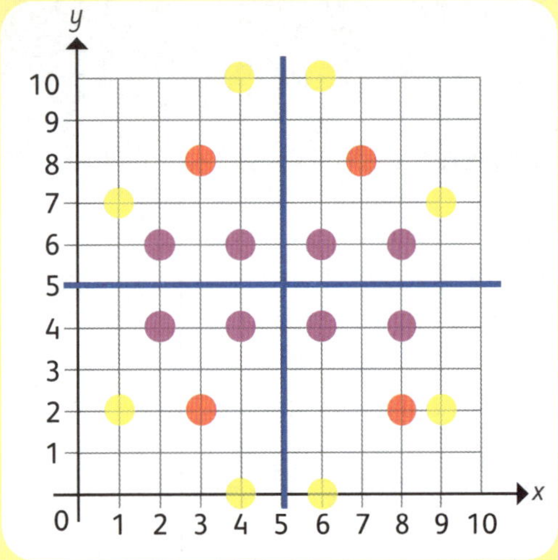

Junior

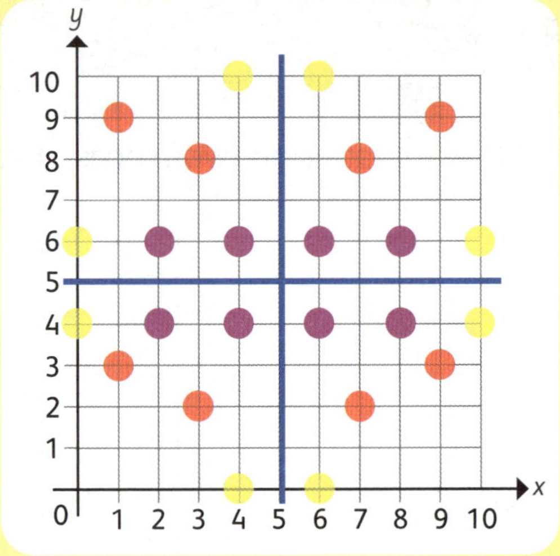

Both Tisha and Junior have made some mistakes.

Find the mistakes and write instructions containing coordinates to show how to correct each mistake.

Example

Tisha

The orange counter at (8, 2) should be moved to (7, 2).

Case 3

1 Do your own research to find out who formalised the use of the Cartesian plane and the coordinate system.

2 Prepare a short slide presentation using information from your research.

What I have learned

Write a paragraph of about five sentences about an object and its image.
The image can be reflected in a vertical line or a horizontal line. Illustrate your answer with a simple diagram.

Practice questions

1 Plot the following points on a 10 × 10 coordinate grid like the one shown here.

(3, 5) (4, 4)
(0, 2) (4, 0)

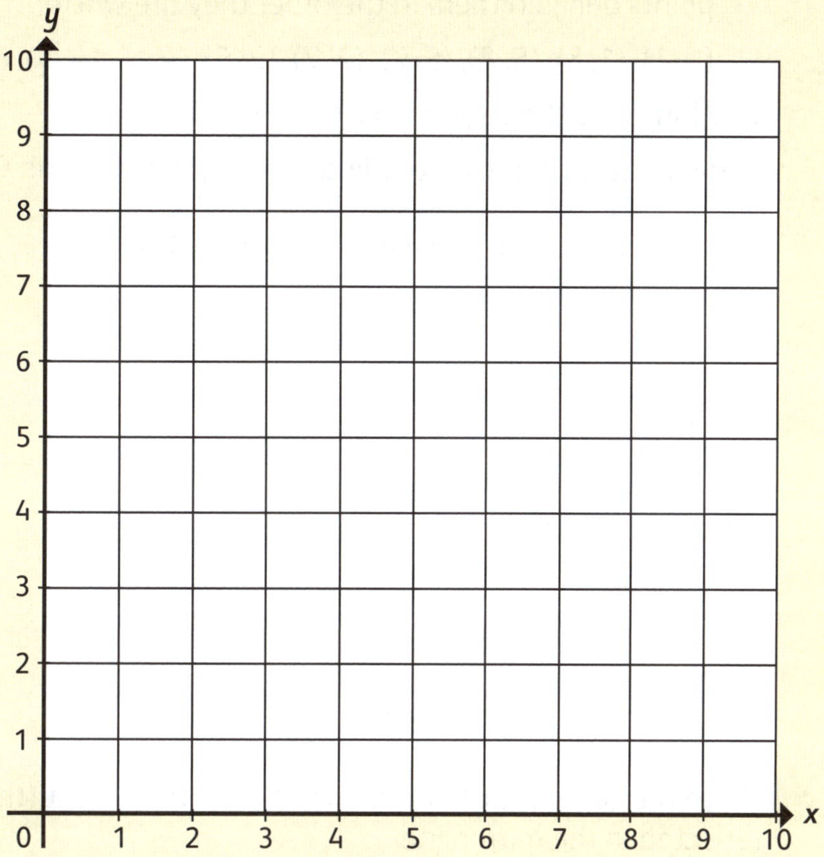

2 What are the coordinates of the vertices of this rectangle?

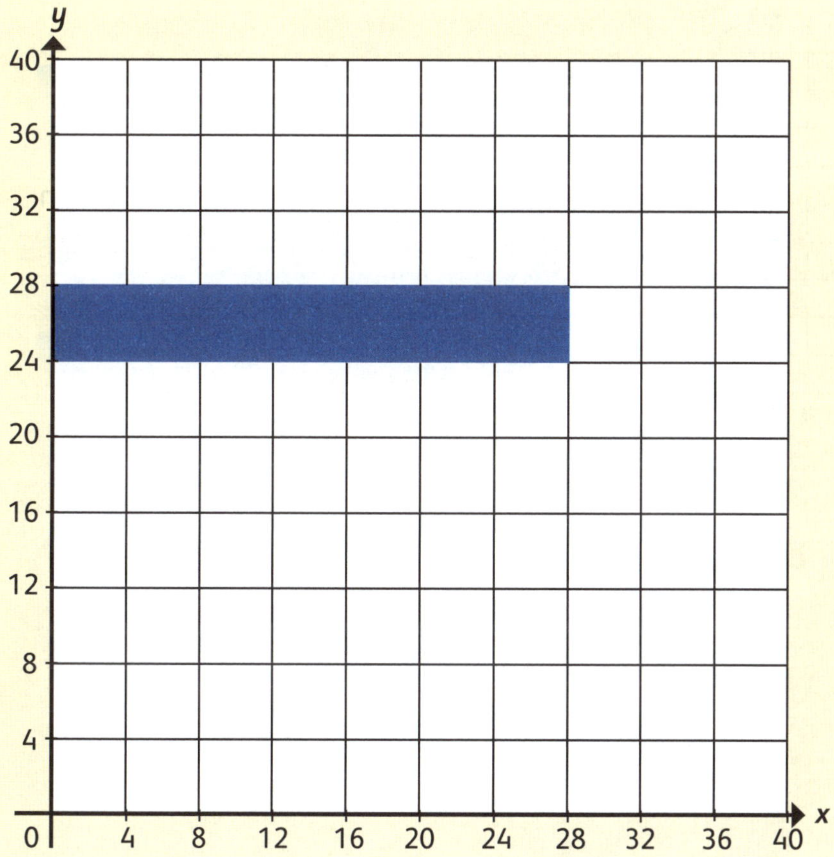

3 **a** Draw x- and y-axes and label each axis from 0 to 10. Now plot the following points and join them in the order they are written.

(1, 1), (1, 8), (5, 8), (5, 7), (2, 7), (2, 6), (4, 6), (4, 5), (2, 5), (2, 1), (1, 1)

b Name the shape you have drawn.

4 Marvin has written the coordinates of points A, B and C as A(1,1), B(2, 7) and C(3, 4).
Mark Marvin's work and correct any mistakes.

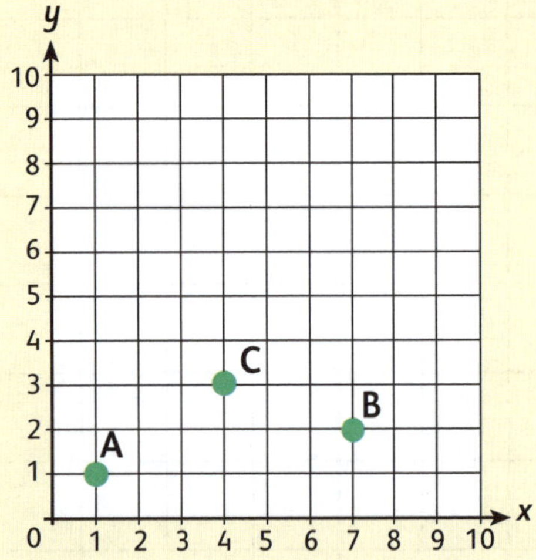

5 Write the coordinates of the image after the object (the triangle) has been reflected in the mirror line.

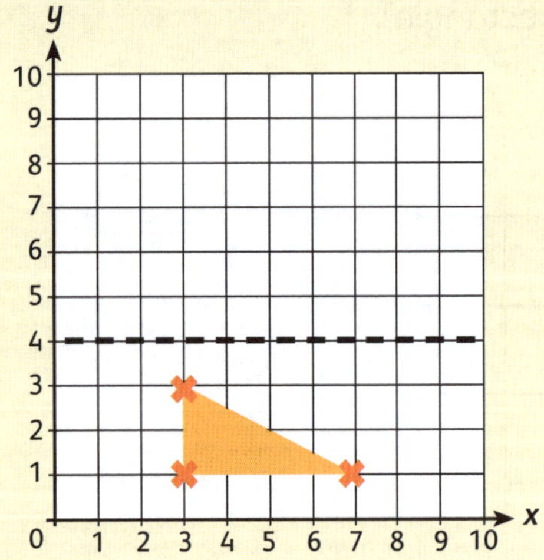

16 Self check

	Check you can do this:	If you have difficulty, try this:
1	Explain and use the concept of reflection within the Cartesian plane.	Revise page 254. Make sure you know what a reflection is, and that you can describe an object and its image in a mirror line.
2	Interpret points on a coordinate system to identify location.	Go back to page 251 and make sure you know what all the key words mean and can locate the x- and y-coordinates on a grid.
3	Use a coordinate grid to plot points accurately.	Revise page 251 to make sure you know how to identify a location on a grid and can use this location to draw a point and label it correctly.
4	Reflect a shape in a horizontal or vertical line on the Cartesian plane.	Go back to page 254 and make sure you know how to plot the reflected vertex points of a shape an equal distance on the other side of the mirror line.

Unit 4: Algebra

Chapter 17 Inequalities, variables and number patterns

In this chapter you will explore these focus questions:

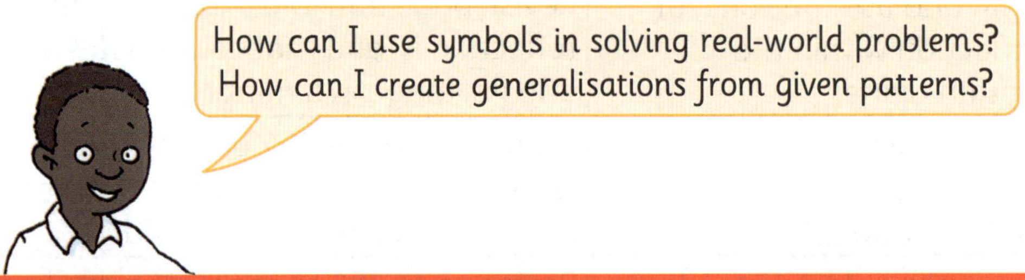

To answer these questions, you will:

- use substitution in formulae, algebraic sentences and inequalities in problem solving with up to two variables
- complete number sentences using the symbols +, −, ÷ and ×
- describe number patterns and predict what comes next in a sequence
- use tables of values to develop algebraic expressions to find given terms in a number pattern.

Starting point

Janette has 20 square red tiles, and 16 square blue tiles. She wants to explore how many different ways she can arrange them in rectangles. She decides to try and work it out using numbers. This is what she does:

$20 + 16 = a \times b$

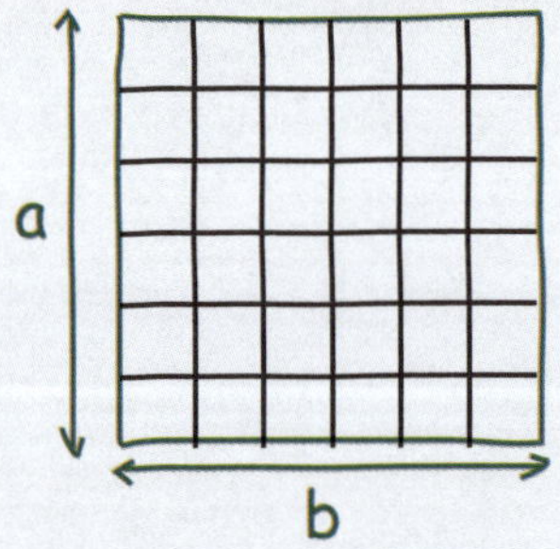

Possible values for a and b:

a	1	2						
b	36	18						

Discuss with your group:

1. How did she come up with her equation?
2. What is a simpler way to write the equation?
3. The diagram of the rectangles shows one arrangement of the tiles. What other ways can you draw it?
4. What does the table show?
5. How can she use the numbers in the table to tell her how many different rectangles are possible?

Using letters and symbols

Key maths idea

Some students are playing 'I think of a number'.

Anna writes this **algebraic expression** to represent the numbers $a + b$.

Then she gives another clue about the numbers.

$a + b = 20$

This number sentence is an **equation**, because it has an equal sign.

An equation is a **statement** that gives expressions or values that are equivalent to each other.

Her classmates **substitute** different values to make the number sentence true:

$5 + 15 = 20$ $8 + 12 = 20$ $19 + 1 = 20$ $11 + 9 = 20$

Anna gives one more clue.

Clue 3

$a - b = 8$

How can Anna's classmates use the last clue to work out the values of a and b?

Key words

algebraic expression
equation
statement
substitute

1 Work in pairs. Use the first equation to work out some different possible values of the variables. Then use the second equation in each pair to help you work out the values of both variables.

- **a** $k + m = 100$ $\frac{1}{2}$ of $k = 24$
- **b** $p - n = 90$ $3 \times p = 360$
- **c** $x + y = 1\,500$ $y \div 4 = 250$

2 A substitute is a replacement. Why do you think we use the word substitution for what we do when we work out the value of a variable?

3 To work out an employee's weekly pay, a company multiplies the number of hours worked (h) by the hourly wage (w). Calculate the pay for the following employees:

- **a** Nell works 30 hours, and has a wage rate of \$250 per hour.
- **b** Kelly works 45 hours, and her wage rate is \$200 per hour.
- **c** Carol works 20 hours and her wage rate is \$315 per hour.

The language of algebra

Key maths idea

Algebra is the part of maths where we use letters and symbols to represent unknown values. An expression can include known values, **variables** and **operators**. Variables are usually letters. They represent an unknown value. Operators tell us which operations to perform on numbers and variables, for example +, –, ÷ or ×.

When you multiply a variable by a value, you always write the value before the variable, like this:

$t \times 4 = 4t$

In this expression, 4 is the **coefficient**. It is the number we are multiplying by the variable. The value of the expression $4t$ will change depending on the value of t.

A **constant** is a value that does not change.

$3 + n$	means 3 more than n, or n plus 3	n is a variable and 3 is a constant
$3n$	means $3 \times n$ or $n \times 3$	3 is the coefficient
$\frac{1}{2}n$ or $\frac{n}{2}$	means half of n or $n \div 2$	$\frac{1}{2}$ is the coefficient (dividing a variable by 2 is the same as multiplying it by $\frac{1}{2}$)

Key words
algebra
variable
operator
coefficient
constant

1 For the expression $3x + 5$, work out the value of the expression if:

a $x = 10$ **b** $x = 35$ **c** $x = \frac{1}{2}$

2 For the expression $2n - 20$, work out the value of the expression if:

a $n = 17$ **b** $n = 48$ **c** $n = 100$

3 I think of a number. I multiply it by 8 and then subtract 14. Write the expression that expresses these instructions.

4 **a** $3p + q = 50$. Work out three different possible sets of values for a and b. You can use a table like this to help you:

p	1	2						
q	47	44						

5 In groups, write a song or an acrostic poem to explain the terms you have learned so far in this chapter. Include these terms in your song or poem: equation, variable, operation, expression, algebra. Perform your song or poem for your class, or collect them to make a class poster or booklet.

Order of operations

Key maths idea

Three students got this mixed operation to solve:
75 × 2 + 90 ÷ 3 = ☐

Look at their work:

Kari:
75 × 2 = 150
150 + 90 = 240
240 ÷ 3 = 80

Alistair:
75 × 2 = 150
90 ÷ 3 = 30
150 + 30 = 180

Megan:
2 + 90 = 92
92 × 75 = 6 900
6 900 ÷ 3 = 2 300

- Which answer is correct? Why?
- What mistakes did the others make?
- How can you rewrite the number sentence to make it true for the other answers?

In mathematics, there is a **convention** called the **order of operations**. A convention is an agreed rule that people follow because they have decided to do things in an agreed way. When a calculation has more than one **operation**, (+, −, ×, ÷), you can use the acronym **BOMDAS** to help you remember which part of the calculation to do first.

What goes first?	Example
Brackets – any parts of the calculation in brackets	(3 × 8) + 3 = 24 + 3 3 × (8 + 3) = 3 × 11
Of – this is usually where you see a fraction of an amount	Once you have calculated what is inside the brackets, you do 'of' calculations: $20 - \frac{1}{2}$ of $8 = 20 - 4$ $17 + \frac{1}{4}$ of $16 + 1 = 17 + 4 + 1$

Key words

convention
order of operations
operation
BOMDAS

Multiplication and **D**ivision	After brackets and 'of' calculations, look for the × and ÷ signs: $450 - 8 \times 10 = 450 - 80$ $130 + 2 \times 10 = 130 + 20$ $200 + 50 \div 10 = 200 \div 5$
Addition and **S**ubtraction	Last, complete the addition and subtraction. $8 + \frac{1}{2}$ of $100 - 7 \times 3$ $= 8 + 50 - 21$ $= 58 - 21$ $= 37$

1 Work out the following number sentences without using a calculator.

a $16 + 8 \times 3$ **b** $(16 + 8) \times 3$ **c** $30 - 4 \times 3$ **d** $(30 - 4) \times 3$

e $36 \div 2 + 4$ **f** $36 \div (2 + 4)$ **g** $56 - 21 \div 7$ **h** $(56 - 21) \div 7$

2 Use the order of operations to calculate these.

a $2 \times 7 + 2 \times 4$ **b** $2 + 7 \times 2 \div 3$ **c** $4 \times 4 - 2 \times 2 \times 2$

d $36 - 2 \times 8$ **e** $3 \times 4 + 4 \times 5 - 14 \div 2$ **f** $4 - 1 \times 2 + 3$

g $3 \times (4 + 4) \times 5 - 14 \div 2$ **h** $(4 - 1) \times (2 + 3)$ **i** $(5 \times 7) \times 2$

j $(20 - 4) \times 3 + 8$ **k** $28 - (4 + 2) \times 2$ **l** $(4 + 4) \times 4 - 4$

3 Copy and complete these number sentences. Use operation symbols to make the sentences true.

a $48 \square 3 \square 4 = 6 \square 6$ **b** $\frac{1}{2} \square 22 + 5 = 4 \square 4$

c $90 - (\frac{1}{4} \square 40) = 100 \square 20$ **d** $8 \square 5 = \frac{1}{2} \square (100 \square 20)$

4 Copy each number sentence, then use brackets to make the sentence true.

a $4 \times 5 + 7 = 48$ **b** $14 \div 10 - 3 = 2$ **c** $2 + 4 \times 3 - 7 = 7$

d $7 \times 4 - 3 + 2 = 27$ **e** $12 \div 3 \times 2 = 8$ **f** $12 + 3 \times 0 = 12$

g $8 + 6 \div 3 + 4 = 14$ **h** $100 - 3 \times 6 + 4 = 970$

5 Some calculators are designed to apply the order of operations, so you can key in a long mixed calculation. Other calculators require you to solve the calculations in steps. Work with a partner. Explore the way your calculator works. Use the calculations from the questions above.

6 In some parts of the world, the order of operations is called PEMDAS. Use the internet to find out what these letters stand for. Why does this acronym also work?

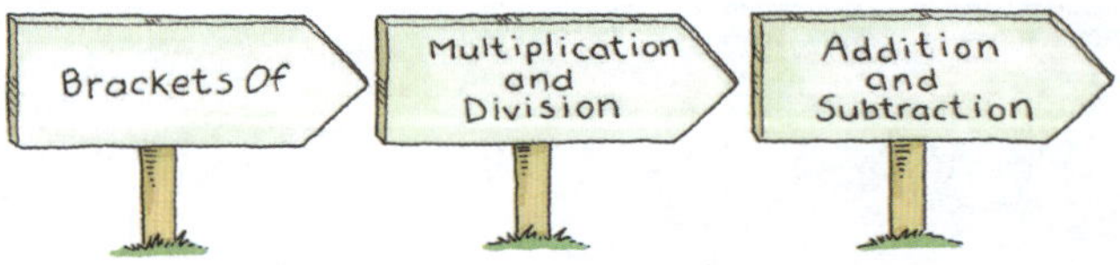

Real-world maths: Working out mobile call costs

Tisha has a mobile phone contract. Her contract says she will pay 33c for a text message and $1.15 per minute (or part thereof) for a call.

In the first month, Tisha sends 55 texts and makes 9 calls of a minute or less.

She wanted to work out how much this would cost. This is her calculation:

9 × 115 + 55 × 33 = □
9 × 115 = 1 035
1 035 + 55 = 1 090
1 090 × 33 = 35 970
35 970 cents = $359.70

Tisha is horrified by this big amount. But she has made a mistake.

1 What did she do wrong?

2 What will she actually have to pay?

3 How could you add brackets to this calculation to make the order of operations clear?

Inequalities

Key maths idea

An **inequality** in a number sentence tells you that the values are not equal to each other.

Example

$3 + 5 \neq 2 \times 3$ This says that 3 + 5 'is not equal to' 2 × 3

We can also use the signs $<$, $>$, $\leq$ and $\geq$ to describe an inequality.

Example

$p < x$ means p is less than 5

$x > 4$ means x is greater than 4

$q \leq 10$ means q is less than or equal to 10

$y \geq 8$ means y is greater than or equal to 8

Key word
inequality

Algebra

Rewrite each of these statements using the correct symbols.

1 x is greater than four

2 x is less than or equal to eleven

3 y is greater than or equal to fifty

4 p is less than nine

5 q is greater or equal to eight

6 5 + 7 is not equal to d

More inequalities

Key maths idea

We can deduce the possible values of a variable from the inequality.

Examples

1 x is a whole number and $x < 5$

The possible values of x are: 1, 2, 3 or 4.

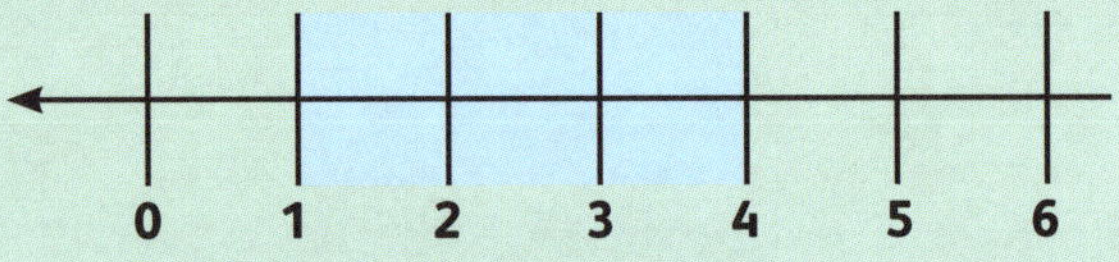

2 y is a whole number, and $0 < y \leq 7$

Which values of y satisfy this inequality?

y is greater than 0 but less than or equal to 7.

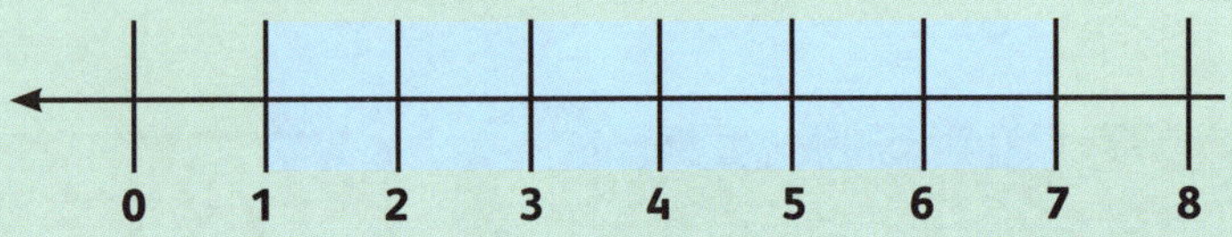

The values of y that satisfy the inequality are: 1, 2, 3, 4, 5, 6 and 7.

Inequalities are examples of **open sentences**. In maths, a sentence is open if it is not known if it true or false. An inequality is an open sentence as the value of the variable can vary. A **closed sentence** will always be true or always be false, such as $3 + 3 = 6$.

We can also write inequalities like this:

$2 < y < 8$　　　　$7 \leq a \leq 10$

We read this as 'y is greater than 2 and less than 8' and 'a is greater than or equal to 7 and less than or equal to 10'.

Key words

open sentence
closed sentence

1 What are the possible values of x that would satisfy these inequalities, if x is always a positive whole number?

a $12 + x < 17$
b $9 - x \geq 6$
c $x \leq 5$
d $x < 4$
e $36 - x > 30$
f $10 + x < 20$

2 Write the possible values that satisfy each of these inequalities. All variables are whole numbers greater than 0.

a x is a prime number; $x \leq 13$
b y is an even number; $4 \leq y \leq 20$
c p is an odd number; $p \leq 17$
d q is a multiple of 4; $q < 24$
e a is a multiple of 10; $50 < a \leq 100$
f b is a factor of 6; $b \leq 8$
g r is a factor of 45; $r \geq 10$
h s is a prime number; $20 < s < 30$

3 Write each inequality in words.

a $4 \leq x \leq 12$
b $1 < y \leq 5$
c $10 < p \leq 25$
d $75 < q < 99$

Real-world maths: Less waste, more profit

Abbey Gardens Farm in Mandeville is a greenhouse farm that produces and sells fruit and vegetables. The farm manager uses inequalities in her daily work. For example:

- she wants wastage to be less than a certain amount
- she want the profits to be greater than the cost of growing the products
- she wants products to be in a certain range, for example, peppers that weigh between 110 and 150 grams
- she wants packs of strawberries to weigh no more than 510 grams and no less than 495 grams.

1 If it costs \$148.50 per kilogram to produce quality cabbage, write an inequality to show what the profit (p) should be.

2 The mass of broccoli heads (m) that the farm sells should be 225 g $< m <$ 260 g. Which of these broccoli heads fall into the range?

A – $\frac{1}{4}$ kg B – 0.24 kg C – 215 gb D – 8 ounces

3 A local supermarket will stock packs of strawberries as long as they weigh at least $\frac{1}{2}$ kg and no more than 1.2 kilograms. Write an inequality using grams to represent this. Use s for the mass of the strawberries. Try to find a way to show this using a number line as well.

Algebra

Expressions, variables and equations

Key maths idea

Remember that an **expression** (or statement) is a sentence of letters and/or numbers linked by an operations symbol.

Example

$3 + 5$ $x \times 5$ $2y$ $\frac{10}{b}$

The letters used in expressions are known as variables. Their value can change.

When we multiply variables together, we do not show the multiplication sign.

So $2x$ simply means $2 \times x$.

An equation is a number sentence with an equal sign.

These are all equations:

$3 + 5 = 8$ $a + b = 9$ $2y = 8$ $\frac{10}{x} = 5$

Examples

$2x = 8$ is an equation; x is the variable and $2x$ is an expression.

If we are given the value of variables, we can calculate the value of expressions.

If $a = 3$ and $b = 4$, we can find the value of these expressions.

1 $a + b = 3 + 4$
$= 7$

2 $3a = 3 \times 3$
$= 9$

3 $b^2 = 4 \times 4$
$= 16$

Key word
expression

1 If $x = 6$ and $y = 3$, calculate the value of these expressions.

a $x - y$ **b** $4x - 2y$ **c** $5x$

d y^2 **e** $2x + 2y$ **f** x^y

2 If $m = 5$ and $n = 2$, find the value of:

a $2m + 3n$ **b** $3m - 5n$ **c** $5m - n$

d mn **e** $\frac{m}{n}$ **f** m^n

3 If $p = 7$ and $q = 4$, find the value of:

a $2p - q$ **b** $3p + 2q$ **c** $p^2 + q$

d $2pq$ **e** $5p - 3q$ **f** q^q

4 a Write in words the meaning of the expression $19 - 2y$.
  b What is the value of the expression $19 - 2y$ when $y = 8$?

5 Nine times a number x and five times another number y add up to sixty-two. Find the values of x and y. Show how you find your answer.

Real-world maths: Converting temperature

Josh and Malik want to visit a resort in the Caribbean.

Josh doesn't like it to be too hot, so he wants somewhere with an average temperature below 27 °C.

Malik doesn't mind the heat, but he doesn't want to go anywhere where the temperature is less than 23 °C.

They find this information about different resorts:

Resort	**Average daytime temperature**
Palm Beach	84.5 °F
Golden Sands	80.2 °F
Sea Breezes	74.5 °F
Blue Mountain View	73 °F

1 Use the formula $C = (F - 32) \times \frac{5}{9}$ to convert the average resort temperatures to degrees Celsius.

2 Which resorts would be acceptable to both Josh and Malik?

Algebra

Number patterns and sequences

Key maths idea

A **sequence** is a list of numbers that follows a **pattern**.

This pattern is an **increasing** sequence, following the **rule** +4. A rule is a way to find the value of **consecutive terms**. One way to do this is to find the **common difference** between each term and the next.

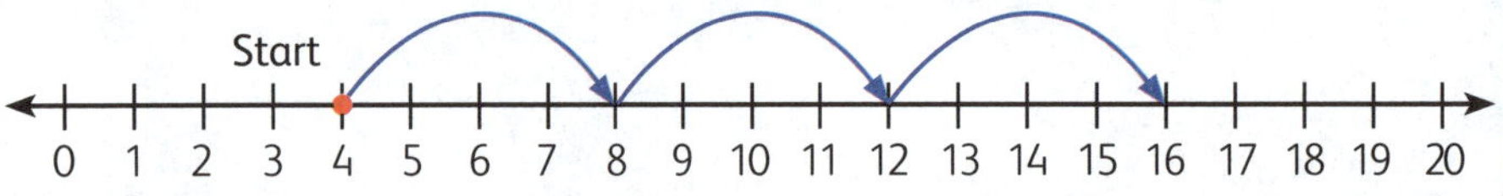

'Consecutive' means 'following each other'.

This pattern is a **decreasing** sequence. What rule does this sequence follow?

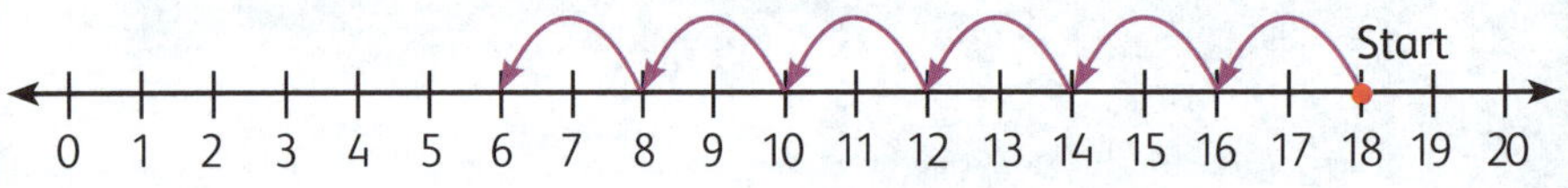

Key words

sequence
pattern
increasing
rule
consecutive terms
common difference
decreasing

1 Here are the first three terms in a sequence.

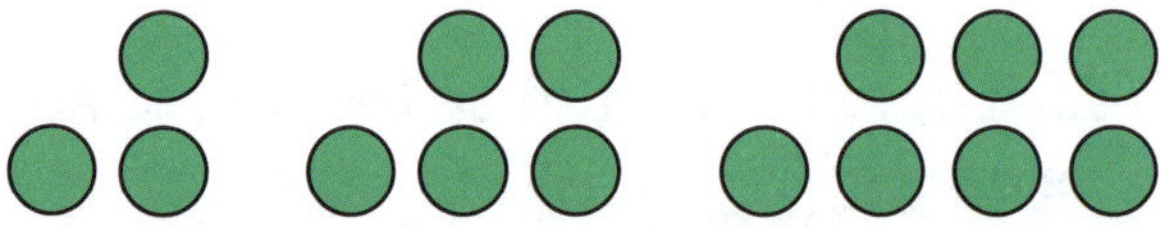

- **a** Draw the next term in the sequence.
- **b** How many circles will make up the 5th term?

2 Copy and complete these sequences. See if you can find the common difference to help you work out the rule.

- **a** 4, 14, 24, ____, ____, ____
- **b** 35, 33, 31, ____, ____, ____
- **c** 7, 11, 15, ____, ____, ____

3 The rule in these patterns is not based on a common difference. With a partner, work out what you have to do to each term to calculate the next term.

- **a** 3, 12, 48, ____, ____, ____
- **b** 1024, 256, 64, ____, ____, ____
- **c** 1, 4, 10, 22, ____, ____, ____

4 Find the next two terms in each of these sequences.

- **a** 51, 47, 43, ____, ____
- **b** 1 500, 2 600, 3 700, ____, ____
- **c** 7.25, 7.45, 7.65, ____, ____

Exploring number patterns and using tables

Key maths idea

Look at this pattern:

We can describe this pattern by writing the number of dots in each shape: 1, 4, 9, 16.

If you look at each array of dots, you will see that we can also describe it using multiplication:

1 row of 1 dot	$1 \times 1 = 1$
2 rows of 2 dots per row	$2 \times 2 = 4$
3 rows of 3 dots per row	$3 \times 3 = 9$

If you multiply a number by itself, the product is a square number. For this reason, multiplying a number by itself is called squaring the number.

3×3 We say: '3 squared' or '3 to the power of 2'. We write it like this: 3^2

3 is the **base**. 2 is the **exponent**.

When we write $n \times n = n^2$, we say the answer as 'n squared'.

Key words
base
exponent

The exponent tells us how many of the base to multiply.

1 Work out the next four numbers in the sequence.

Term	1	2	3	4	5	6	7	8
Number of dots	1	4	9	16				

2 Is 150 a square number? How can you work it out?

3 Triangular numbers can be organised into triangles.

Work out the next five triangular numbers.

Term	1	2	3	4	5	6	7	8
Number of dots	1	3	6					

Think of patterns, including square or triangular numbers, in other subjects at school. What patterns can you think of in Physical Education, Social Studies, Music, Visual Arts and Language Arts?

4 This pattern is made from matchsticks.

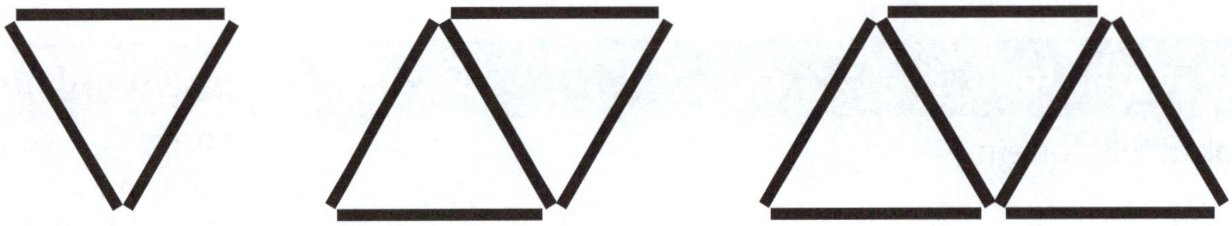

a Record the pattern in a table like this.

Term			
Number of matchsticks			

b How many matchsticks are needed for the next two terms?
Show by drawing how you get your answer.

5 Use matchsticks or cotton earbuds to create your own shape pattern with a partner. Then record your pattern in a table to show the rule.

Real-world maths: Choose the best packaging

Moira sells packaging boxes in sets like this:

The boxes are cubes.

The smallest box has sides 8 cm long.

Each box has sides 1.5 cm longer than the smaller box that fits into it.

1 Work out the side lengths of each box. Use a table to help you record your work.

2 Abbie wants to pack four boxes of volume 686 cm^3 into one of these boxes with as little space as possible left over. Which box should she use?

The n-th term

Key maths idea

Sometimes we might need to find more than the 5th or 10th term of a sequence. We might be asked to find the 100th term and that could take a lot of time!

We can find the value of the **n-th term** to help us find any term we are asked for.

Look at this sequence: 4, 8, 12, 16, 20, …

We can write it in table form:

Term number	1	2	3	4	5	6	n
Term	4	8	12	16	20	24	4

We can describe this sequence in different ways:

The rule is +4.

Multiply the term number by 4.

If the term is n and the sequence is formed by multiplying the term by 4, $n \times 4$,
then we write this as $4n$ in algebra.

So the algebraic expression for this sequence is $4n$.

Once we know this, we can calculate any given term in the sequence.

What is the 100th term of our sequence?

Our n is 100, so $4 \times n = 400$.

The 100th term of this sequence would be 400.

Key word

n-th term

1 For each table, write the terms that are missing.

a Write term 4, term 6 and n term for this number pattern.

Term	1	2	3	4	5	6	n
Number	5	10	15		25		

b Write term 5, term 6 and term n for this number pattern.

Term	1	2	3	4	5	6	n
Number	0	7	14	21			

c Write term 3, term 6 and term n for this number pattern.

Term	1	2	3	4	5	6	n
Number	100	99		97	96		

2 For each table in Exercise 1, write an algebraic expression for the rule.

3 Calculate the 100th term for each of the three patterns in Exercise 1.

Maths detective: Algebraic patterns

Use what you know about patterns and algebra to solve these problems.

Case 1

1 Make a list of what happens when you add one even number to another even number.

2 Make a list of what happens when you add one odd number to another odd number.

3 Make a list of what happens when you add one even number to an odd number.

4 Draw up a table like this one to record what you notice.

	Patterns I noticed
Even + Even	
Odd + Odd	
Odd + Even	

Case 2

Select any two numbers from the list to make calculations with answers that are as large as possible for each operation.

$\frac{1}{2}$ 4 8 40 80 200

1 $\square + \square =$ ____

2 $\square - \square =$ ____

3 $\square \times \square =$ ____

4 $\square \div \square =$ ____

Case 3

Andrea wears a watch that tracks how far she walks and how long it takes.

Hours walked (h)	**Kilometres walked** (m)
2.5	8.75
4	14

1 Which of these equations best represents this relationship?

a $h = m + 10$ b $h = 3.5\,m$ c $m = h + 10$ d $m = 3.5\,h$

2 If Andrea walks for 6 hours and 30 minutes, how many kilometres will she walk at this rate? Show how you got your answer.

Case 4

Letters A to N of the alphabet are assigned a value as shown in the table.

A	B	C	D	E	F	G	H	I	J	K	L	M	N
1	2	3	4	5	6	7	8	9	10	11	12	13	14

Find the value of each expression.

1 The sum of D and H.

2 E is increased by three times C.

3 The product of B and F.

4 Two times L reduced by K.

5 Four times N all divided by G.

6 The difference between M and J.

Hint: there is more than one possible answer here.

Real-world maths: Tiling a floor

Salim covers a floor area using square tiles that form a pattern. The diagram shows the first three parts of the pattern.

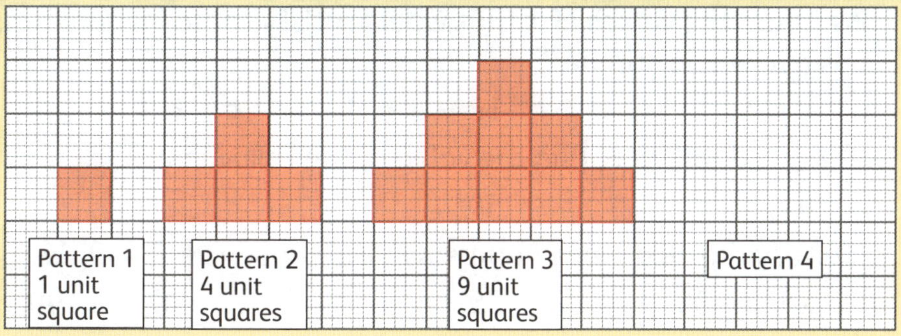

1 Draw pattern 4 in your notebook.

2 How many square tiles will there be in pattern 5?

3 Explain how you figured this out.

4 Salim represents the tiling pattern algebraically in the table below. Study the number pattern and complete the rows for patterns 4 and 5.

Pattern	Area of pattern (square units)	Perimeter of pattern (units)
1	1	4
2	4	10
3	9	16
4		
5		

5 a Each square tile costs $250.00. Salim says the cost for tiles for pattern 7 will be $35 000.00. Explain why his calculation is incorrect.

b What will the correct cost be for the tiles for pattern 7?

Algebra

What I have learned

Copy this diagram into your notebook. Write short notes to explain and summarise the most important information about each concept.

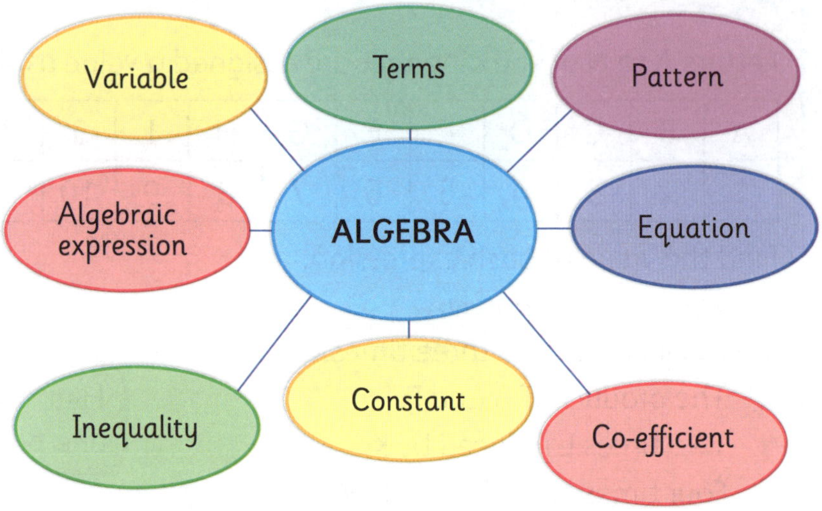

Practice questions

1 Calculate without using a calculator.

a $11 \times (25 + 6) - 18$ b $75 - 15 \times 4 + 28$

c $26 - (17 - 3) \div 2$ d $100 \div 4 \times (8 \times 4)$

2 What are the values of x that would satisfy these inequalities, if x is always a positive whole number?

a $x < 5$ b $12 + x < 17$ c $x \leq 4$ d $x - 10 \geq 24$

3 If $a = 5$ and $b = 6$, find the value of the following:

a $b - a$ b $5b$ c a^2 d $2a + 2b$

4 Identify the next three numbers in the sequence: 1, 4, 9, 16, …

a 24, 29, 36 b 25, 36, 49 c 30, 50, 100 d 36, 49, 64

5 Find the rule and give the next three terms for each sequence.

a 320, 400, 480, 560, 640, …

b 5 650, 5 550, 5 450, 5 350, 5 250, …

6 a What shape will come next?

b What shape will the 25th bead be?

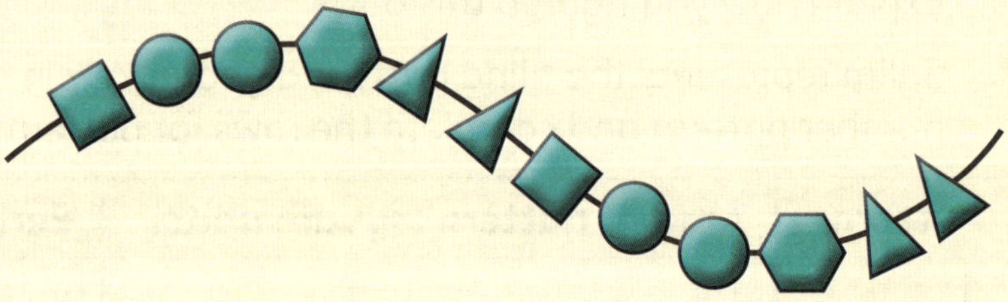

7 Look at this sequence:

7, 14, 21, 28, 35, …

What is the algebraic expression for the n-th term of this sequence?

8 Over four years, the number of students enrolled at the local primary school increased as shown.

Year	2018	2019	2020	2021
Number of students	110	140	170	200

This pattern is expected to continue for each year after 2021.

a How many students can be expected in the year 2022?

b How many students can be expected in the year 2024?

c How many students can be expected in the year 2026?

17 Self check

Check you can do this:	If you have difficulty, try this:
1 Use substitution in a formula to solve a problem.	Go over page 263 and remind yourself how we substitute different values into number sentences.
2 Use substitution in algebraic sentences and inequalities to solve a problem.	Go over the activities you did on page 264 and also revise page 267.
3 Use the symbols for the four mathematical operations to complete number sentences.	Revise page 265. Make sure that you know the four operator symbols and can use them to make true number statements.
4 Substitute a number for a variable in a mathematical sentence with up to two variables.	Revise the activities from page 270.
5 Calculate the common difference between consecutive terms in a pattern, and use the common difference you found to predict further terms in the pattern.	Go over page 272.
6 Investigate tables of values to identify the algebraic expression that represents any term in a simple numeric pattern.	Revise page 273.
7 Use the algebraic expressions you identified to make predictions about further terms in the pattern.	Go over page 275.

TERM 3

UNIT 1: NUMBER
p281

Problem solving

Money matters

UNIT 2: MEASUREMENT
p313

Speed, time and distance

Circles

UNIT 3: GEOMETRY
p338

Angles

UNIT 4: ALGEBRA
p350

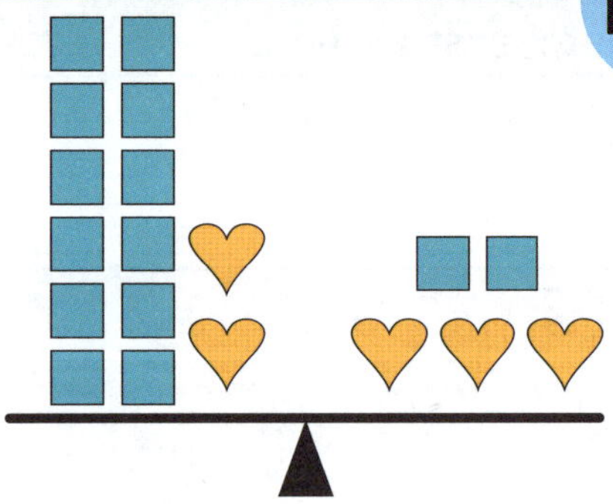

Solving simple equations

UNIT 5: STATISTICS AND PROBABILITY
p361

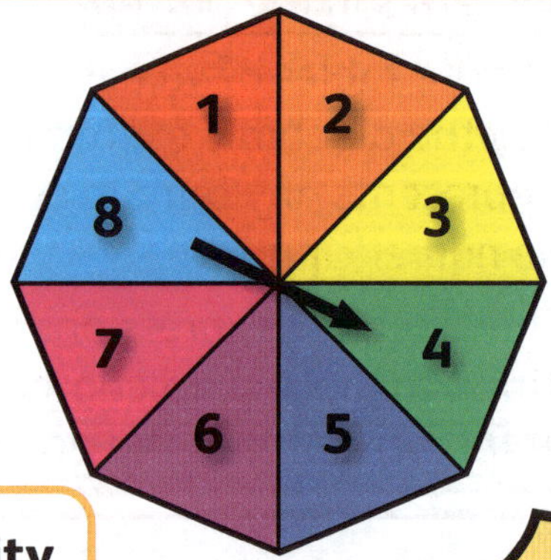

Probability

Practice test p373

Unit 1: Number

Chapter 18 Money matters

In this chapter you will explore this focus question:

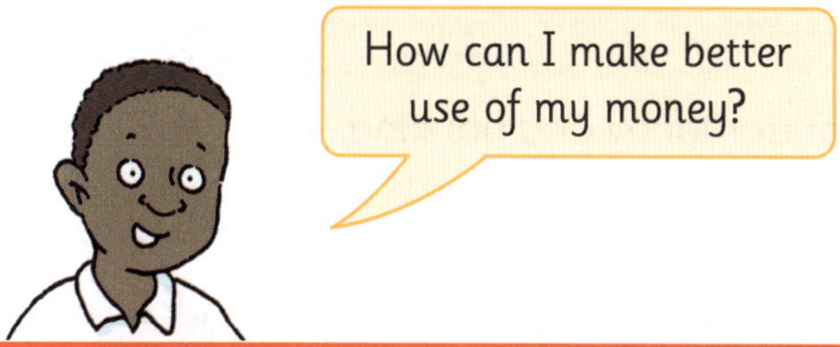

To answer this question, you will:

- show knowledge of different financial institutions
- explain the functions of different financial service providers
- identify types of financial institutions
- complete bank deposit and withdrawal slips
- calculate simple interest
- calculate profit and loss.

Starting point

1 The Jordans' 3-year-old washing machine has broken down. They need to decide which option is the best use of their money: replace the machine with a new one, replace the machine with a used one, or repair the machine they have.

Use this information to help them decide:

a List the advantages and disadvantages for each of the three options.

b What would you advise the family to choose and why? Share your thoughts with the class.

Talk about

Discuss as a class why the option you have chosen (replace with new, replace with used, repair) might not always be the best option for other breakdowns, for example, a car, games console or computer.

2 Here are some terms people use to talk about money and banking. Match each term to a suitable definition.

1 Account number
2 Cash machine
3 Transaction
4 Balance
5 Contactless payments
6 Debit card
7 Online banking
8 PIN
9 Withdrawal

A Taking money out of your account, either at a bank or an ATM, or making a transfer to another account.
B A Personal Identification Number is a private number you need to know when you want to withdraw cash or pay for goods with a card.
C A plastic bank card that can be used instead of cash when buying goods. It withdraws the money directly from your bank account.
D The amount of money you have in your account.
E Machine from which you can withdraw cash from your bank account; also called an ATM (automated teller machine) or cashpoint.
F Moving money somewhere else by buying or selling something or transferring money to another account.
G A way of accessing your bank account using the internet on your phone, tablet or computer.
H Paying by bank card without typing in your PIN.
I A special reference number for your bank account.

The financial sector

Key maths idea

The economy of any country is made up of different sectors, for example, the tourism sector, the agricultural sector and the manufacturing sector. The financial sector is also part of the economy. The diagram shows some of the key parts of this sector.

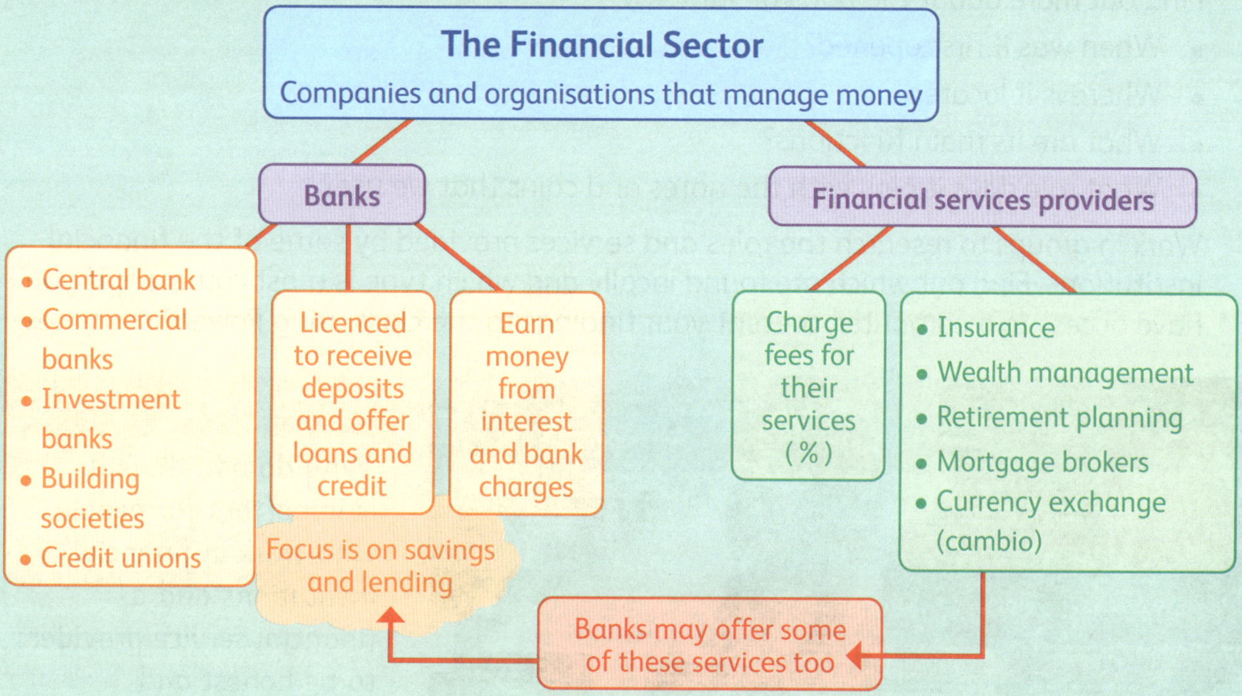

Financial institutions are organisations that process monetary transactions, including business and private **loans**, deposits, and **investments**. They include **commercial banks**, **credit unions** and **building societies**.

Financial institutions in Jamaica are governed by the **Bank of Jamaica**, which is Jamaica's **central bank**. It acts as the government's banker. This means that it monitors and administers the country's money reserves and currency.

Key words

loan
invest/investment
commercial bank
credit union
building society
Bank of Jamaica
central bank
insurance company

1 Choose the correct term to complete each sentence.

a A credit union is a type of (financial institution/commercial bank).

b When you put money in a bank account, this is called making a (deposit/ loan/ withdrawal).

c When you borrow money from a bank, you take out a (loan/withdrawal).

d The Bank of Jamaica is Jamaica's (commercial bank/central bank).

2 Find out more about the Bank of Jamaica:

- When was it first opened?
- Where is it located?
- What are its main functions?
- What role does it play with the notes and coins that we use?

3 Work in groups to research the roles and services provided by some of the financial institutions. Find out which are found locally and which type is most common. If you have access to a computer, present your findings to the class using PowerPoint slides.

Talk about

Why do you think it is important for people who work in financial institutions and as financial service providers to be honest and trustworthy? Why is it important that we are all honest and trustworthy?

Real-world maths: Financial service providers people use

Conduct a survey to find out how people in your community use financial service providers, for example, **insurance companies**. What is the most common service used? What is the least common? Present your findings to the class.

Maths detective: Budget allocations

Use your knowledge of numbers, percentages, general calculation skills and a calculator to solve these problems.

Case 1

This graph shows the proportion of the national budget allocated to different branches of government.

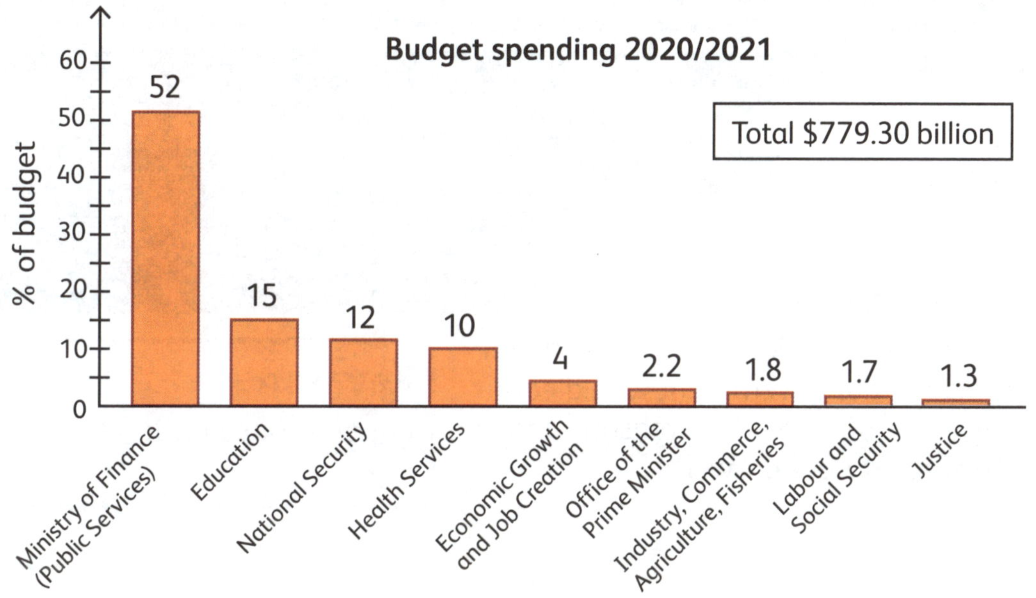

1 What does this graph tell you?

2 How much is the total budget?

3 Which branch of government receives more than half the annual budget?

4 What percentage of the budget is allocated to Health Services?

5 Work out how many billion dollars are allocated to Education.

Case 2

When he made his budget speech, the prime minister said that only 28 % of senior citizens were eligible for a national insurance scheme pension. To support senior citizens he announced an overhaul of the system, so that all persons who needed financial support could access it.

1 There were 392 930 senior citizens in Jamaica in 2020. How many of these persons were not eligible for the pension scheme?

2 In July 2021, the government aimed to pay 8 000 senior citizens a pension of $ 3 400. How much is this in total?

Case 3

In April 2020, the rate of GCT was reduced from 16.5 % to 15 %. The government estimated that this would mean a loss of income of almost $14 billion.

General consumption tax or GCT is a tax that is added to goods or services in Jamaica and is paid to the government. The **tax** is usually already included in the price you pay.

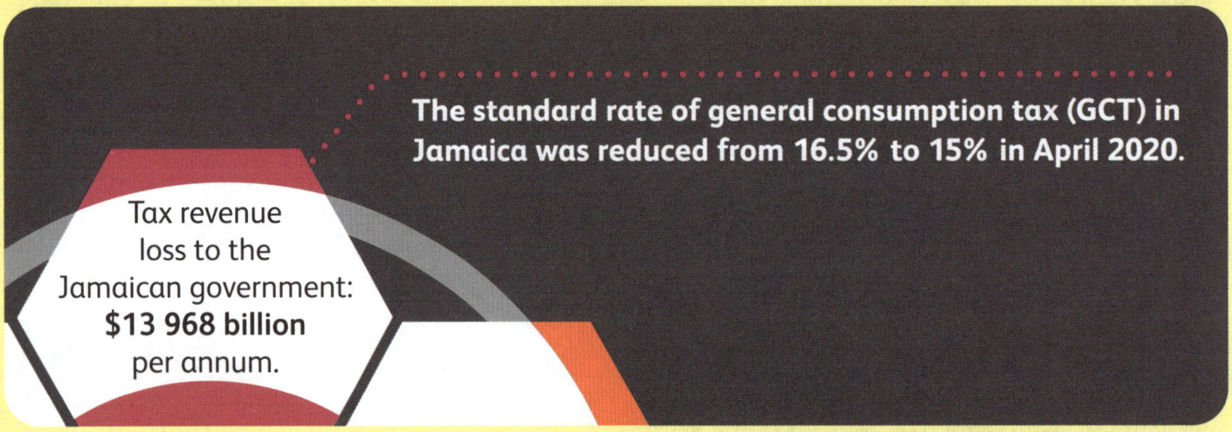

Is it possible to use this information to work out how much income the government would earn from GCT at a rate of 15 %? Give reasons and examples to support your answer.

Key words

general consumption tax (GCT)

tax

Deposits and withdrawals

Key maths idea

Kelvin saved these notes and coins on his birthday:

He wants to **invest** the money, so he will transfer it into his **savings account**, because the bank pays him **interest** of 3 % per year. Kelvin needs to complete a **deposit** slip at the bank. This is a record of how much money you are paying into your account.

Key words

invest

savings account

interest

deposit

withdrawal

finances

What else does Kelvin need to include on the deposit slip? How much does he deposit in total?

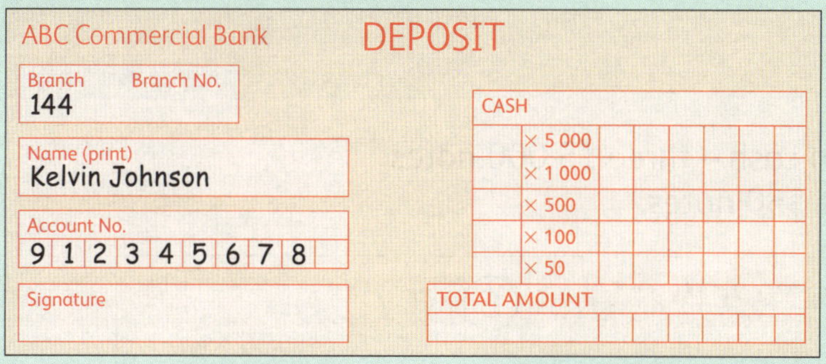

Kelvin's sister, Hattie, also has a saving account. She would like to take \$2 500 out of her account, so she needs to complete a **withdrawal** slip.
What else does Hattie need to include on the withdrawal slip? What different combinations of notes could she request?

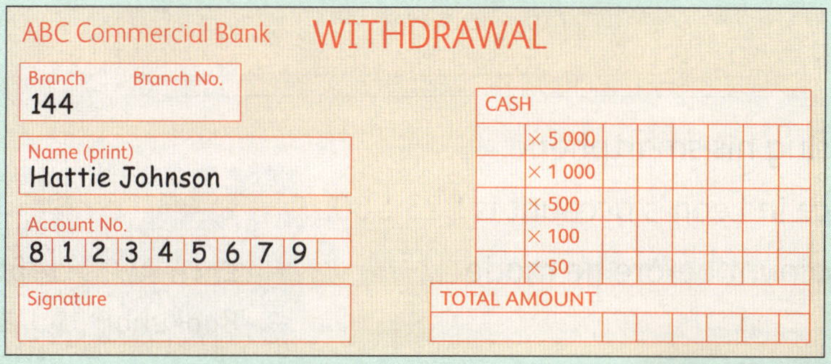

Before people used credit and debit cards, they used cheques when they wanted to pay larger amounts. Some businesses still accept cheques, but many have stopped using them as it is not as safe as electronic money transfers.
Instead of using cash or cheques to make payments, we can today use debit and credit cards.

- When you use a debit card, the money is taken immediately from your bank account.
- When you use a credit card, you receive a bill at the end of the month to pay for everything you purchased. But be careful: you will pay interest on the amount you used, so the purchase will be more expensive.

When you pay by card, you will see the details of each transaction in your bank account. This includes the date, the person or company you have paid (payee), the amount you paid and your balance (the amount of money left in your account).

Many people use e-banking to keep track of their **finances**.

You can also withdraw money at an ATM (automated teller machine) or at a cashpoint by using your bank card. The machines give you cash from your bank account. In larger city bank branches they can also be used to deposit cheques and cash.

1 Use this information to complete a copy of the bank deposit slip below:

- your name
- your signature
- today's date
- account number: 123456
- amount: Jamaican dollars in cash – three $5 000 notes, two $1 000 notes and three $50 notes.

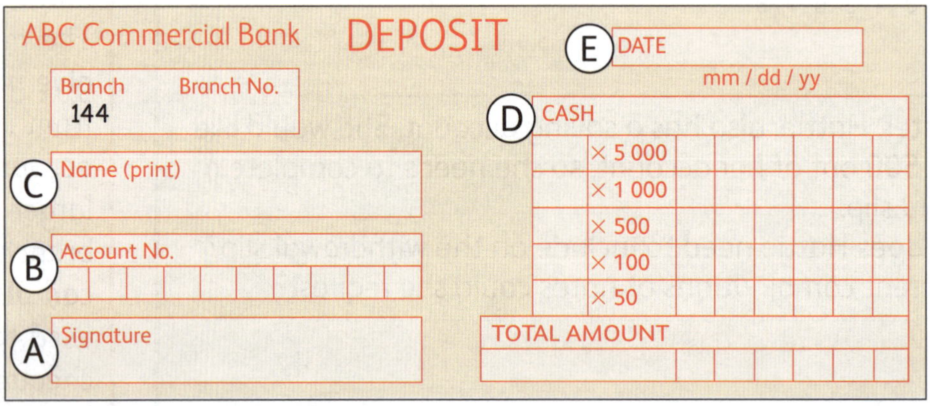

2 Leon makes some payments using his smartphone:

After the payments, the balance in Leon's account is $23 560.

What was the balance in his account before he made the payments?

Café	$450
Bookshop	$3 200
Kiosk	$195

3 Work in a small group to role play depositing and withdrawing money at the bank. Present this to the class.

4 Use a computer program, such as PowerPoint, to make a presentation about the different ways that people can make payments. Include any advantages and disadvantages. Be sure to use examples of payments made and the balance left in an account.

Interest

Key maths idea

You can earn **interest** from the bank, but the bank also charges you interest to borrow money from it, for example, to **finance** the purchase of a car. When you borrow money from the bank (a loan) you have to pay the bank a fee for this. A large loan used to buy a home is called a **mortgage**.

Example

Mikayla wants to buy a car. The price of the car is $2 500 000. She has $2 000 000 and decides to borrow $500 000 from the bank for three years.

Key words

interest
finance
mortgage
rate
principal

The bank charges simple interest on the loan at a **rate** of 2 % per year. This is called simple interest.

We calculate interest by using this formula:

Simple interest = principal × rate × time

$$SI = P \times R \times t$$

How much interest will Mikayla pay on her loan?

The **principal** (loan amount) is $500 000. The rate is 2 %.
The terms are 3 years.

$500 000 × 2 % × 3
= $500 000 × 0.02 × 3
= $30 000 simple interest

What will be the cost of Mikayla's monthly repayments?

$500 000 + $30 000 = $530 000

There are 36 months in 3 years (12 × 3 = 36)

$530 000 ÷ 36 = $14 722.22 per month (to the nearest cent).

1 Calculate the interest on each amount in the table. Write the solutions a to e in your notebook.

Principal amount	Time	Rate	Interest payable
$1 000	1 year	20 %	a
$10 000	2 years	10 %	b
$12 000	1 year	4 %	c
$18 000	3 years	5 %	d
$18 000	5 years	5 %	e

2 A bank charges 5 % interest on a loan. How much will Anthony have to pay back in total for a loan of $80 000 over three years?

3 Mrs Lamone has $12 000 to invest in a new savings account. She will not withdraw any money in the first year. She finds out about the interest rates and offers on some different accounts.

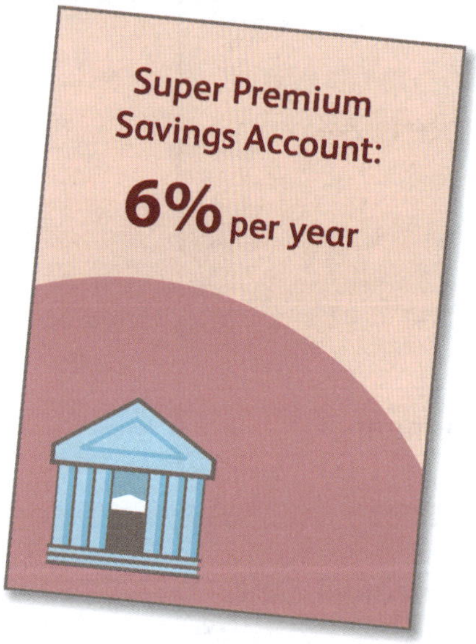

- **a** Calculate how much interest Mrs Lamone will earn each year in the different savings accounts.
- **b** Explain to Mrs Lamone which savings account she should open and how much money she will have in total at the end of the year.

> We describe interest per year as 'per annum'. This is often abbreviated as p.a., so the Standard Savings Account, for example, gives interest of 4% p.a. (or per annum).

4 A company wants to purchase a new printer. The cash price of the printer is $20 000. A deposit of 15 % is required. The remaining loan amount will be paid off over 24 months at an interest rate of 10 %.

- **a** How much is the deposit for the printer?
- **b** How much is the remaining loan before interest?
- **c** How much is the remaining loan after interest?
- **d** How much are the monthly repayments?

Profit and loss

Key maths idea

If you buy or make something and sell it for more than it cost you, you make a **profit**. If you have to sell it for less than it cost you, you make a **loss**.

You can calculate the percentage of profit or loss if you know the cost price and the selling price.

Example

Jayden bought a car for $950 000 and sold it three years later for $750 000.

To calculate the profit or loss Jayden made:

He sold it for less than he paid, so he made a loss.

The loss is $950 000 − $750 000 = $200 000.

$$\frac{\text{profit or loss}}{\text{original amount}} \times 100 \longrightarrow \frac{\$\ 200\,000}{\$\ 950\,000} = 21\,\%$$ (rounded to the nearest whole number)

Jayden made a loss of 21 %.

Key words
profit
loss

1 Brianna's mother bakes cakes to earn an extra income. Work out the profit she makes on the different cakes she bakes. Write the solutions from a to e in your notebook.

Item	Cost price	Selling price	Profit
Chocolate cake	$1 257.50	$1 650.00	a
Toto	$714.00	$950.00	b
Plantain tarts	$894.30	$1 250.00	c
Fruit cake	$1 554.20	$1 850.00	d
Coconut macaroons (each)	$245.55	$325.00	e

2 Sean bought some toys and resold them. Decide if there is profit or loss, and then calculate how much. Write your answers in the last column.

	Item	Cost price	Selling price	Profit or Loss
a	Pen	$85.00	$99.00	$14 profit
b	Ball	$155.50	$185.00	
c	Game	$255.99	$265.00	
d	Train set	$1 805.75	$1 550.00	
e	Skipping rope	$555.85	$500.00	
f	Pack of cards	$266.50	$300.00	

3 Hakeem bought a second-hand phone for $19 280. He sold it online and made 10 % profit. What price did he sell it for?

4 The bar graph shows the profit and loss for the first part of a company's fiscal year.

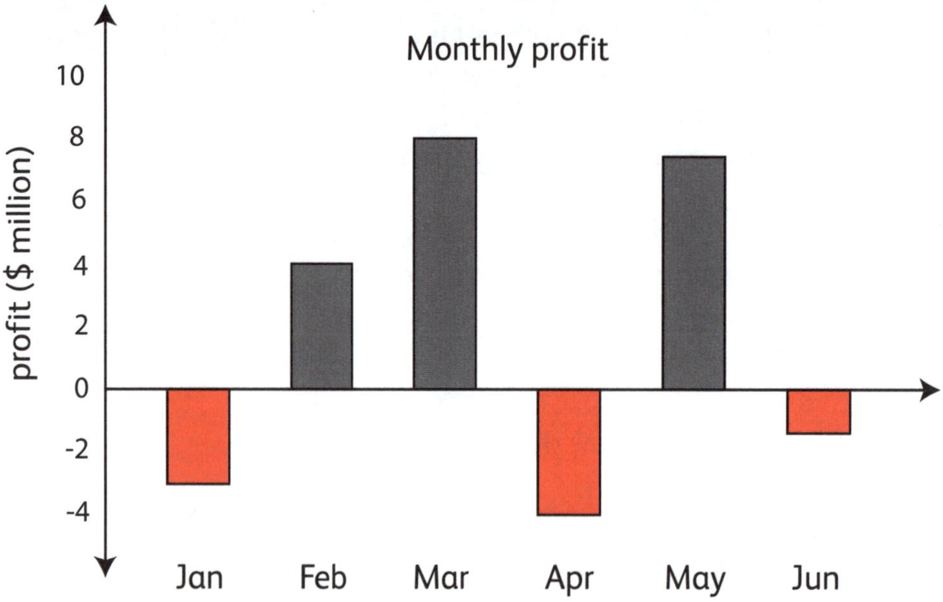

- **a** In which months did the company report a profit of over $7 000 000?
- **b** How does the bar graph show a loss?
- **c** In which month was the greatest loss made?
- **d** Do you think this company is in a good financial situation or should it be worried? Explain your answer.

5 Jackson exchanges Jamaican dollars for 500 USD for his trip to the United States.

- **a** How many Jamaican dollars does Jackson exchange?
- **b** During his trip, Jackson paid for everything by bank card so he came home with the 500 USD. He exchanges the currency at the bureau. The bureau paid 135 JMD for every 1 USD. How many Jamaican dollars does Jackson get back?
- **c** Was this a profit or loss? By how much?

Foreign currency exchange rates	
1 USD	149 JMD
1 CAD	118 JMD
1 GBP	210 JMD
1 EUR	180 JMD

Real-world maths: Profit or loss?

Mrs Smith runs a shop in Falmouth. She bought 16 cartons of pineapple juice at a cost price of \$150 a carton and sold ten cartons in her shop for \$200 each. Since she needed to sell the cartons before their expiry date, she sold the rest of the cartons at a 50% discount.

1 What was the total cost price of the pineapple juice cartons?

2 How many cartons did Mrs Smith discount?

3 What was the total selling price of the discounted cartons of pineapple juice?

4 What was the total selling price of all the cartons of pineapple juice?

5 Decide whether Mrs Smith made a profit or a loss overall, and how much it was as both an amount of money and as a percentage.

Talk about

Do some internet research about import and export between countries. How do countries profit financially from what and how much they export? Discuss why countries that import more than they export could face difficulties.

Maths detective: Run a profitable coffee shop

Use your knowledge of money and finance to solve these cases.

Case 1

You want to run a coffee shop. You are keen to find one that is already set up and the current owner wants to sell on. You find three possible coffee shops and ask the current owner about their running costs. The table shows what you found out:

	Coffee Shop A	Coffee Shop B	Coffee Shop C
Floor area available for tables and chairs	30 m² (6 m × 5 m)	48 m² (8 m × 6 m)	63 m² (9 m × 7 m)
Monthly lease of shop:	\$100 000	\$135 000	\$170 000
Monthly running costs for the building, for example, insurance, water, electricity	\$51 000	\$64 000	\$105 000
Number of staff needed	2	3	4

From the information above, decide which coffee shop you will choose. Explain why.

Case 2

You want to make a minimum 70 % profit on each of the coffees you sell. You have worked out the cost value for each of the following coffee options. What is the minimum you should you charge the customer for each cup? Round your answers up to the nearest 5 dollars.

Coffee type	Americano		Latte		Cappuccino		Flat white	
Size of cup (small/large)	S	L	S	L	S	L	S	L
Cost per cup	\$235	\$294	\$282	\$360	\$264	\$350	\$260	\$350
Selling price								
Profit on each cup in dollars								

Case 3

You want to make the best use of the floor area for chairs and tables to maximise the number of customers.

A table for 4 people is 1 m² and needs 1 m space around it in each direction, for example:

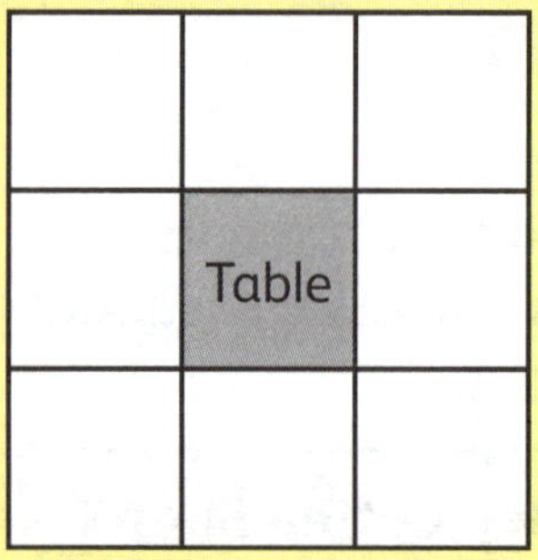

A table for 6 covers is 2 m² and also needs 1 m space around it in each direction, for example:

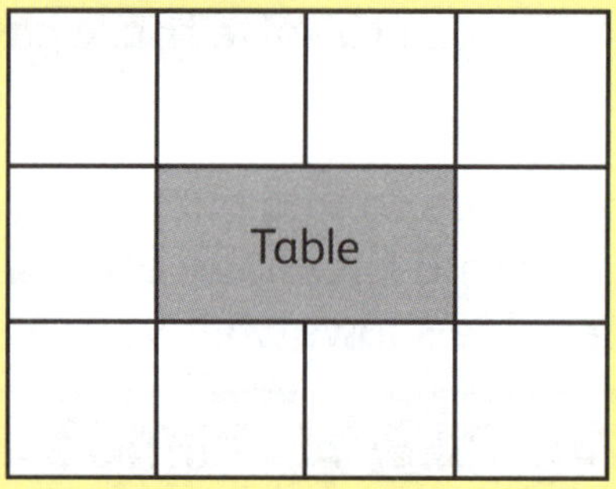

Tables can go on the edge of the floor area but you cannot put a chair on this edge, for example, the following table will only seat 3 people.

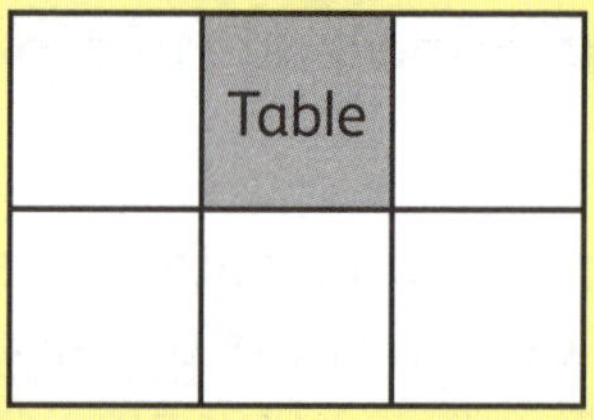

Use the dimensions for the floor area of the coffee shop you have chosen (see table in Case 1) and use cubes or squared paper to represent it. Let each cube or square represent 1 m².

Design the layout of your coffee shop so you can get the greatest number of people sitting at tables.

Case 4

You want to open the coffee shop for an average of 25 days each month. Assume that:

- all tables in the coffee shop will be full 5 times a day
- each person at a table will buy one coffee.

Calculate the mean profit for the coffees you will sell so you know the average profit for one cup of coffee sold.

1 Use a computer spreadsheet to work out your profit from the sale of coffees for a month (use 25 days).

2 Work out if your profits will cover the monthly running costs of your building and staff salaries ($70 000 per month for a full-time member of staff).

3 Will there be enough left over each month to pay the building lease each year?

If your coffee shop is running at a loss, think about these questions.

4 How can you change your table layout to provide extra seating?

5 How many take-away coffees will you need to sell?

6 Adapt your spreadsheet to show any changes you have made.

What I have learned

Make a brochure like this one to give people advice on making better use of their money. Use these headings and add any other sections you need.

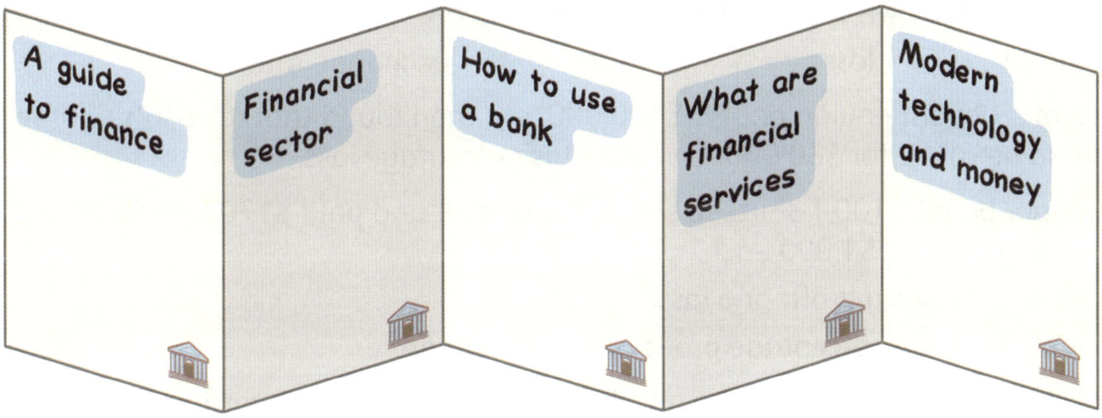

Practice questions

1 Look at this list of companies and organisations in the financial sector.

Insurance company	Bank of Jamaica	Commercial bank

Which one is described each time?

a This is where most people do their banking. This type of organisation accepts deposits, and offers loans and saving accounts to individuals and small businesses.

b They offer services where customers pay to be protected in case something unexpected happens, for example, they become ill, or their car is damaged.

c This is where the country's money reserves and currency are monitored and administered.

2 Mrs Redmond is paying in a deposit into her bank account. Spot two mistakes she has made.

ABC Commercial Bank **DEPOSIT**

DATE 02/30/23
mm / dd / yy

Branch 144 Branch No.
MAIN 788

Name (print)
A.F. REDMOND

Account No.
3 4 1 2 6 8 9

Signature

CASH						
2	× 5 000	1	0	0	0	0
2	× 1 000		2	0	0	0
0	× 500					0
0	× 100					0
4	× 50			2	0	0
TOTAL AMOUNT		1	2	2	0	0

3 Iris bought a table for $100 000 and two chairs for $40 000 each. She later sold them as a set for $200 000. What was her profit?

4 How much money will Anthony save if he buys a $70 000 bicycle and gets a discount of 15 %?

5 Explain in your own words what these terms mean.

a profit b loss c deposit d exchange rate

6 A farmer bought equipment for $100 000, but then found that it wasn't right for him, so he sold it for $80 000. Calculate his percentage loss.

7 A vendor bought 100 T-shirts to sell at the market and paid $120 000 for them. She sold them for $1 000 each.

a Did she make a profit or a loss?

b Calculate the percentage profit or loss she made.

8 Akiel wants to buy a car, but he only has $500 000. This is a quarter of the price of the car. He borrows the rest from the bank for three years, at a rate of 15 %.

a What is the cost of the car?

b How much interest was paid each year?

c Calculate the total amount of interest Akiel will pay over the three years.

d How much would Akiel have saved if he bought the car at a cash price?

9 Shericka wants to borrow $150 000 from the bank. Her bank has different interest rates for different terms:

- If you borrow money over one year, you pay 2 % interest.
- If you borrow over two years, you pay 3 % interest.
- If you borrow over five years, you pay 5 % interest.

a Calculate Shericka's monthly payment, including interest, if she borrows the money over two years and pays it back in equal monthly amounts.

b Calculate Shericka's monthly payment, including interest, if she borrows the money over five years and pays it back in equal monthly amounts.

c Why do you think Shericka might choose to borrow the money over five years?

18 Self check

Check if you can do this:	If you have difficulty, try this:
1 Name the different financial institutions in Jamaica and explain their functions.	Go back to page 283. Look back at the research you did about their roles and the services they provide.
2 Name some of the different financial service providers and explain what they offer.	Think about the service providers you found out about in your survey, and which service providers are used by the people you surveyed.
3 Complete a bank deposit slip.	Go back to pages 286–287. Remind yourself about the information that needs to be included on a bank deposit slip.
4 Complete a bank withdrawal slip.	Go back to the section on pages 286–287. Remind yourself about the information that needs to be included on a bank withdrawal slip.
5 Calculate simple interest on a sum of money.	Remember to think about your work on finding percentages of amounts to help you calculate simple interest.
6 Calculate a profit or loss.	Go back to page 291 and remind yourself about the steps you need to take to help you calculate.

Chapter 19 Problem solving

In this chapter you will explore this focus question:

To answer this question, you will:

- use a variety of strategies to solve problems involving decimals and percentages, and explain your decisions
- write story problems for calculations involving decimals and the four operations
- identify the 'hidden question' in a two-step problem
- write and solve mathematical sentences for a two-step problem
- generate number patterns and identify their rules
- solve problems requiring the use of percentages.

Starting point

A conference venue has square tables that can seat four people.

The tables can be moved around to seat different numbers of people like this:

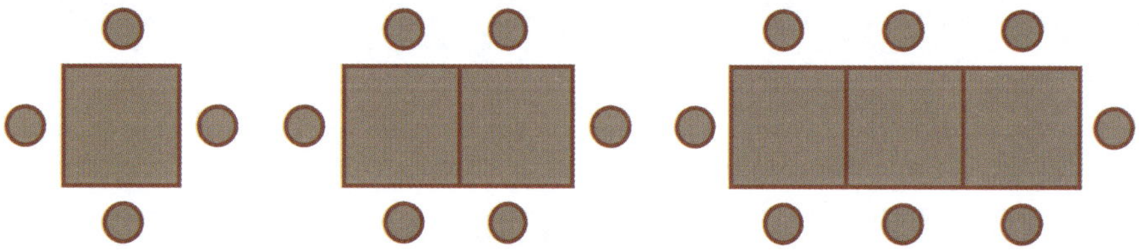

1 Work in a group to find out the following information.

a How many seats are there when five tables are arranged in this way?

b How many people can be seated if ten tables are arranged in this way?

c Write a number pattern to show the number of seats for the first five arrangements.

d What is the rule for finding the next **term** in the sequence?

Key words

term

algebraic expression

operation

problem solving

e What is the rule for working out the number of seats you will need for *n* number of tables? Try to write this as an **algebraic expression**.

f The maximum length of a row in the conference centre is 23 tables. How many seats will be needed for that row?

g Explain why this arrangement makes it impossible to have an odd number of seats.

In a problem,
- the phrase 'more than' means you have to use addition
- the word 'double' or 'doubled' means you have to multiply by 2
- the word 'share' or 'shared' means you have to divide equally.

2 Make up some word problems that use 'more than', 'double' or 'share' and can be solved in the way the children describe.

3 Now think about some word problems that use 'more than', 'double' or 'share' but are not solved in the way described – in fact, you need to do the opposite and use the inverse **operation**.

4 Make a list of other words and phrases used in **problem solving** for the four operations. Make up some different word problems and include examples where you need to do the opposite to solve them.

More about two-step problems

Key maths idea

Sometimes when we solve problems we need to work out extra information before we can find the final answer. So we need to identify a 'hidden' question to help find a solution.

We call these problems two-step problems. To solve a two-step problem, we need to use two of the four arithmetic operations.

Look at this problem. We can use the three-step problem-solving approach (understand, plan, solve) to help us.

Example
There are 30 students in a class.
The teacher has 7 litres of mango juice.
She uses 215 millilitres of mango juice for every student.
How many millilitres of mango juice is left over?

<table>
<tr><th>UNDERSTAND:
What do you need to find? What do you know?</th><th>PLAN:
How can you solve the problem?</th></tr>
<tr><td>We know that the teacher has 7 litres of mango juice and she uses 215 ml of juice for each of the 30 students. We need to find out how much juice is left over.</td><td>Identify the hidden question as ‘How much juice was used altogether for 30 glasses of 215 ml?’</td></tr>
<tr><td><table><tr><th colspan="2">7 litres</th></tr><tr><td>juice used</td><td>left over</td></tr></table></td><td>We can represent the problem with the mathematical sentence 7 000 − 215 × 30, as multiplication has priority over subtraction.

We will complete the problem in two steps: multiplication to find out how much juice the teacher uses altogether for the students, and then a subtraction to find the difference between the amount of juice used and the original 7 litres.</td></tr>
</table>

SOLVE:
Find the solution or solutions

Step 1

Find out the amount of mango juice used for the 30 glasses of 215 ml:

215 ml × 30 = 200 × 30 + 10 × 30 + 5 × 30

= 6 000 + 300 + 150 Remember the order of operations

= 6 450 ml

Why could you have also calculated this as 215 × 3 × 10?

Step 2

Find the difference between 7 litres and 6 450 millilitres to find out how much juice is left. Remember to convert 7 litres to 7 000 millilitres so we are working with the same units of measurement.

<table><tr><th colspan="2">7 000 ml</th></tr><tr><td>6 450 ml</td><td>left over</td></tr></table>

7 000 − 6 450 = 550

So, there is 550 ml of mango juice left over.

Finally, make sure to look at the original question again. Did you answer it? Check the answer on a calculator.

1 Lisa makes bead jewellery. Bracelets have 37 beads. Necklaces have 74 beads. Lisa makes 28 bracelets and 81 necklaces. How many beads does she use altogether?

2 A field measures 15 m by 20 m. The next field measures 18 m by 23 m. What is the difference in area between the two fields?

3 A bakery has 4 apple pies left for sale. A customer buys 3 slices of apple pie. Each slice is $\frac{1}{5}$ of a pie. How much pie does the bakery have left to sell? Write your answer as a mixed number.

4 Grandma equally shares $6 000 among her 5 grandchildren. One of the grandchildren spends $475 of their share on a notebook. How much does this grandchild have left?

5 Look at the diagram shown here.

a What is the difference between the perimeters of the shapes?

b What is the area of the blue portion?

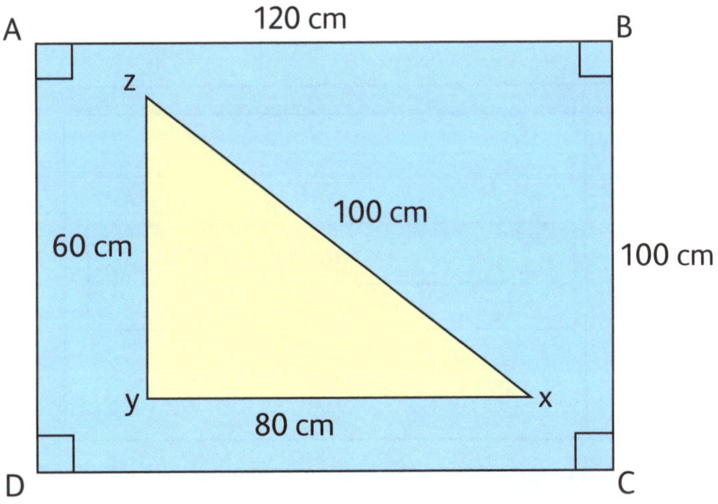

6 Kaci solves a two-step problem. The answer to the 'hidden question' is $3\frac{1}{2}$ hours. She writes the answer to the problem as 'The train journey took 45 minutes longer.' Write a word problem that Kaci could have been solving.

Problems involving decimals

Key maths idea

When we solve real-life problems involving **decimals**, they are usually problems about money or measures.

Example
Patrick has a 10-metre roll of electrical tape. He cuts off five pieces that are each 1.25 metres in length. What length of tape is left on the roll?

A **bar model** diagram can help us understand the problem:

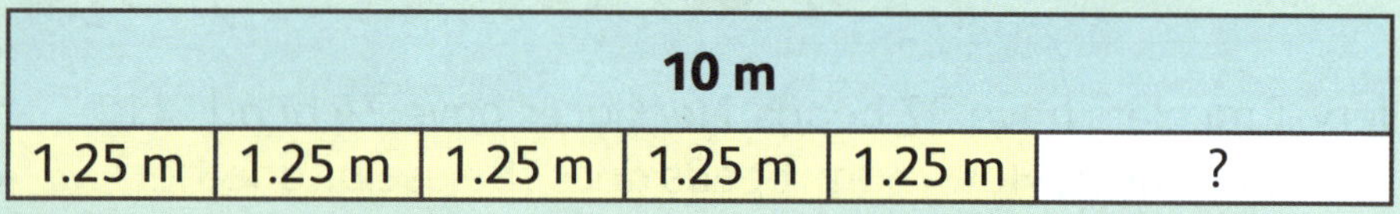

Explain to your partner how the diagram helps you decide which mathematical sentence you will write to solve the problem.

This is a two-step problem, so a 'hidden question' must be answered first.
In this example we need to first answer: How much electrical tape does Patrick cut off the roll altogether?

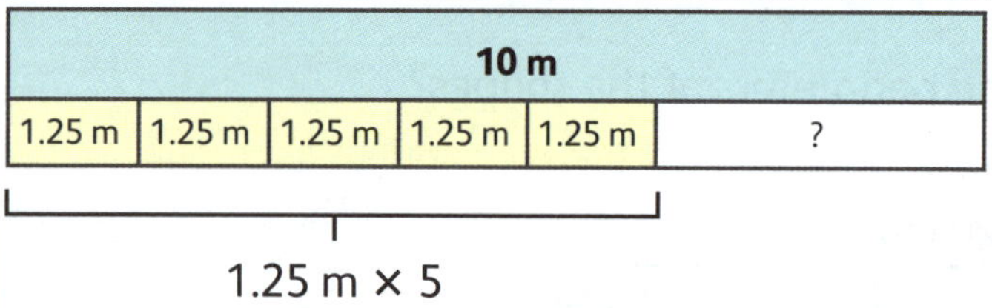

1.25 m × 5
= 6.25 m

Then we subtract 6.25 m from 10 metres to find the difference.

10 − 6.25 = 3.75

There is 3.75 m of tape left on the roll.

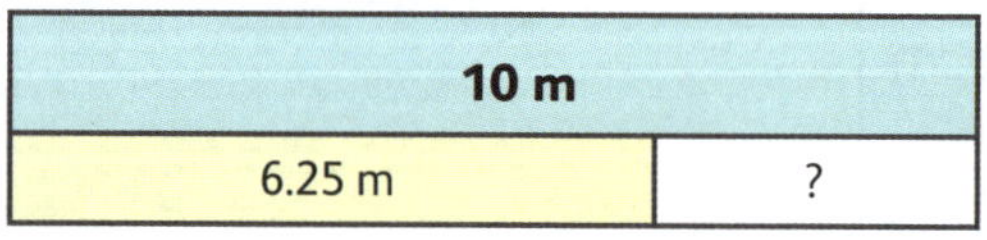

Imagine that the answer to another problem is also 3.75 metres, but this time it is in the context of height. What could the problem be? Share your ideas with the class.

Key words
decimal
bar model
measure
convert

Be careful! Sometimes the units of **measure** in problems differ and you might need to **convert** them!

1 Solve these one-step problems. Write the mathematical sentence you will use to solve each one.

- **a** A parcel has a mass of 950 g. A box has a mass of 2.65 kg. What is the total mass of the parcel and box?
- **b** Leon is 1.65 m tall. Malcolm is 0.13 m taller. How tall is Malcolm?
- **c** Daniella is 1.59 m tall. She is 0.45 m taller than her brother. How tall is Daniella's brother?
- **d** A drink costs $295.50. What is the cost of three of these drinks?

2 Malik solves a problem by carrying out the calculation 15.75×10. What could Malik's problem be? Make up an example in the context of:

a mass **b** capacity **c** distance

3 Anthony studied for a total of 12.4 hours over a period of four days. What is the mean of the hours Anthony studied each day?

4 Solve these two-step problems. Think about the 'hidden question' that you need to answer first. Write the mathematical sentence you will use to solve each problem.

- **a** Hattie is carrying 13 jars of sugar on a tray. The total mass she is carrying is 11.2 kg. The mass of each jar of sugar is 0.8 kg. What is the mass of the tray?
- **b** Jayden is making fruit punch for his party. The bowl for the fruit punch will hold 4.75 litres. The local shop only sells fruit juice in 150 ml cartons. How many cartons does Jayden need to fill the bowl?
- **c** A metal tag is 2.4 cm long. Each tag is sold for $99.99. The tags are cut from a strip of metal that is 36 cm long. How much money in total is taken when all of these tags are sold?

> Remember that bar model diagrams are useful – you may find it a good idea to draw one for each step.

5 Here is a plan of a garden. Calculate the perimeter.

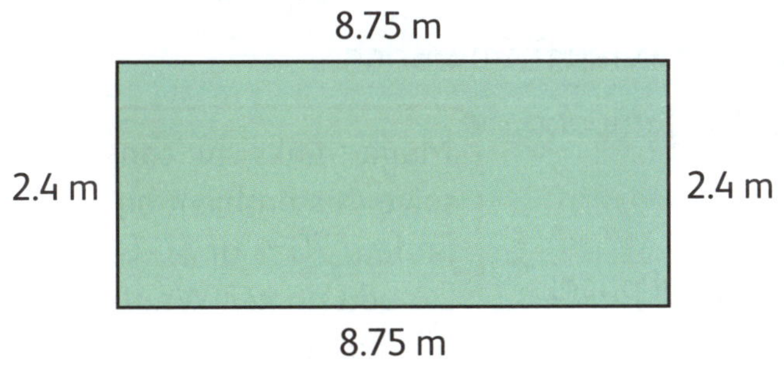

6 A group of friends earn $9 000 by mowing lawns. They share the money equally. They get $562.50 each. How many friends are there in the group?

7 A swimming pool is 10.8 metres wide and has 5 equal lanes marked out for swimming. Four lane dividers are used, each with a width of 10 cm. Approximately how wide is each of the lanes for swimming?

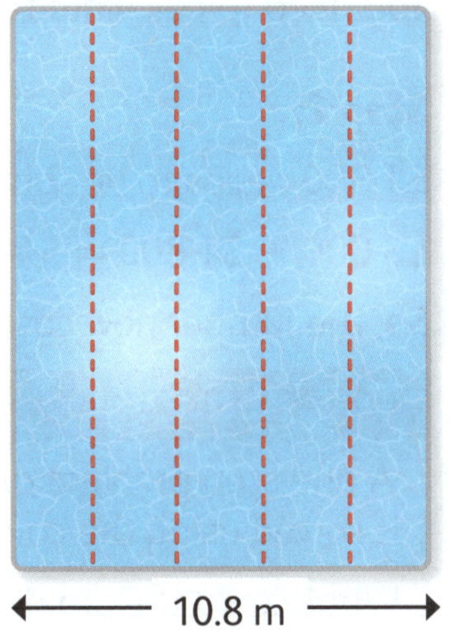

Problems involving percentages

Key maths idea

You know how to find percentages of quantities, but when they are used in word problems you will need to read carefully to check what the problem requires you to do with them.

Example

A jumper in a shop window is priced at $6 000.

The next day the jumper is reduced in price by 30 %.

What is the reduced cost of the jumper?

Think about the three-step problem-solving approach: understand, plan, solve.

It is useful to represent the problem in a diagram that shows what you know and what you need to find out. A diagram also helps you plan how to solve the problem.

Full cost: $6 000	
Reduced cost	30 % reduction

Explain to a partner how the bar model diagram represents the problem.

What is the 'hidden question' that you need to answer first?

Now let us solve it.

Step 1: Find 30 % of $6 000.

You can do this by finding 10 % of $6 000. $600

Then multiply this by 3 to find 30 %. $1 800

You have found 30 %, but you have not yet answered the question.

You need to calculate the reduced price of the jumper.

Step 2:

$1 800 needs to be subtracted from the original price:

$6 000 − $1 800 = $4 200

So, the jumper now costs $4 200.

Maya thinks she can solve this problem by finding 70% of $6 000. Do you agree? What is the 'hidden question' when you use this strategy?

Talk about

A one-step problem about percentages is solved using the calculation 56 ÷ 4.

What could the problem be? Try to think of more than one possibility.

Think about how you will solve these problems. Try to represent them first.

1 A supermarket has 75 employees. 20 % of the employees work part-time. How many part-time employees does the supermarket have?

2 At the sewing store, Mrs Butler bought a bag of mixed buttons. The bag included 65 buttons, of which 60 % were wooden. How many buttons are not wooden?

3 A company meeting is attended by 56 people. 75 % of them are managers. How many people in the meeting are not managers?

4 Joshua takes chocolate and vanilla cupcakes to school for his birthday. 20 students take a cupcake. 40 % of them picked vanilla. How many students picked a chocolate cupcake?

5 Last Wednesday, students could choose egg or cheese sandwiches for lunch. The pie chart shows the proportions of the different sandwiches the canteen staff made. How many cheese sandwiches did they make?

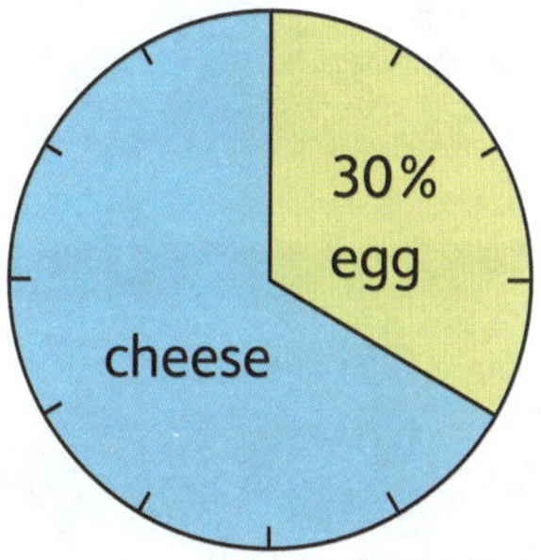

240 sandwiches made

6 A golf club has 500 members. 58 % of the members are over 60 years old. 20 % of the younger members are children.

a How many 'over 60' members are in the golf club?

b How many children are members of the golf club?

7 1 million people visited the Dunn's River Falls last year. 15 % of the people visited on a Monday. 40 % of the people visited on a Saturday. How many people visited during the rest of the week?

8 The answer to a percentage problem is $450.
What could the problem be if it is:

a a one-step problem

b a two-step problem

c a problem with more than two steps?

Talk about

Exercise 7 can be solved in a different number of steps. Talk about the different ways you could solve this problem and how many steps you need for each one.

Real-world maths: Timetables

This timetable shows flight arrival and departure times at Norman Manley International Airport in Kingston.

ARRIVALS

From	Arrival time	Status
Montego Bay	13:10	delayed
Ochos Rios	14:15	landed
Miami	16:15	expected on time

DEPARTURES

To	Departs at	Arrival time at destination
Montego Bay	14:10	14:55
Miami	14:45	17:40
Port of Spain	15:15	19:40 (via Miami)
London	18:00	08:55
Ochos Rios	18:30	19:05

1 a Lloyd arrived at the airport 1 hour 45 minutes before his departure to Miami. The plane departed 55 minutes late. How long was he waiting at the airport before departure?

b Brianna took the 13:10 flight from Montego Bay to Kingston. The flight continues on to London. If it landed 25 minutes late in Kingston, how long will the passengers have to wait before they take off for London?

c The gates close 25 minutes before the departure time and if passengers are not in time, they may not board. What is the latest time that Andre could get to the gate for his flight to Miami?

d The flight distance from Kingston to Montego Bay is 135 km. Use the formula speed = distance ÷ time to work out the average speed of the plane for this flight.

2 Write two word problems of your own using the flight timetables. Each problem should involve at least two steps. Exchange problems with a partner and solve each other's problems. Check the answers together.

Using patterns and rules to solve problems

Key maths idea

Junelle creates a **pattern** with counters.

Term 1 Term 2 Term 3

So far, there are three **terms** in the pattern.
Talk to a partner about what changes and what stays the same as you move from term to term.

We can create a table to help look for relationships within the pattern and think about how it will continue, for example, What will term 4 look like? How many counters will it have?

Term	1	2	3	4	5	6
Number of counters	1	5	9			

Use counters or sketch circles to show term 4, term 5 and then term 6 in the pattern.

We need to find the rule for working out other terms in the sequence. We could use a term-to-term rule to describe this pattern as next term = previous term + 4

but often we are asked to find the value of a term much later in the pattern, for example, the 50th term, and are unlikely to know the value of the previous term.

Note that in the pattern, each term has one central counter and layers of 4 counters are then added – in fact, 1 fewer layer than its position in the pattern. So, term 2 is in the second position in the pattern and is made up of 1 central counter and 1 layer of 4 counters (term number subtract 1, so $2 - 1 = 1$ layer).

The letter n is used to represent any term in a pattern. We can then write an **algebraic expression** to help us find the value of any term. We call this finding a rule for the n-th term.

The rule is: number of counters $= 1 + 4(n - 1)$
Use the rule to now find the value of the 50th term.
What about the 65th or the 100th term?

Key words

pattern
term
algebraic expression

1 Look at this pattern of match sticks.

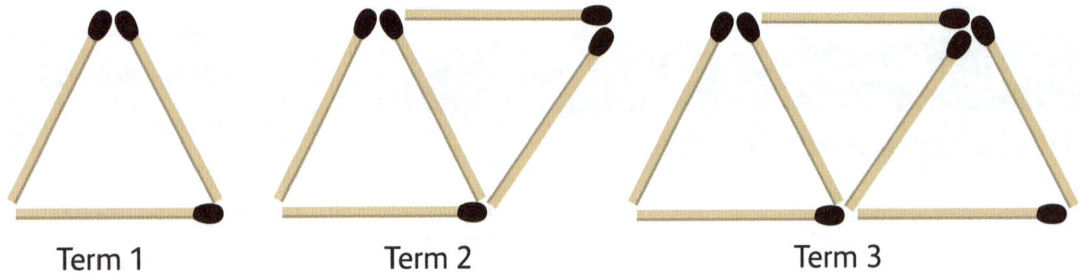

a Recreate the pattern and continue it up to the 6th term.

b Record your findings in a table.

c Write the rule for finding the number of match sticks for the next term in the pattern.

d Write the rule for finding the number of match sticks for the *n*-th term.

e Use your rule to find the number of match sticks for the 15th term and the 100th term.

2 Look at this pattern of numbers:

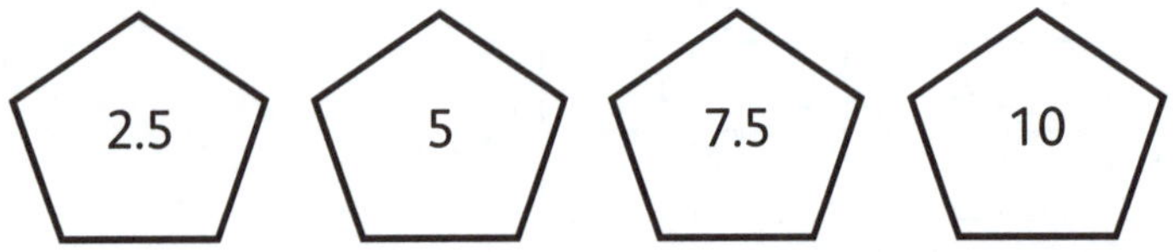

a What number will come next in the pattern?

b Write the rule for finding the value of the next term in the number pattern.

c Write the rule for finding the *n*-th term.

d Use your rule to find the value of the 12th term and the 31st term.

3 Malcolm and Kaci are making a pattern with fudge sticks. They use a table to record the number of fudge sticks they use for each term in the pattern.

Term	1	2	3	4	5	6
Number of fudge sticks	4	7	10	13		

a Create a possible pattern of fudge sticks to match the table. Think about what stays the same and what changes each time.

b Write the term-to-term rule for finding the next term in the pattern.

c Record the number of fudge sticks for the 5th and 6th term.

d Write the rule for finding the number of fudge sticks for *n*-th term.

e Use your rule to find the number of fudge sticks for the 25th term and the 60th term.

Real-world maths: A sunflower garden

Amira's class planted sunflowers in the school garden as part of a science project. Amira measured her plant at the end of each week. She got these measurements:

Week 1	Week 2	Week 3
4.5 cm	6.7 cm	8.9 cm

1 If the plant continues to grow at this rate, how tall will it be at the end of week 4 and week 5?

2 Write a rule for finding the height of the plant at the end of n weeks.

3 Use the rule to work out how tall the plant will be after 15 weeks.

4 Amira used the rule to work out that the sunflower will be 222.3 cm high after 100 weeks. Explain why this might not happen in real life.

Maths detective: Investigate pressure and weight

Use what you know about formulae and what you have learned in science about pressure and weight to solve these problems.

Case 1

Elaine and Tisha weigh the same (55 kilograms).

1 Elaine wears trainers and Tisha wears high heels. They both walk across a grass field, but only Tisha sinks into the grass. Explain why.

When a force pushes down on a **surface**, it exerts pressure. The amount of pressure depends on the size of the force (weight) and the area it acts on (area).

You can work out the pressure exerted on a surface using the formula

Pressure (bars) = Weight (kg) ÷ Area (cm^2) or

$$P = \frac{W}{A}$$

2 How would you rearrange this formula to calculate weight?

3 If you know the pressure and the weight, how would you work out the area?

4 Elaine's shoe covers a **surface area** of 28 x 11 cm and Tisha's shoe covers an **area** of (1 × 1 cm) + (8 × 8 cm). Work out the area on which each person will exert pressure. Remember they are wearing a pair of shoes!

5 Use the formula to work out how much pressure each person exerts on the ground.

Key words
- pressure
- weight
- surface
- surface area
- area
- tyre footprint

Case 2

Use your own footprints to work out how much pressure you exert when you are standing on the ground.

Case 3

How can you use tyre pressure and the area of the tyre in contact with the ground to work out the weight of a car?

a The amount of tyre that touches the ground is called the contact patch or the **tyre footprint**. The contact patch of each tyre on this car is 114 cm^2. What is the total contact area of the four tyres?

b If the tyres are inflated to a pressure of 2.2 bars, use the formula Weight = pressure × force to work out the approximate weight of the car in kilograms.

c A car has a weight of 1 250 kilograms and each tyre has a contact patch of 120 cm^2. What pressure does each tyre exert on the ground?

Case 4

A car has a weight of 1 250 kilograms. Each tyre has a contact patch of 118 cm^2. If there are four passengers with a mean weight of 72 kg in the car, how much pressure is exerted on each wheel when the car is parked?

What I have learned

Draw a flow diagram like this one to summarise the steps you follow to understand and solve problems. Add more boxes if you need them.

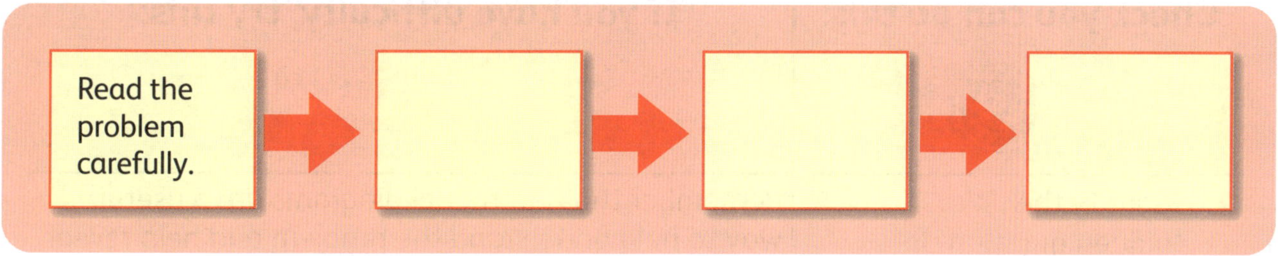

Practice questions

1 Mr Lewis runs 7.7 km on Monday, 6 500 m on Tuesday and 9.35 km on Wednesday. Find the total of his runs. Give your answers in kilometres.

2 A teaspoon contains 5.26 ml of cough medicine. Mikayla takes 5 teaspoons full every day. How many millilitres of cough medicine does Mikayla take each day?

3 The answer to a story problem about temperature is –2°C.

   a What could the problem be?

   b Write two different examples. Try to make one problem involve both an addition and a subtraction.

4 A racing car can travel at 255 kilometres per hour. If it continues at this same speed, how far does it travel in 90 minutes?

5 Jasper's Diner offers a choice of regular and diet soda. Last night, the diner served 50 sodas in all, 80% of which were regular. How many regular sodas did the Jasper's Diner serve?

6 Look at this number pattern:

   7, 14, 21, 28, 35, …

   What is the algebraic expression for the *n*-th term for this pattern?

7 Mr Barker is planting flowers. He plants 3 flowers in the first row, 5 flowers in the second and 7 flowers in the third row. He notices a pattern and continues it. Mr Barker plants his favourite flower in the 25th row. Each flower cost $500. How much does Mr Barker spend on his favourite flowers for this row?

> Think about how you can use a rule for the pattern to help you.

19 Self check

Check you can do this:	If you have difficulty, try this:
1 Identify the 'hidden question' in a two-step problem.	Remember that bar model diagrams are a useful way to help understand the problem and help reveal the steps you need to take, including the 'hidden question' you need to answer first.
2 Write mathematical sentences for a two-step problem.	Go back to pages 299–300 and recap the mathematical sentences you used to solve the problems. Revisit the order of operations to help make decisions.
3 Solve two-step problems.	Go back to pages 299–300 and remind yourself how to use the three-step problem-solving approach to help you understand, plan and then solve the problem.
4 Use different strategies to solve problems with decimals.	Remember that you can use similar calculation strategies to solve problems with whole numbers or with decimal numbers. Use place value knowledge and known facts to help you.
5 Use different strategies to solve problems with percentages.	Remember that a percentage is a proportion of a whole, so it is important to think carefully about the information you know and what you need to find out. Bar models are a useful way to think about parts and the whole.
6 Extend and generate number patterns.	Go back to page 307. It is important to look at how the value of the terms in the pattern changes. You can use a rule to continue the pattern.
7 Write an algebraic expression for a number pattern.	Go back to page 307. Check that you know how to find the rule to work out the next term, and the rule to work out the *n*-th term for any value in the pattern.
8 Write my own story problems to calculate decimals and percentages.	You can sketch bar models to represent calculations and use these to help create the steps in your story problems.
9 Use the four operations in my story problems.	Recap the different problems you worked on in this chapter. First practise changing only the numbers in a problem, then think about only changing the context. Once you are confident with this, challenge yourself to think of different contexts and different numbers.

Unit 2: Measurement

Chapter 20 Circles

In this chapter you will explore this focus question:

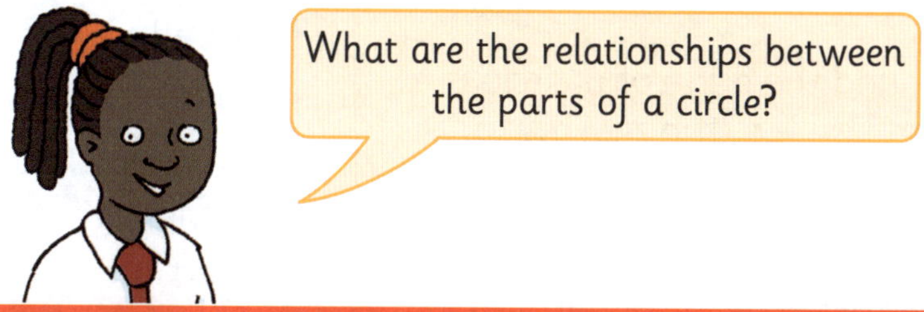

To answer this question, you will:

- use a compass (pair of compasses) to draw circles to create designs
- name parts of a circle: radius, diameter, centre, circumference, sector
- investigate numerical relationships between the circumference of a circle and its radius and diameter
- solve problems involving the estimation and calculation of the circumference of a circle.

Starting point

This design is called the flower of life. It is made by drawing overlapping circles.

Talk about how the design works.

Use a circle stencil or a pair of compasses to test your ideas and see if you can copy the design.

Look at the triangles that have been formed. Use a **protractor** to help you name them correctly.

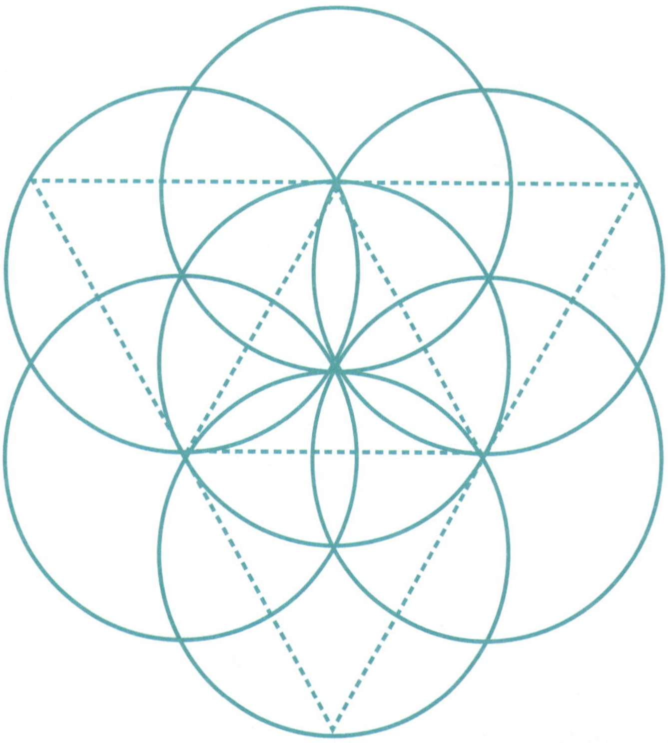

Key word
protractor

Parts of a circle

Key maths idea

A **circle** is a closed plane shape, with all points on its edge the same distance from its **centre**.

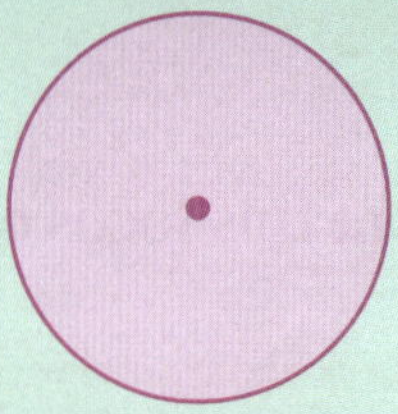

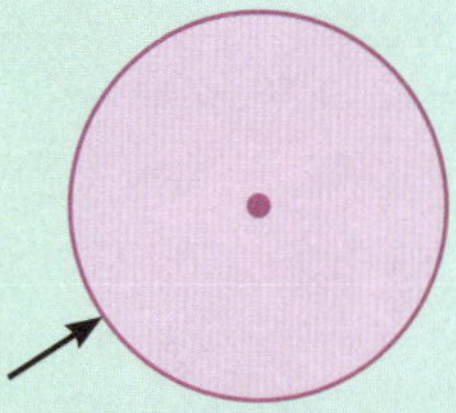

The distance around the edge of a circle is called the **circumference** (the **perimeter** of the circle).

The **diameter** of a circle cuts the circle exactly in half and goes through the centre.

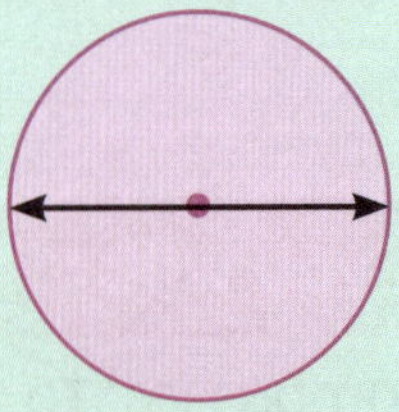

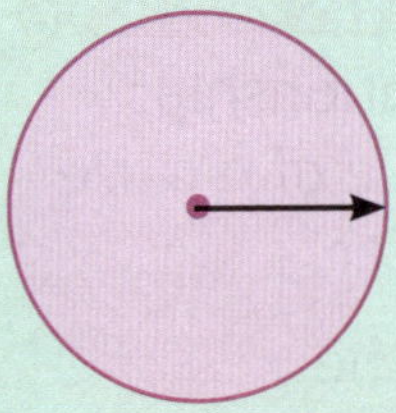

The **radius** of a circle joins a point on the circumference to the centre of the circle.

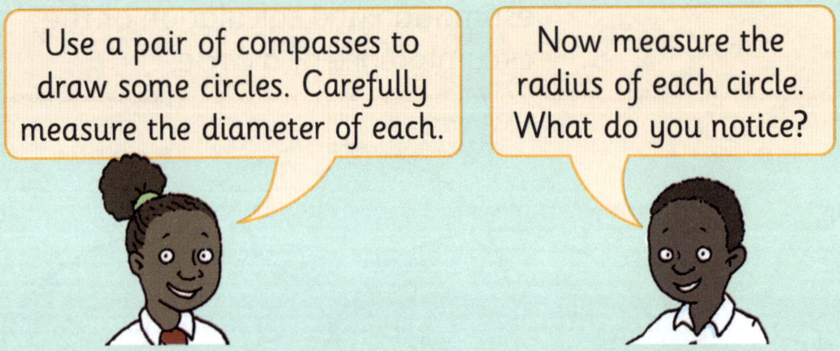

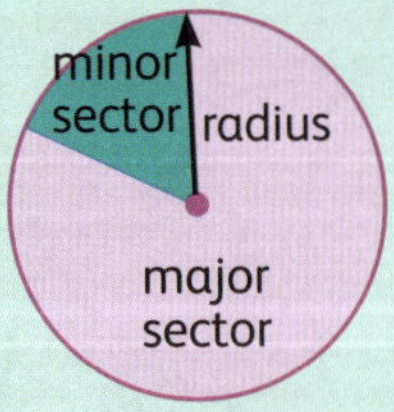

The **sector** of a circle is the area made with 2 radii (radius) and a section of the circumference.

It looks like a slice of cake or pizza.

We call the smaller part a minor sector.

We call the larger part the major sector.

A **segment** of a circle is formed when a straight line (called a **chord**) is drawn to divide a circle into two parts. You can draw a chord by joining two points on the circumference of a circle. We call the smaller part the minor segment. We call the larger part the major segment. A section of the circumference of a circle is called an **arc**.

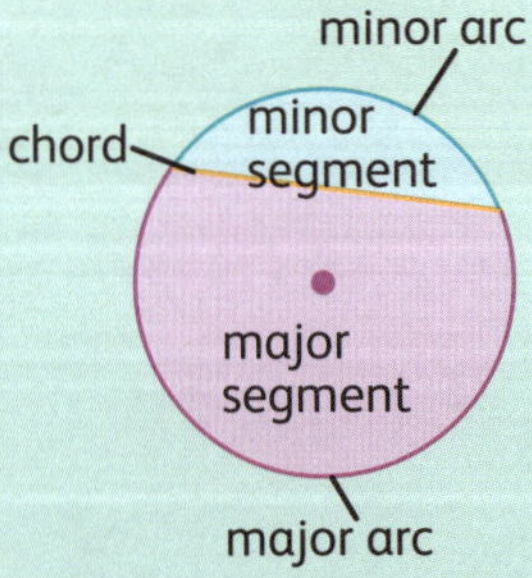

Key words

circle
centre
circumference
perimeter
diameter
radius
sector
segment
chord
arc

1 Which is the odd one out? Explain why each time.

a

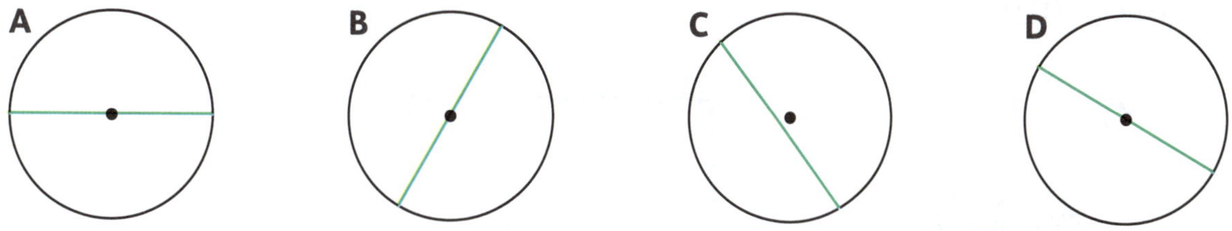

b

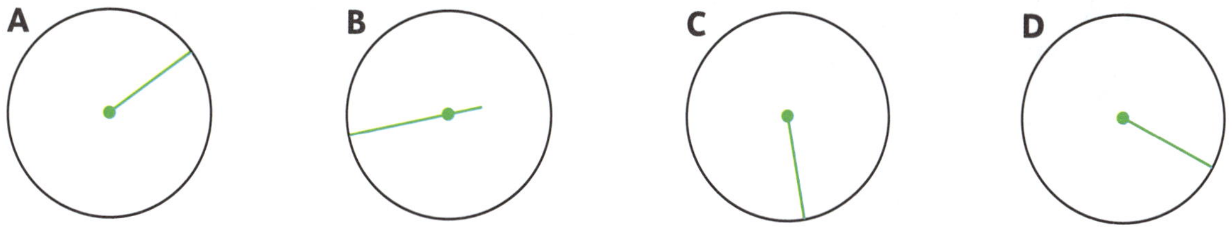

c

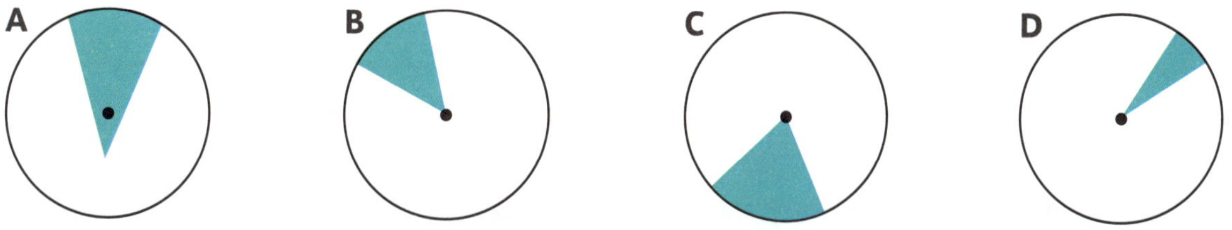

2 True or false? Correct any that are false.

a The radius of a circle is half the length of the diameter.

b The perimeter of a circle is usually known as the chord.

c When two points on the circumference are joined, we call this line a chord.

d Segments are formed when a chord is drawn to divide the circle into two parts.

e The blue part of the circle on the right is called the major segment.

3 Look at this circle.

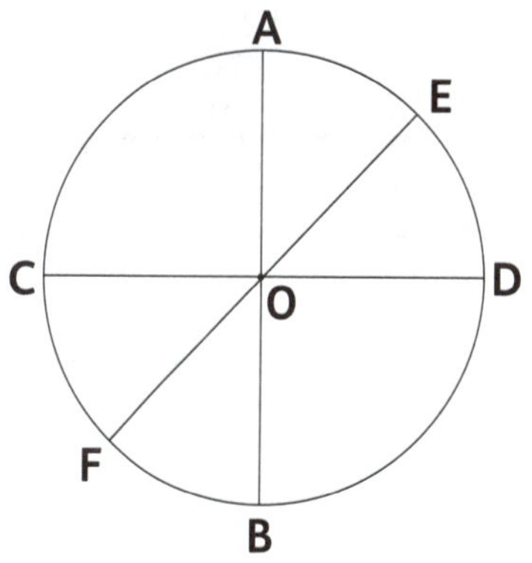

a Where is the centre of this circle?

b Line segment AB shows a diameter of the circle. What other line segments are a diameter?

c Line segment FO shows a radius of the circle What other line segments show a radius?

d Choose three letters that make a sector of the circle.

e If you drew a line from point C to point E, what would it be called?

4 Draw a circle and use different colours to show:

- the centre
- the radius
- the diameter
- the circumference
- a sector.

Label the parts of the circle clearly.

5 Work in groups to find examples of circles we use in real life. Here are some to get you started.

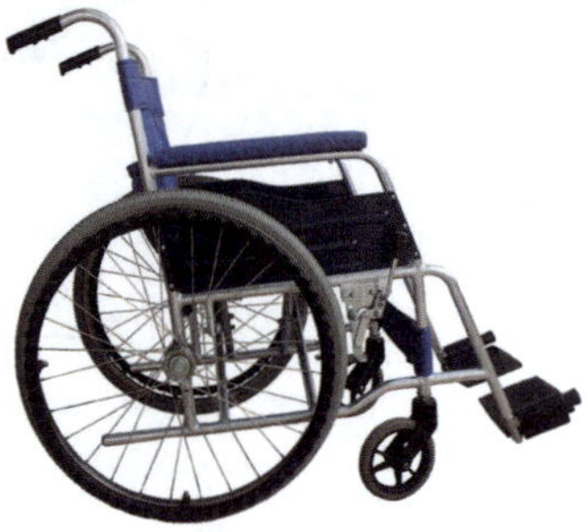

a Use a camera to take pictures of the examples you find.

b Prepare a short slide show to show the class where you found circles used in real life.

Drawing circles

Key maths idea

Sometimes we need to draw circles with a given radius or diameter.

Example

Use a pair of compasses to draw a circle of radius 4 cm.

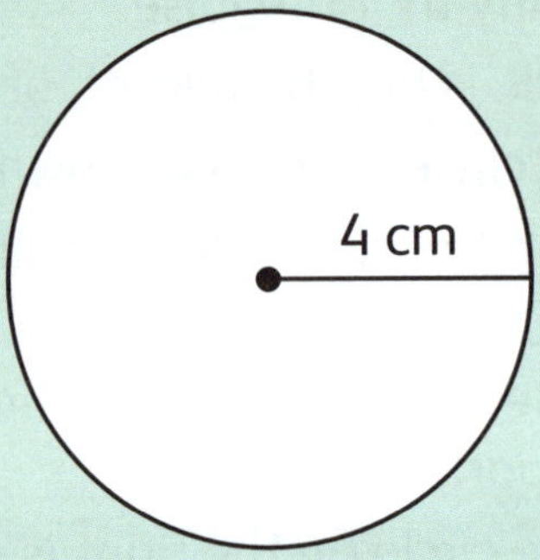

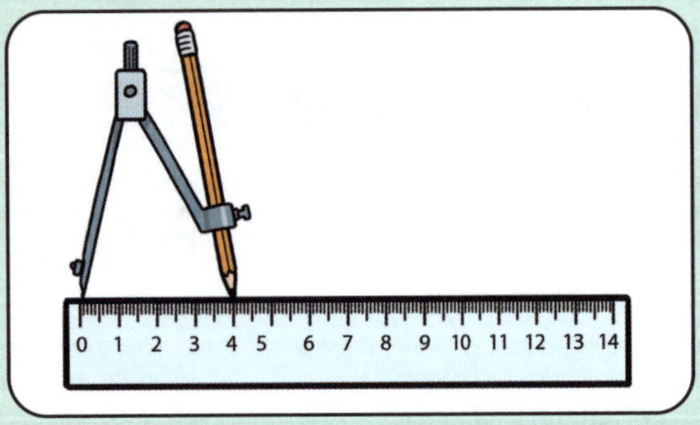

Step 1: Use a ruler to set the distance from the point of the pair of compasses to the pencil's lead at 4 cm.

Step 2: Place the point of the **pair of compasses** at the centre of the circle.

Step 3: Draw the circle by turning the pair of compasses through 360°.

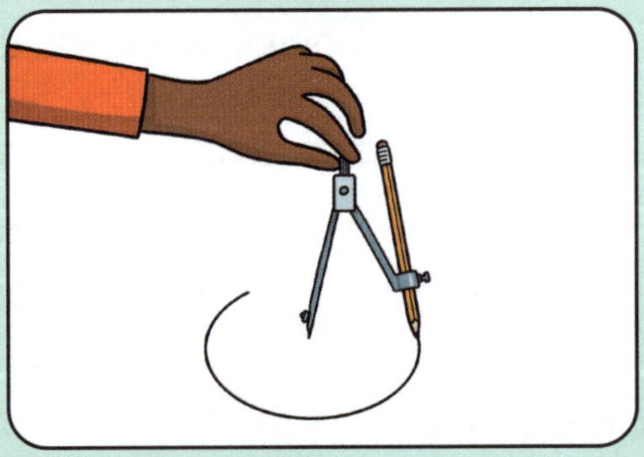

Key word

pair of compasses

1 Use a pair of compasses to draw a circle of radius 5 cm.

2 Use a pair of compasses to draw a circle of diameter 12 cm.

3 **a** Use a pair of compasses to draw a circle of radius 4.5 cm.

b Draw the diameter of the circle. Then use a ruler to measure the length of the diameter.

c Write an equation to represent the relation between the radius, *r*, and the diameter, *d*.

4 Start by drawing a circle with a radius of 6 cm. Use the same centre and draw another circle with a radius that is a little shorter, for example, $\frac{1}{2}$ cm or 1 cm shorter. Continue to reduce the length of the radius each time so you have at least 6 circles in your pattern. Find out what we call this arrangement of circles.

Real-world maths: Circles in design

Many artists and designers use circles in their work. You can find circle designs on fabrics, floor tiles, wallpaper and business logos.

1 Use circles, lines and colour to design a round or rectangular table mat.

2 Jewellery designers often use geometric circle designs in their work. Look at these ideas for wire pendants.

Choose one design and try to reproduce it on paper.

3 Samira Mian is an artist who makes beautiful geometric designs and patterns. Do some internet research to find out more about her work. She also produced a series of YouTube tutorials with instructions for drawing geometric designs. Try them out!

Exploring circumference and diameter

Key maths idea

These students are exploring the length of the diameter and the circumference of some different circles.

Work with a partner. Carefully draw some different sized circles and measure the diameter and circumference of each as accurately as possible. What do you notice?

You can also use a flexible ruler instead of string.

The circumference of a circle is a little over 3 times as long as the diameter. In fact, a circle with a diameter of 1 unit has a circumference of 3.14159265... units.

We call this number **pi** (pronounced 'pie'). We use the symbol π to represent it.

In geometry π is the ratio of a circle's circumference to its diameter.

$$\frac{\text{circumference}}{\text{diameter}} = \pi$$

The value of pi is $\frac{22}{7}$ or approximately 3.14.

If you know the diameter (or radius) of a circle, then you can calculate the circumference using the formula:

$\pi \times \text{diameter} = \text{circumference}$

$2 \times \pi \times \text{radius} = \text{circumference}$

So, for the children's circle with a diameter of 10 cm we can find the circumference as:

$\pi \times 10 = 3.14 \times 10 = 31.4$ cm (We take π as 3.14)

Or, if they had measured the radius as 5 cm:

$2 \times \pi \times 5 = 2 \times 3.14 \times 5 = 31.4$ cm

Key word

pi (π)

Use a calculator to check the circumference of each of your circles. How close were you with your measuring? Use 3.14 for π.

1 Calculate the diameter of each of these circles.

a radius = 10 cm
b radius = 20 mm
c radius = 3 m
d radius = 4.5 cm

2 Calculate the radius of each of these circles.

a diameter = 12 cm
b diameter = 10 mm
c diameter = 20 m
d diameter = 15 cm

You will need a calculator for Exercises 3 to 5. Use $\pi = 3.14$.

3 Calculate the circumference of each of these circles.

a diameter = 8 cm
b diameter = 12 cm
c radius = 3 cm
d radius = 4.5 cm

4 A boy rides his bicycle around a circular flower bed with a radius of 13 metres. How far does the boy ride if he goes around the circle:

a once
b four times
c ten times?

5 The diameter of a circular clock face is 56 cm. What is the circumference?

Maths detective: Dendrochronology

You are going to use your measuring skills, knowledge of circles, formulae and graphs to solve these problems.

Case 1

Dendrochronology is the science of working out how old a tree is by counting its rings.

1. Each dark ring on this tree stump represents a year of growth. Start at the centre and count the rings till you reach the bark. How old was this tree when it was chopped down?
2. When the dark rings are further apart, it means the tree grew quickly that year. When did this tree grow the fastest?
3. What does it suggest when the dark rings are very close together?
4. Find out what cross-dating is and how scientists use it to work out the age of wooden objects. Share what you find out with your group.

Case 2

You can only count the rings on a tree if it has been cut down, but you can use maths to work out the approximate age of a tree without damaging it.

1. Use a tape measure or a piece of string and a ruler to measure the circumference of three trees in the school garden. One should be a young tree, one a medium tree and the other the oldest tree you can find. Take the measurement about one metre above ground level.
2. Use the formula diameter $= \frac{\text{circumference}}{\pi}$ to find the diameter of the trees. Round your answer to the nearest tenth (one decimal place). Record your data in a table like this one:

	Circumference (cm)	Diameter (cm)
Tree 1		
Tree 2		
Tree 3		

3. Is there a relationship between the diameter and the age of a tree? Explain your answer.

Measurement

4 To work out the age of a tree you can use the formula:

Age = diameter × growth factor

This table gives the growth factor for some different species of tree. The higher the growth factor, the faster the tree grows.

Tree species	Growth factor
Rubber tree	3.0
Spiked pepper	3.2
Scotch pine	3.5
Sweetgum	4.0
Black walnut	5.0
Jamaican cherry	5.1
Ironwood	7.0
Horse chestnut	8.0

This formula uses inches and not centimetres. If you measure in cm you need to convert to inches before you work out the age of the tree. You can divide the centimetre length by 2.5 to get an approximate length in inches.

Sharmayne measured the circumference of some trees in a botanical garden and recorded this data:

Tree	Scotch pine	Jamaican cherry	Horse chestnut	Rubber tree	Sweetgum
Circumference in inches	63	36	51	114	120

Calculate the diameter of each tree (in inches) and use the growth factor from the table to calculate its age.

Case 3

Junior found a Colorado blue spruce tree in a national park while he was on holiday in the United States. He worked out its age to the nearest year like this:

diameter × growth factor
= 18.46 in × 4.5 = 83 years.

Junior saw this poster at the Forestry Office in the national park:

a What do you think this means? Share your ideas with the class.

b Use Junior's calculations to approximate the diameter of the spruce tree 20 years ago.

c If you planted a black walnut tree today, what would its diameter be in 20 years' time?

Case 4

1 This graph shows how the diameter of a tree increased over a 45-year period.

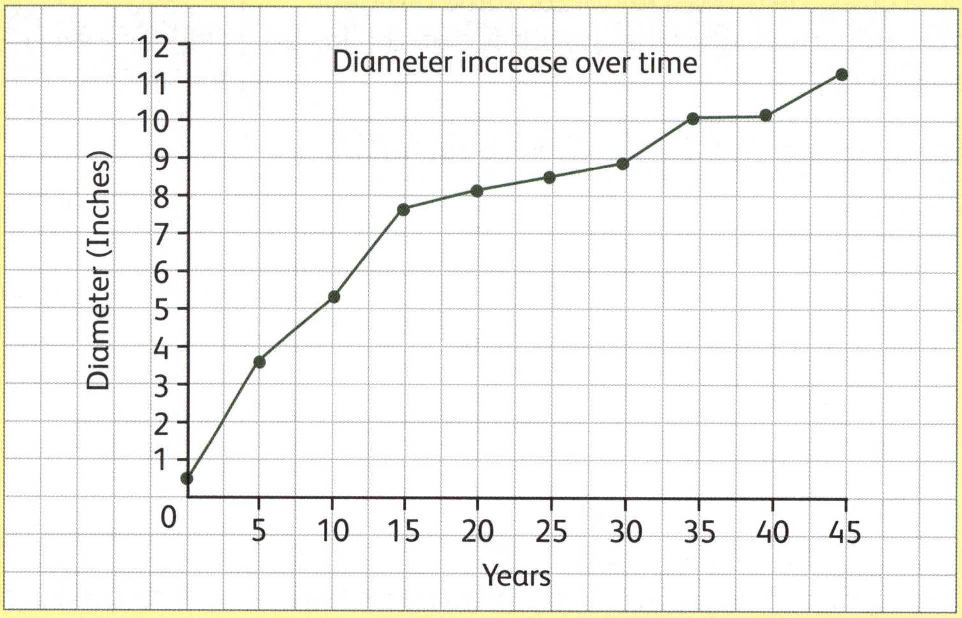

a What was the diameter when the tree was 5 years old?

b By how much did the diameter increase from year 0 to year 10?

c Describe the change in the diameter from year 35 to 40. Suggest a reason for this.

d When did the tree grow at the fastest rate?

e Estimate the tree's circumference in inches at the end of year 45.

f Give two reasons why this graph is not a straight line.

2 The table gives the mean diameter (in centimetres) of live oak trees of different ages in Cinchona Botanical Gardens.

Diameter (cm)	6.4	10.2	15.2	20.3	22.9	24.1	31.8	39.5
Age (years)	15	24	32	56	49	76	90	89

a Draw a line graph to show this data set.

b Use your graph to estimate the age of an oak tree if its diameter is 45 cm.

What I have learned

Write an email to a classmate who has been absent from school to summarise what you learned in this chapter. Answer these questions in your email.

- What is a circle?
- What are the correct names for parts of a circle?
- How do you draw a circle if you are given the radius or diameter?
- What is pi (π) and how does it help you calculate the circumference of a circle?
- How are circles used in real life?
- What was the most enjoyable and interesting thing you learned?

Practice questions

1 Use the words from the box to label the parts of the circle.

centre diameter radius circumference

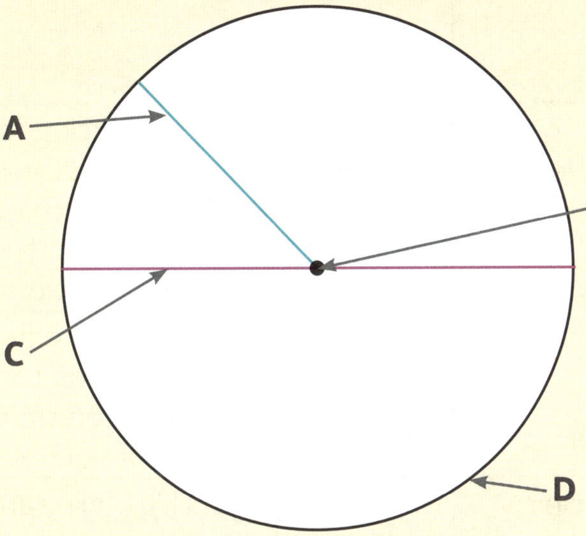

2 If a circle has a radius of 4 cm, what is its diameter?

3 If a circle has a diameter of 10 cm, what is its radius?

4 Calculate the circumference of each of these circles. You may use a calculator. (Take π as 3.14.)

a diameter = 15 cm b diameter = 8 cm c radius = 3.5 cm

5 Draw a circle of radius 5 cm.

6 Draw a circle of diameter 12 cm.

7 In the diagram, the diameter of a large circle is 8 cm.

Find out:

a the circumference of a large circle

b the diameter of a small circle

c the perimeter of the large blue rectangle.

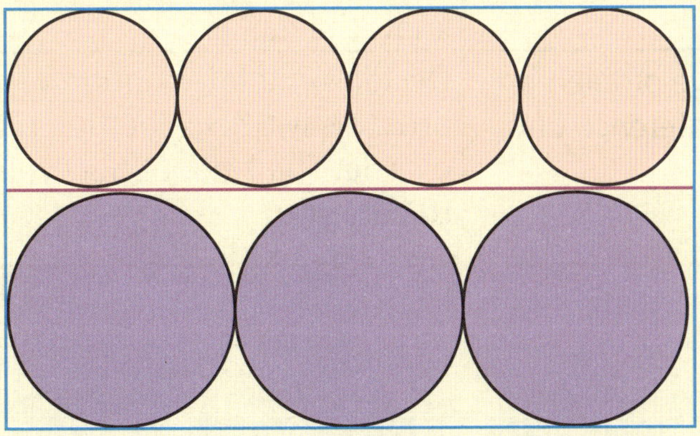

20 Self check

Check you can do this:	If you have difficulty, try this:
1 Use a compass (pair of compasses) to draw circles.	Practise drawing large circles, for example, with a radius of 4 cm. Check that your pencil is secure and the point is lined up with that of the pair of compasses. Hold it tightly at the knob at the top and turn it to draw your circle.
2 Create a circle design.	Look carefully at the designs in this chapter and at designs around you or on the internet. Think about how they have been created. Now create some simple designs of your own.
3 Name the parts of a circle: radius, diameter, centre, circumference, sector, segment.	Go back to page 314. Make sure you know what each part is called.
4 Calculate the diameter of a circle if the radius is given.	Remember that the radius is only half of the length of the diameter, because it joins the centre to a point on the circumference.
5 Calculate the radius of a circle if the diameter is given.	Remember that the diameter of a circle goes through the centre and cuts the circle in half. It is double the length of the radius.
6 Calculate the circumference of a circle if the radius or diameter is given.	Go back to page 318. Remember that pi (π) represents the relationship between the circumference and diameter and that you can take π as $\frac{22}{7}$ or 3.14 when you calculate circumference. Look carefully at the formula you can use when you know the diameter of a circle and what you need to do differently when you know the radius.
7 Estimate and solve problems to find the circumference of a circle.	Decide what information you already know and how this will help you find the circumference of a circle. Think about how you can use a formula to help you.

Chapter 21 Speed, time and distance

In this chapter you will explore this focus question:

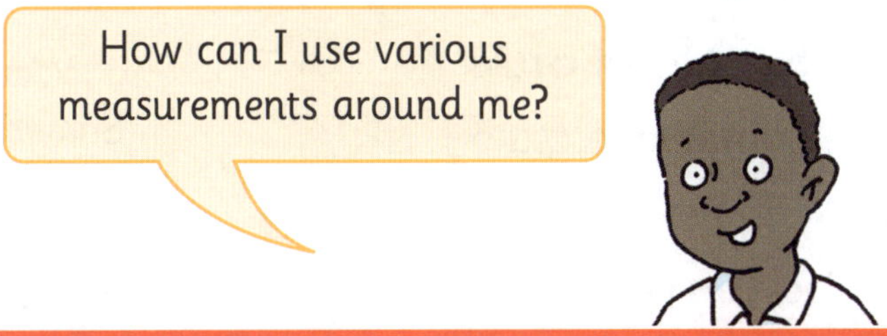

To answer this question, you will:

- use what you know about units of measure and measuring instruments to work with distance, time, volume/capacity and mass
- calculate any one of the measures of distance, time and speed when given the measures of the other two
- understand and use the ideas of distance, speed and time in the context of road safety.

Starting point

Look at these **road signs**.

- What does each one tell you?
- Where would you find them?
- For whom are the road signs?

a

b

c

d

e

1 Match the units of measurement to the things that need to be measured.

Key words
road sign
measurement

Unit of measurement	Object or thing to be measured
1 kilogram (kg)	**A** The width of a ruler
2 gram (g)	**B** Room temperature
3 pound (lb)	**C** Your mass
4 hour (h)	**D** How quickly you can write your name
5 minute (min)	**E** How long your pencil is
6 second (s)	**F** How long it takes to drive from your home to the airport
7 kilometre (km)	**G** How much cola there is in a can
8 metre (m)	**H** How far it is to New York
9 centimetre (cm)	**I** How long it takes to complete your multiplication tables to 12 x 12
10 millimetre (mm)	**J** How long a room is
11 mile (mi)	**K** The mass of an apple
12 feet (ft)	**L** How much water a bucket hold
13 inch (in)	
14 litre (l)	
15 millilitre (ml)	
16 pint (pt)	
17 gallon (gal)	
18 degrees Celsius (°C)	

2 Work in groups. Write the following headings on a sheet of paper:

Distance/length	Mass	Capacity
Volume	Time	Temperature

List of as many measuring instruments as you can for each type of measurement.

Speed, distance and time

Key maths idea

The children are talking about the time they should meet at the cinema and when they should each leave home to get there.

I live furthest, so I need longest to get there.

I'm walking, so I need longest to get there.

I live closest to the cinema, so it will take me no time at all!

Leon

Rhianna

Malcolm

Discuss the children's statements as a group. Think of examples where each child will be correct. Then think of examples where this may not be true for each child. Present your ideas to the class.

When we travel, we think about the distance, the time it takes and how fast we have gone (our speed).

Speed is the **rate** of **distance** per unit of **time**.

Example

A car is driving at 60 kilometres per hour or 60 km/h. This means that if the car continues at the same speed for one hour, it will cover 60 kilometres.

We can use a formula to calculate the distance covered if we know the **average speed** and the time taken:

distance = average speed × time

Rhianna took half an hour to get to the cinema. She walked at an average speed of 5 km/h. We can calculate the distance she travelled as:

$$\text{Distance} = 5 \times \frac{1}{2} \text{ (or } 5 \times 0.5)$$
$$= 2\frac{1}{2} \text{ km (or 2.5 km)}$$

So, Rhianna travelled a distance of $2\frac{1}{2}$ km.

Average speed is also called the **rate of travel**.

Key words

speed
rate
distance
time
average speed (rate of travel)

When you work with time, distance and speed, you need to take note of the units. Speed can also be given in metres per second.

1 Solve these problems. Use the formula to calculate the distance travelled.

a Alison took 2 hours to complete her morning run. Her average speed was 8 km/h. What distance did she run?

b Lamar travels at 60 km/h. How far does he travel in:

- **a** 1 hour
- **b** 2 hours
- **c** 3 hours
- **d** 5 hours
- **e** $\frac{1}{2}$ hour
- **f** $\frac{1}{4}$ hour?

2 How far did each of the friends travel to the cinema?

	Distance	Average speed	Time taken
Leroy		48 km/h	$\frac{1}{4}$ hour
Rhianna		5 km/h	$\frac{1}{2}$ hour
Malcolm		6 km/h	10 minutes

3 Mr Johnson spent 3 hours on a coach. The average speed for the first half of the journey was 70 km/h. For the remainder of the journey, the coach travelled at an average speed of 80 km/h. What was the total distance travelled?

Real-world maths: Speed of sound

Sound travels at approximately 350 m/s. We can use what we know about calculating distance to find out how far away lightning struck during a storm. If it take 10 seconds to hear thunder after the flash of lightning, then lightning struck about 3.5 km away.

How far away did the lightning strike when thunder was heard:

- 5 seconds later
- 2 seconds later
- 12 seconds later?

Calculating average speed and time taken

Key maths idea

Look at this diagram. It shows the relationship between speed, distance and time.

Remember that when we multiply speed by the time taken, we can find the distance travelled:

distance = speed × time

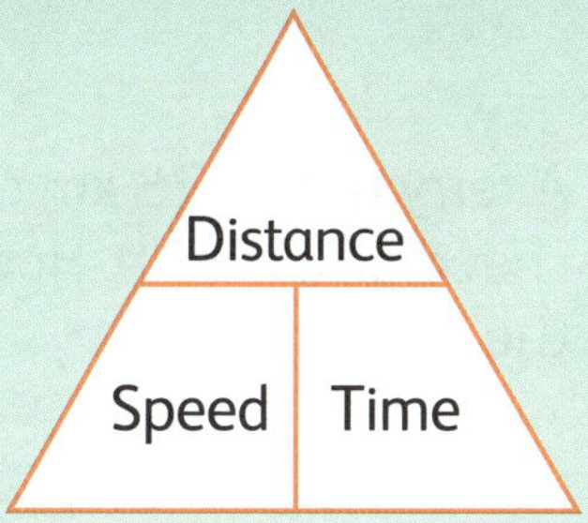

How do you think you can use the diagram to find the speed, when you know the distance covered and the time taken?

How do you think you can use the diagram to find the time taken, when you know the distance covered and the speed?

Explain to your partner how the diagram represents this.

Try out your ideas with the following problems:

- If an athlete runs 200 metres in 30 seconds, what is their average speed in metres per second (m/s)?
- If you cycle 15 km at an average speed of 20 km/h, how long did you cycle?

1 Use the formula to calculate the average speed each time.

Average speed $= \frac{\text{distance}}{\text{time}}$

We need to divide the distance by the time taken to calculate the average speed.

a Iris cycled 8 km. It took her 1 hour. How fast did she cycle?

b If it took Tanja $\frac{1}{4}$ hour less time to cycle 8 km, what was her average speed?

c In 2009, Usain Bolt broke the world record for the 100 m sprint. He ran the race in 9.58 seconds. What was his average speed in m/s? Give your answer rounded to 2 decimal places.

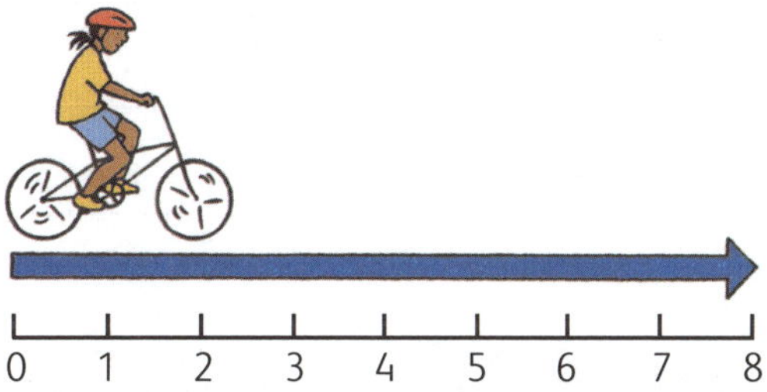

2 What is the average speed of a plane that travels 240 km in 2 hours?

3 Use the formula time = $\frac{\text{distance}}{\text{speed}}$ to calculate the time taken.

a A ferry travels 75 km at an average speed of 25 km/h. How long does the journey take?

b A taxi travels for 150 km at an average speed of 60 km/h. How long does the journey take?

4 Candace walks a distance of 500 metres at an average speed of 1.5 m/s. How long did it take her? Give your answer in seconds and also in minutes.

5 Complete this table. Write only the letter and the answer.

Average speed	Distance	Time
20 km/h	40 km	a
36 km/h	b	$1\frac{1}{2}$ hours
c	64 km	1 hour and 20 minutes

Real-world maths: Speed and braking distances

Look at the graph shown here.

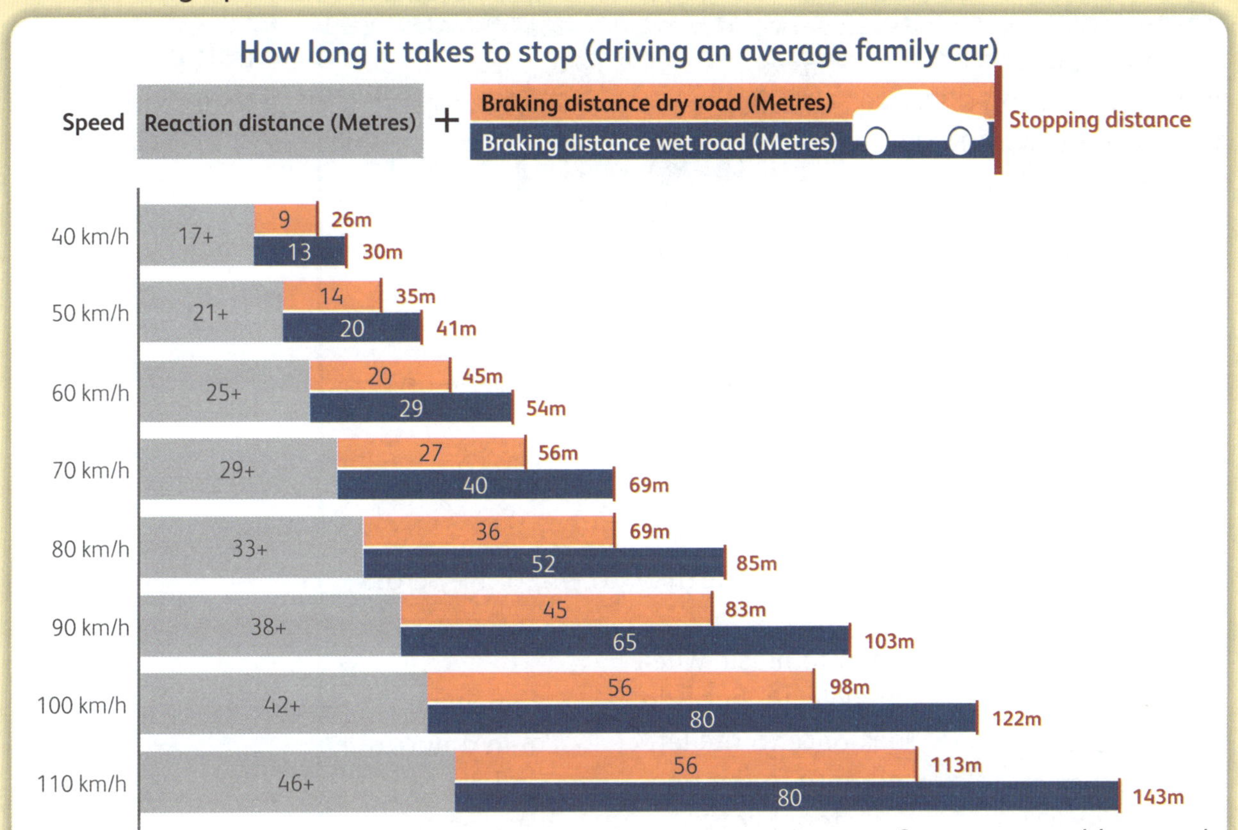

Source: www.qld.gov.au/

The reaction distance in the graph shows how far the car travels while the driver is just thinking about stopping. The braking distance is the distance the car travels once the driver starts to brake.

1 Explain what this graph shows.

2 Refer to the graph to answer these questions:

- **a** What do you notice about the pattern (trend) in the data? Describe the relationship between speed and braking distance.
- **b** About how many metres does a vehicle travelling at 80 km/h travel before it comes to a stop?
- **c** If a person is driving at 70 km/h how much greater is the reaction distance than if they were driving at 30 km/h?
- **d** A car travelling at 110 km/h on the highway will travel about 113 m before stopping in dry conditions. How much further will it travel before stopping when it is raining?

3 Why should a driver keep a few car lengths between their car and the car in front when they are driving? Support your answer with data from the graph.

4 If a driver is tired, distracted, on the phone, or under the influence of alcohol or medication, their thinking distance increases because their reactions are slower. Why is it important for drivers to react quickly? What rules are in place to try and make sure drivers stay focused and can react quickly?

Road safety

Key maths idea

You found out earlier about the braking distance for vehicles travelling at different speeds on the roads, but it is important for all road users to know the **traffic** rules and take special care.

Pedestrians who cross a road have to think carefully about:

- the width of the road
- the speed at which they can safely cross the road
- the speed of the vehicles on the road.

These children are talking about what they do when they cross a road:

Who follows the rules to cross the road safely? Who does not cross safely and is at risk of being in an accident? Explain your reasons. What else do you think is important for the children to know about road safety?

Key words
traffic
pedestrian

1 True or false?

a It will take longer for a pedestrian to cross a wider road than a narrower road when they are walking at the same speed.

b When a pedestrian crosses a 20 metre wide road in 5 seconds, they are walking at an average speed of 5 m/s.

c If a car is travelling at 60 km/h on a dry road, you have plenty of time to cross when it is at a distance of 30 metres.

2 Look at the paths these children used to cross the road:

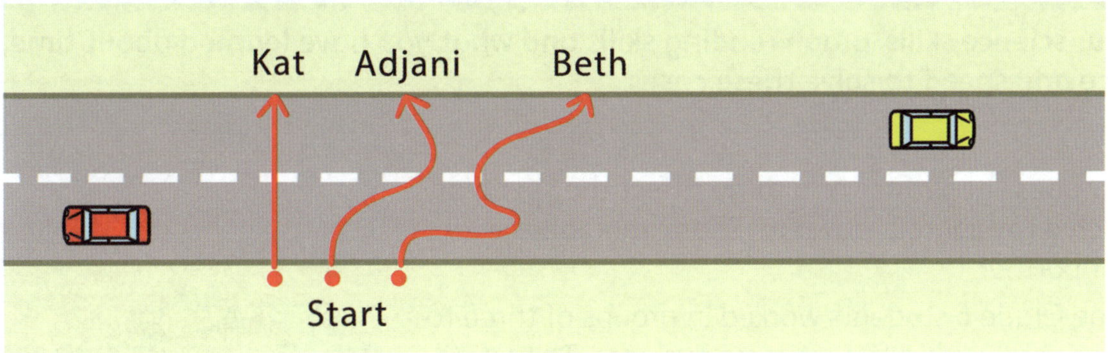

Write the children's names in order to show the safest path to the least safe path used to cross the road. Explain the decision you made.

3 Describe three conditions that make roads more dangerous to cross. What would you do in each case?

4 Leroy wants to cross a double carriageway in a 30 km/h speed limit area. He can see several cars coming.

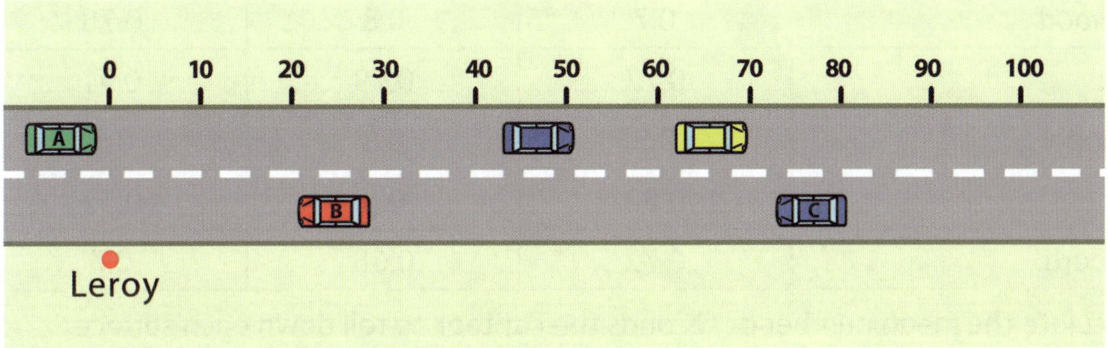

After which car, A, B or C, will it be safest for Leroy to cross the road? Use the diagram to help you explain your reasons.

Real-world maths: Average speed

Mr George left home at 5 to 10 in the morning and he arrived at his destination at 12:55. At the start of his journey, his car's odometer showed 65 463 km. At the end it showed 65 718 km.

1 For Mr George's journey, work out:

a the total time taken

b the distance he travelled

c his average speed in kilometres per hour.

2 On the return journey, Mr George's average speed was 90 km/h. He says this isn't possible, as he did not travel 90 km/h at any point on this journey and that he travelled much slower through the towns and much faster than that on the highway. How could you convince Mr George that 90 km/h was indeed his average speed?

Maths detective: Smooth and rough surfaces

Use your science skills, graph reading skills and what you have learned about time, distance and speed to solve these cases.

Case 1

1 How does the surface of a ramp affect the speed at which a toy car rolls down it? Share your ideas with your partner.

Some Grade 6 students worked in groups of three to experiment with different ramp surfaces. They used the same 1.2 m long ramp for each surface and the same car. They let the car run down each surface three times and used a stopwatch to record the times. They recorded their data in a table.

Surface	Time in seconds		
	1st roll	2nd roll	3rd roll
Plain wood	0.7	0.8	0.75
Carpet	0.77	0.79	0.91
Sandpaper	0.9	0.8	0.9
Plastic	0.68	0.5	0.52
Cardboard	0.72	0.68	0.7

2 Calculate the mean number of seconds the car took to roll down each surface.

3 Use the mean time for each surface to calculate the average speed of the car in metres per second.

Case 2

Ayla recorded how long it took the same car to roll down a 5 m ramp covered with different materials.

Surface	Time (s)
Smooth plastic	4.1
A	8.3
B	6.9
C	5.0

1 What could the surfaces A, B and C be?

2 Explain how you decided.

Case 3

Elaine was watching her brother playing with a toy car.

She wondered if different toy cars would travel at the same speed on a smooth floor.

She set up an experiment in which she rolled a small car and a larger car across a total distance of 5 metres, recording the time taken for every metre travelled. She repeated each test twice and worked out the average times for each car.

Here are her data.

Small car

Distance (m)	1	2	3	4	5
Average time (s)	0.5	1	1.5	2.5	5

Large car

Distance (m)	1	2	3	4	5
Average time (s)	1	2	3	5	8

1 Calculate the average speed of each car in metres per second for the whole distance.

2 Draw a line graph to show both sets of data on the same axes.

3 Write a few sentences to explain what the graph shows.

Real-world maths: Converting rates

Ms Marshall saw this sign while she was driving.

1 What is the speed limit in this area?

2 Ms Marshall's car gives her speed in kilometres per hour. What is the fastest speed in kilometres that she should be going in this area? (Take 1 mile as approximately 1.6 km.)

3 If you drive at this speed for $\frac{1}{2}$ an hour, how far will you get?

4 The speed sign applies to a road that is 12 km long. How long will it take you to cover this distance at that speed?

What I have learned

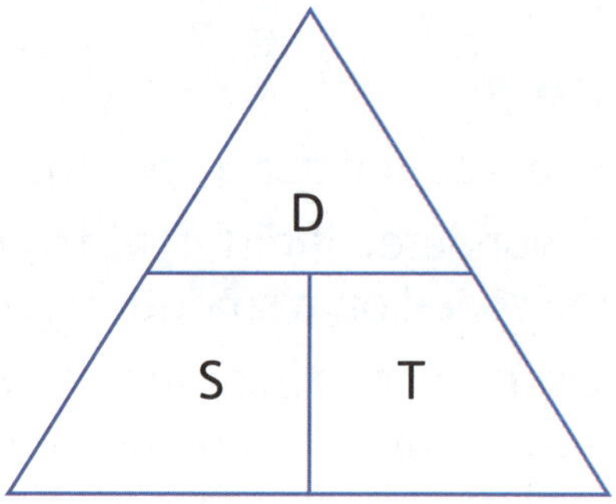

Copy this distance, speed and time triangle into your notebook.

Write short notes to explain to someone who missed class how to use the triangle to find the formula for each measurement – distance, time and speed.

Practice questions

1 What is the average speed of a train that travels 340 km in 5 hours?

2 Mrs Nelson drove 450 kilometres in 5 hours.

- a What was her average speed in km/h?
- b How far will she drive in $7\frac{1}{2}$ hours if she keeps the same speed?
- c How long will it take her to drive 1 080 km at this speed?

3 What distance would you travel if you drove for 4 hours at 80 km/h?

4 How long does it take to cycle 35 km at an average speed of 7 km/h?

5 Malaki drove his car at an average speed of 80 km/h.

How many kilometres did he travel in:

- a 4 hours
- b $5\frac{1}{2}$ hours
- c $2\frac{1}{4}$ hours?

6 A truck leaves a shop at 4:10 p.m. and arrives at the warehouse at 5:40 p.m. The warehouse is 84 km away from the shop. Did the truck driver exceed an average speed of 60 km/h? Explain how you know.

7 Candace walked from her home to her grandparents' house at a steady speed of 5 km/h. She walked for three hours. How far did she walk?

21 Self check

Check you can do this:	If you have difficulty, try this:
1 Calculate distance, time and average speed when you have two of the three measures.	Remind yourself that you can use the distance, speed and time triangle to help you work out which formula to use. Also remember to check the units of measurement.
2 Explain the meaning of the most important road signs.	Go back to page 325 and recap the most important road signs.
3 Explain the important traffic safety rules and comply with them.	Go back to page 332 and remind yourself of the rules for crossing roads safely.
4 Apply the principles of measurement to road safety.	Remember to think about how speed, distance and time relates to road safety and braking distances.
5 Use the road safely.	Make sure to: • find a safe place to cross, preferably a pedestrian crossing, and stop • use your eyes and ears: look right, left, right and listen and look out for vehicles • wait until it is safe: if traffic is coming, wait for it to pass, then look right, left and right again • when there is no traffic near, walk straight across the road • keep looking and listening out for traffic while you cross.

Unit 3: Geometry

Chapter 22 Angles

In this chapter you will explore this focus question:

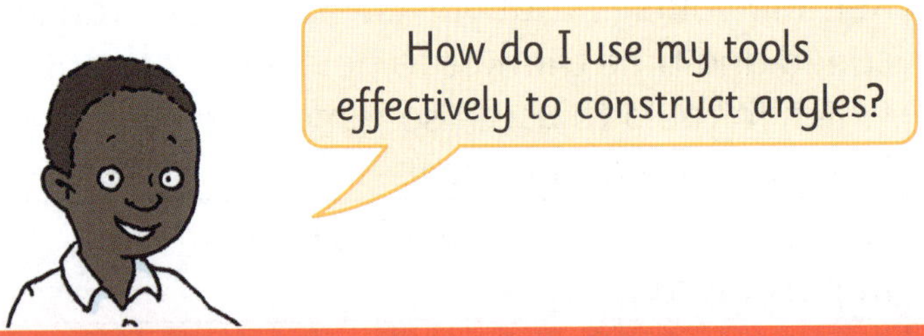

To answer this question, you will:

- explore concepts of angle formation, drawing, naming and measuring
- draw and measure angles using the protractor.

Starting point

This photograph was taken at the port in Falmouth.

1 You can see different kinds of angles in the picture. Identify examples of:

  a right angles

  b angles that are smaller than a right angle.

2 Look at the hands on the clock face.

  a What angle do you see between the hour hand (short hand) and the minute hand (long hand)?

  b What kind of angle do you see if you imagine the turn going in the other direction?

3 Look at the frame around the clock. What angles do you notice on that shape?

4 Choose one of the angles in the picture. Draw an angle made up of two line segments to show an angle the same size.

Classifying angles

Key maths idea

An **angle** is formed by two **rays** or **line segments** that meet at a point. The meeting point is called the **vertex**. The sides of the angle are known as its **arms**.
In this diagram, BA and BC meet at point B. They form angle ABC. We write it as $\angle ABC$.

As point B is the vertex of the angle, it is in the middle of the letters. You can also refer to the angle by the vertex on its own as $\angle B$.

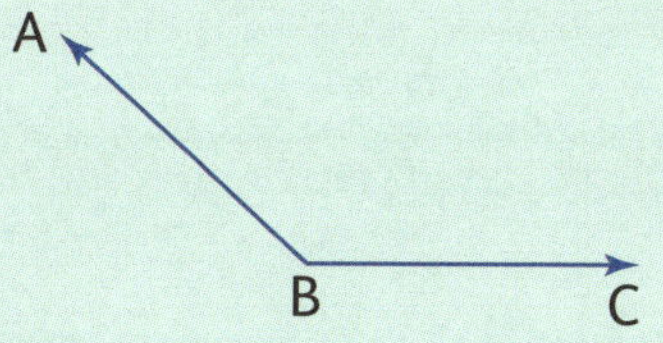

We measure angles in **degrees** (°). You already know that a right angle measures 90°. It looks like a square corner. When you mark a right angle on a diagram, you use a corner rather than a curve.

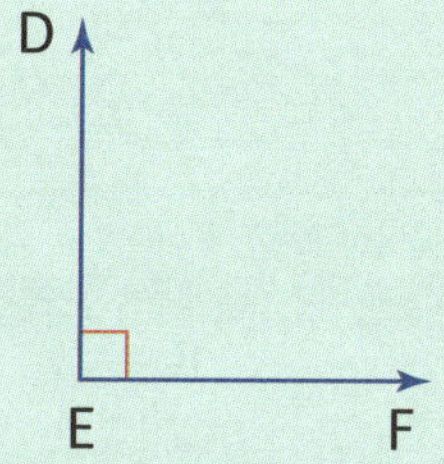

An **acute** angle is smaller than a right angle. It measures between 0 and 90°.

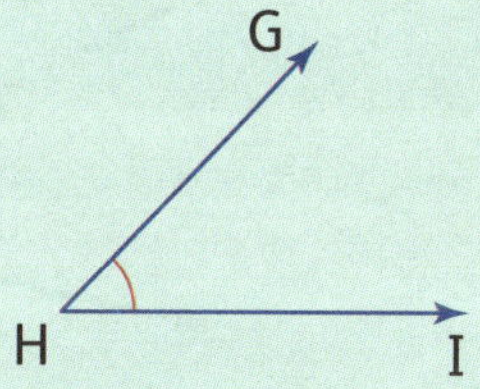

An angle that measures between 90° and 180° is called an **obtuse** angle.

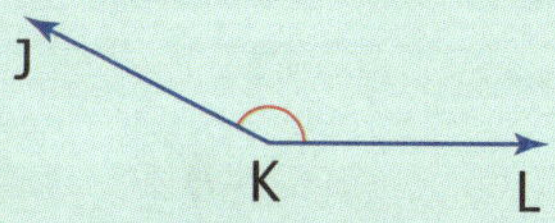

An angle that measures exactly 180° and looks like a straight line is called a **straight line angle**. You can also think of this as a **half turn** around a point.

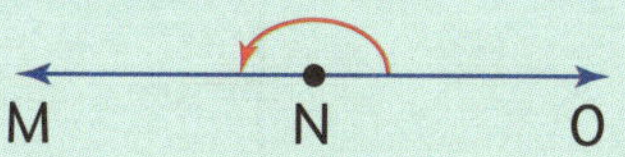

A **full turn** around a point measures 360°. Another word for a full turn is a **revolution**.

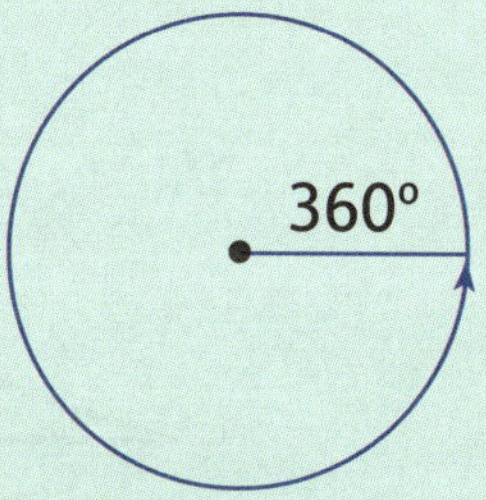

An angle that measures between 180° and 360° is called a **reflex** angle.

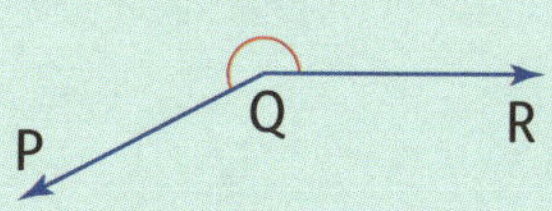

Key words

angle
ray
line segment
vertex
arm
degree
acute
obtuse
straight line angle
half turn
full turn
revolution
reflex

Classify each of these angles as acute, obtuse, right angle, straight line, reflex or revolution.

1

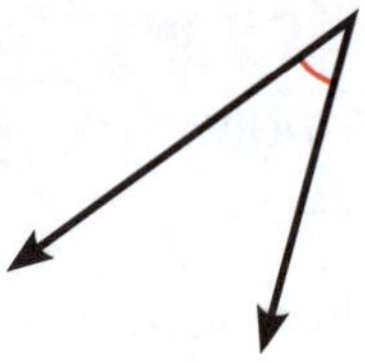

2

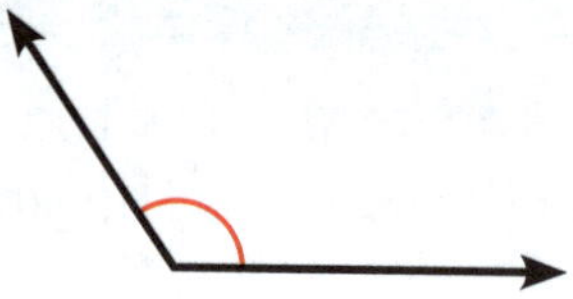

3

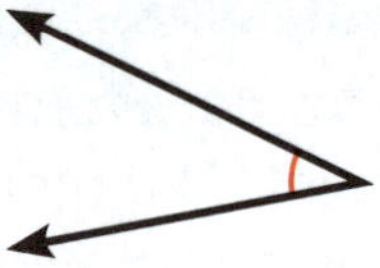

4

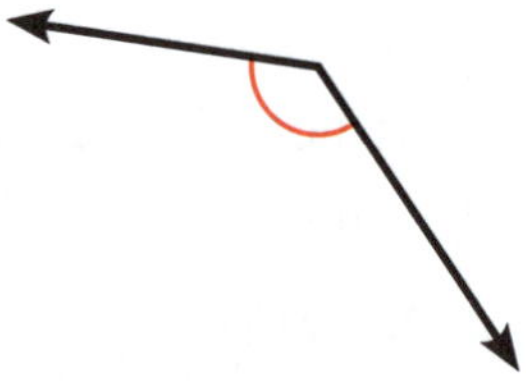

5

6

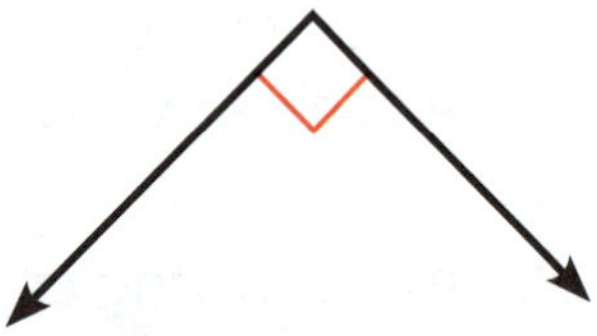

7

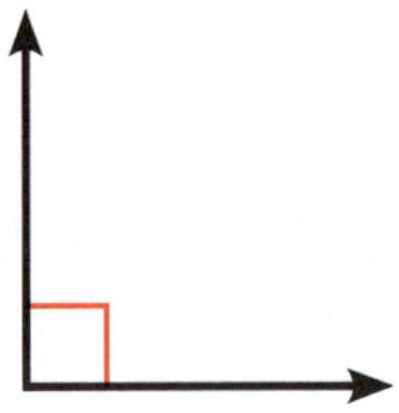

8

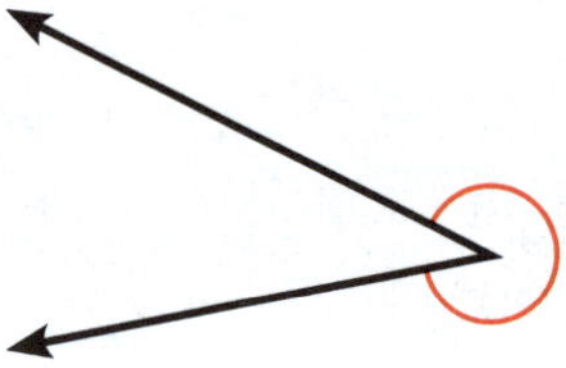

9

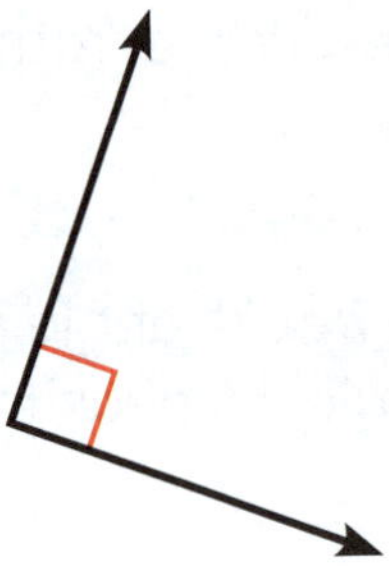

10

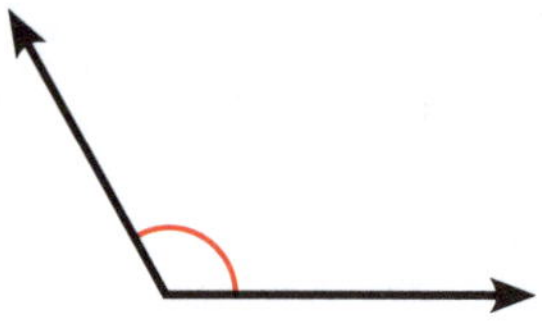

11

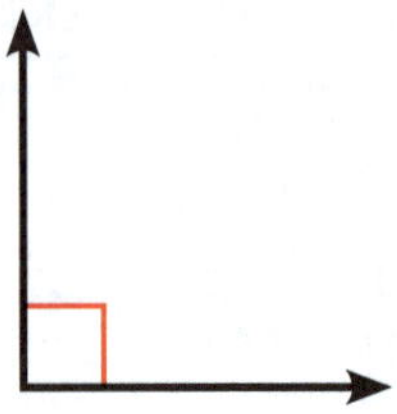

12

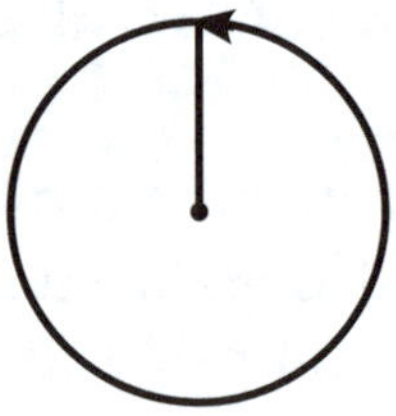

13

14

15

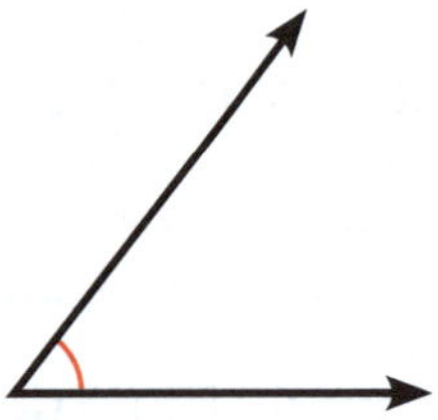

More angles

Key maths idea

Angles inside a shape are called **interior angles**.

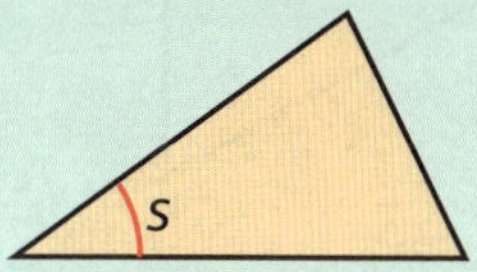

interior angle

Angles on the outside of a shape are called **exterior angles**.

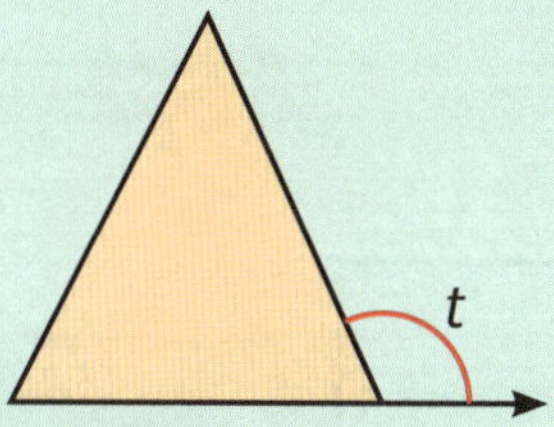

exterior angle

Angles on a straight line add up to 180°. They are called **supplementary angles**.

Two angles that share the same vertex and lie on opposite sides of a common arm are called **adjacent angles**. OB is the common arm of adjacent angles AOB and BOC.

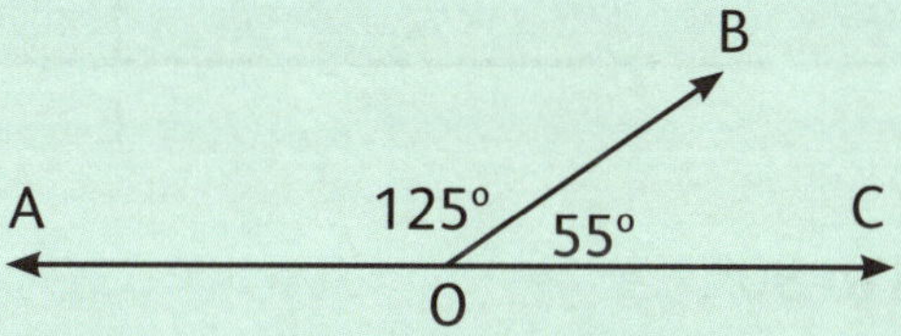

Angles that make up a right angle are called **complementary angles**. They add up to 90°.

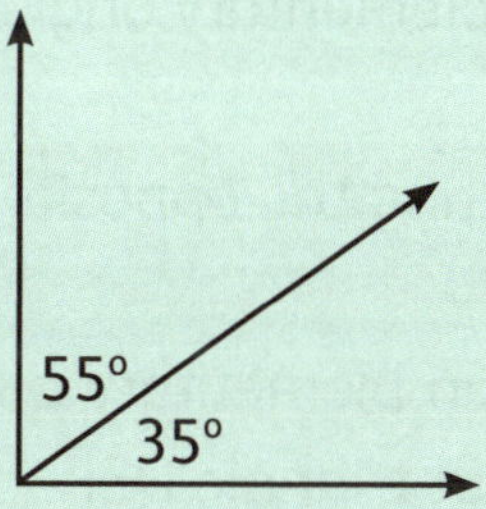

Key words

interior angles
exterior angles
supplementary angles
adjacent angles
complementary angles

Geometry

1 **a** Name two pairs of supplementary angles on line segment TW.

b Name three exterior angles of triangle SQR.

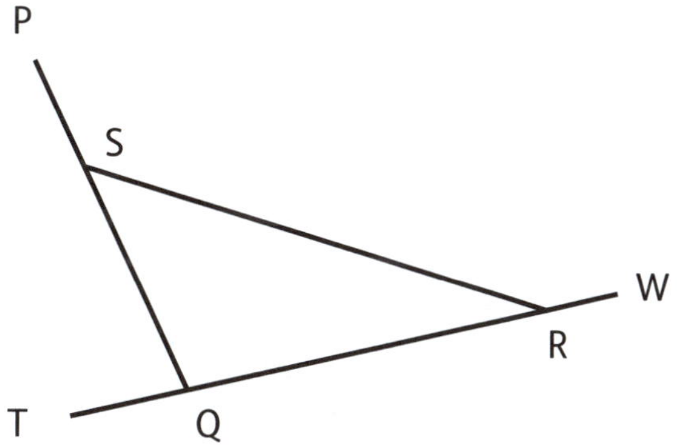

2 **a** Name three interior angles of triangle DCE.

b Name three pairs of supplementary angles.

c Name two angles that are exterior angles to triangle DCE.

d Name one angle that is an exterior angle to triangle ABC.

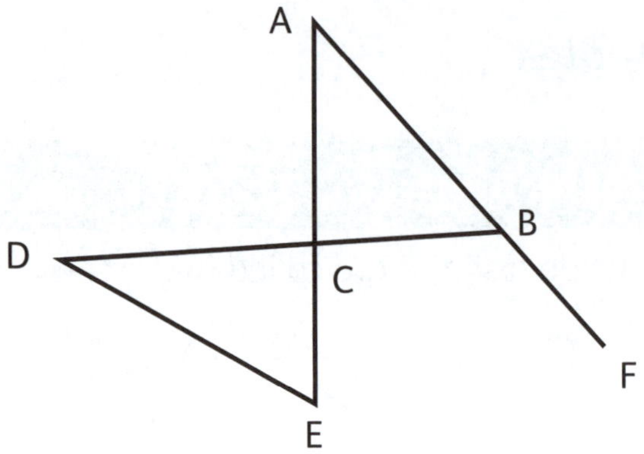

3 In this diagram, we know that ∠JLM is a right angle, and that HK is the same length as JL and HJ is the same length as KL. For each statement, say whether it is true or false, and give a reason.

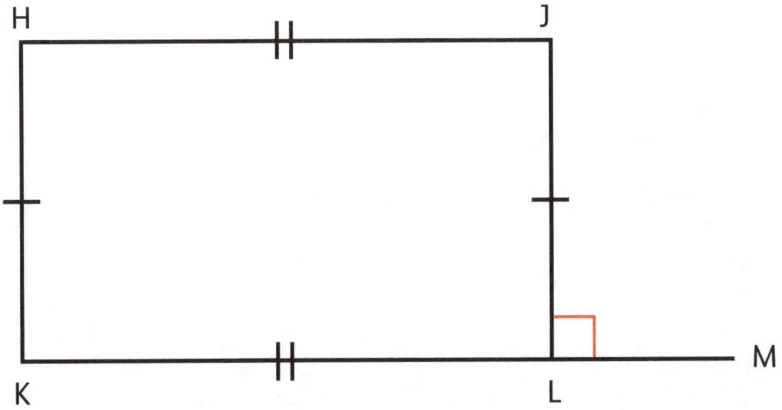

a ∠JLK must be equal to 90°.

b ∠HKL could be an acute or obtuse angle.

c ∠KLJ and ∠JLM are complementary angles.

d HJLK must be a rectangle.

e If you draw line JK connecting the opposite corners of HJLK, this would create two congruent triangles.

4 In triangle DEF, a and b are complementary angles.

a What does this tell us about triangle DEF?

b If $a = 55°$, what is the size of b?

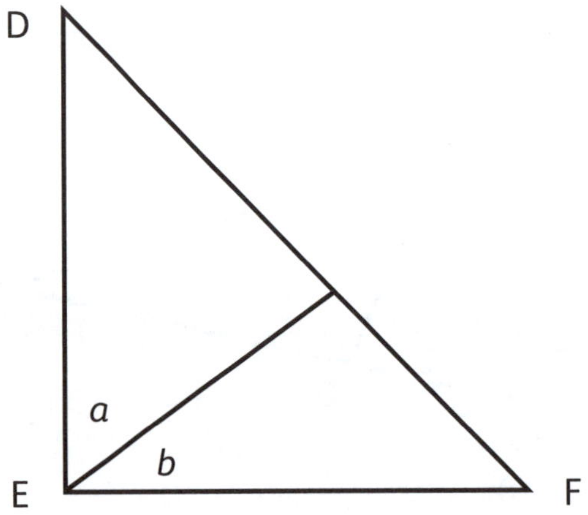

Maths detective: More angles

Use the internet to help you find out what these types of angles are.

1 **corresponding** angles
2 **alternate** angles
3 **vertically opposite** angles.

Draw sketches to show what you found out.

Key words
corresponding
alternate
vertically opposite

Measuring angles with a protractor

Key maths idea

We measure angles with a **protractor**.

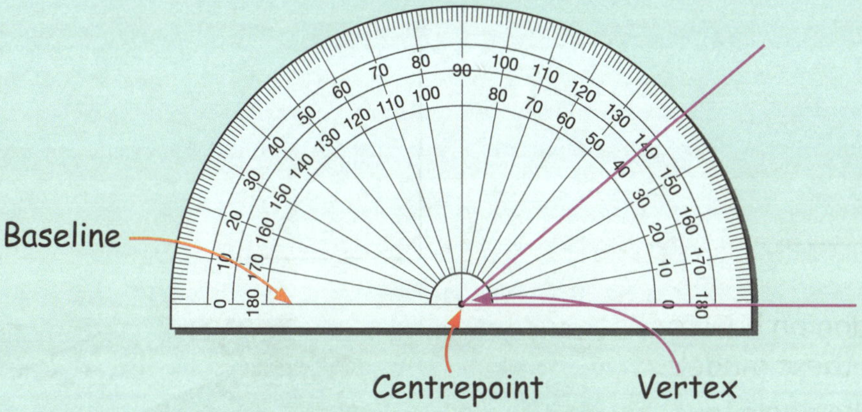

To measure an angle using a protractor:

- Before you measure, think about whether the angle is acute, right, obtuse, straight line or reflex. This will help you to estimate your answer and check it later.
- To measure, place the centrepoint of the protractor on the vertex of the angle.
- One ray of the angle must lie on the baseline of the protractor.
- The other ray will fall on a degree greater than 0°.
- Read off the measurement where the second ray meets or crosses the scale of the protractor.

This angle is 40°.

Key words
protractor
pair of compasses

1 Measure each of these angles using a protractor.

a

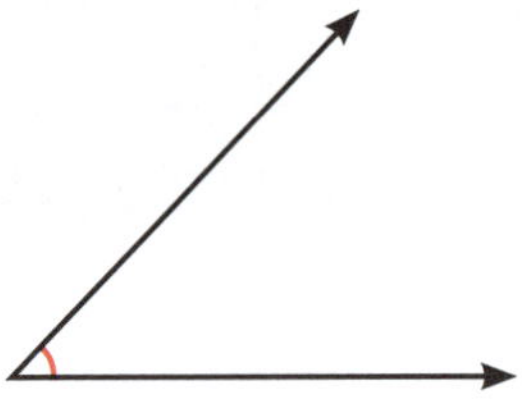

b

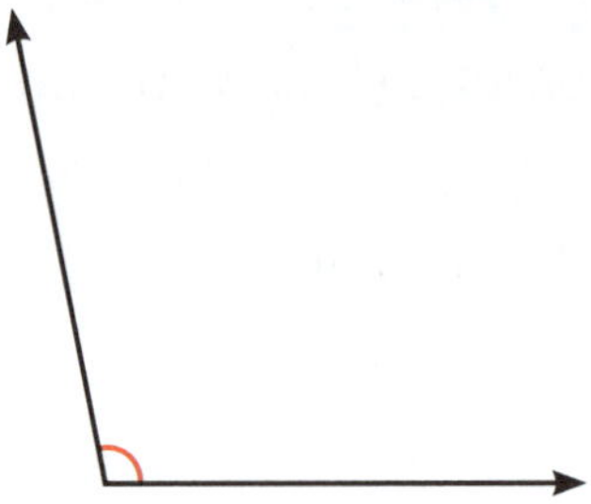

c

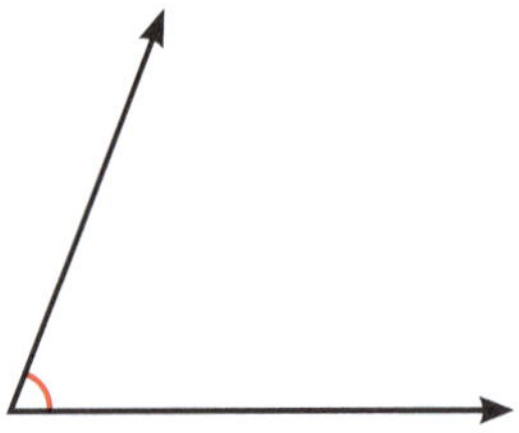

d

e

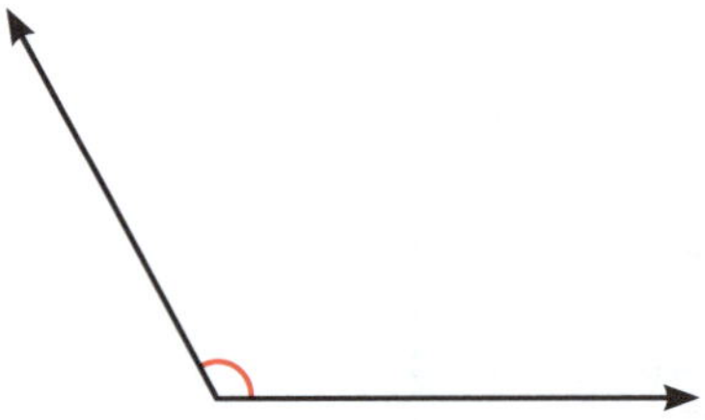

f

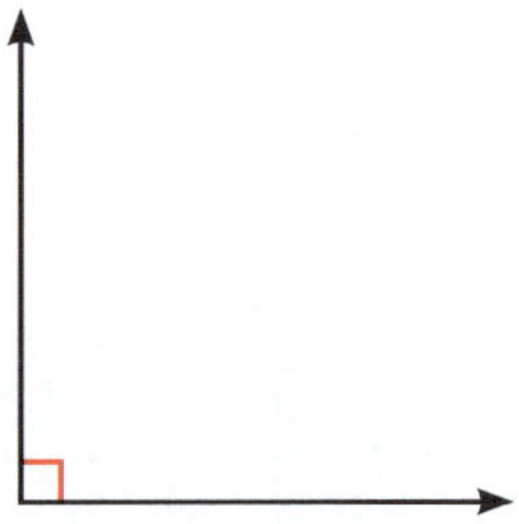

2 Zara is doing a presentation on how to use pie charts. She made these models.

a Estimate the size of the angle for each different coloured sector on her models.

b Use the formula (size of angle ÷ 360) × 100 to work out what percentage of the whole each sector represents.

3 Use a pair of compasses to draw some circles. Draw lines from the centre of the circle to the circumference. Use your circles to make sketches that illustrate the different kinds of angles you have learned about in this lesson.

> You learned in Chapter 20 how to use a **pair of compasses** to draw circles. Look back at page 316 to remind yourself how to use it.

Drawing angles with a protractor

Key maths idea

You can use a protractor to draw angles of a particular size.

For example, draw an angle, x, measuring 140°.

Step 1: Draw one ray as a horizontal line.
Mark where you want the vertex to be.

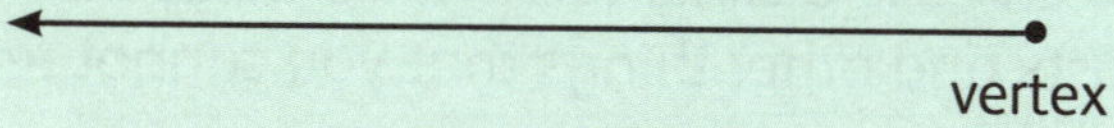

Step 2: Place the protractor on the ray with the centrepoint on the vertex.

Step 3: Mark the point where the protractor shows 140° from the baseline.

Remember 140° is an obtuse angle, so you know the angle is larger than 90°.

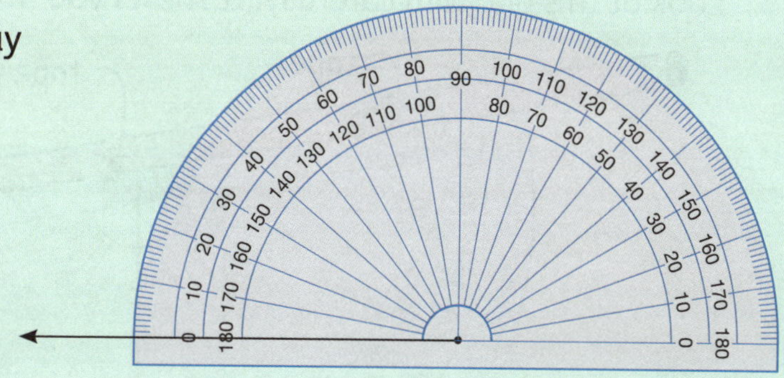

Step 4: Remove the protractor and draw a straight line from the vertex through the point.

Step 5: Label your angle x.

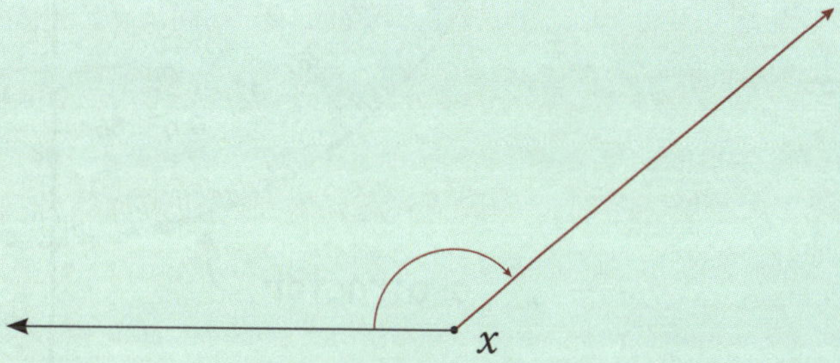

Draw these angles accurately.

1 30° **2** 70° **3** 90°

4 15° **5** 160° **6** 200°

Real-world maths: Angles in structures

Look at this photograph of a Ferris wheel.

1 What is the angle between the cars? Show how you work this out.

2 The green metal post is perpendicular to the ground. Estimate the size of the angle between the yellow post and the ground. How could you check this?

3 Use your compasses to draw a circle. Mark where you would put the cars on a Ferris wheel with only 15 cars. Remember the cars need to be the same distance apart.

Maths detective: A clinometer

You are going to make a simple instrument and use angles and right-angled triangles to solve problems involving measuring the height of tall things.

Case 1

A clinometer is a piece of equipment used to measure the angles that are formed when you look up (or down) at objects.

You can use a clinometer to work out the height of tall buildings, trees, mobile phone masts and other things that you cannot reach to measure directly.

1 Look at this diagram carefully. It shows you how to make your own clinometer.

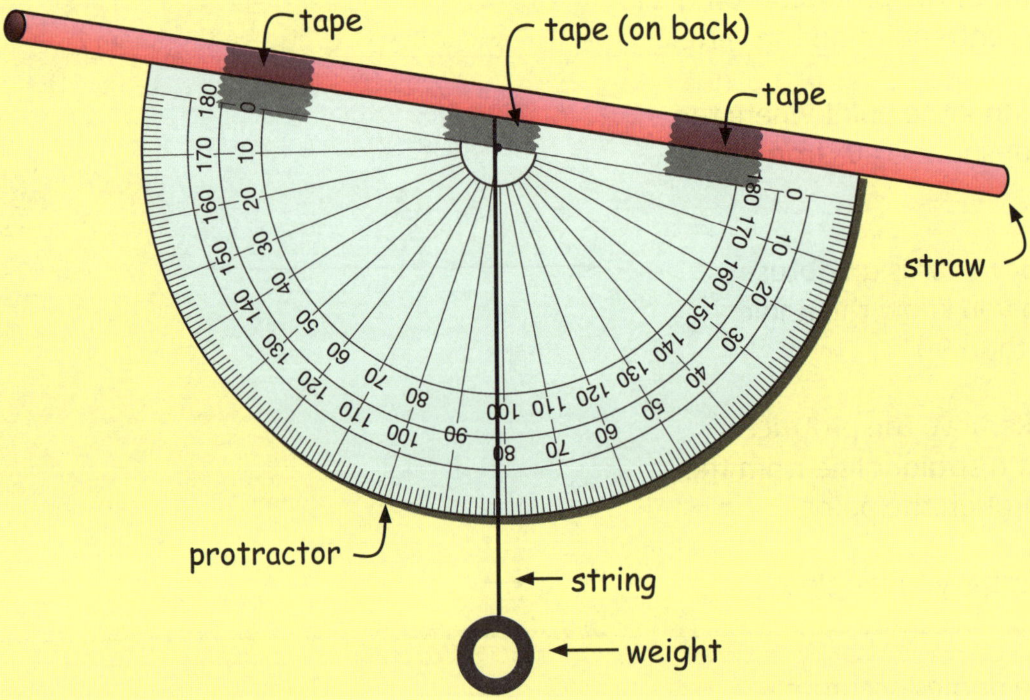

a List the equipment you need to make a clinometer.

b If there is anything on the list that you do not have, suggest what you have that you can use instead.

c Make the clinometer.

2 Read these instructions for using a clinometer and then work with a partner to try yours out.

How to use your clinometer

- Work in pairs. You will need a tape measure or metre stick.
- Go out into the school yard and find a tall tree, flagpole, basketball hoop or building. You need to be able to measure the distance from yourself to the object and you must have space to move closer to it, or further from it.
- Hold the clinometer with the straw on top. Look through the straw and find the top of the thing you are measuring.
- Ask your partner to look where the string is touching the protractor scale and read the angle on the clinometer.
- Move back (or forwards) till the angle on the clinometer is 45°.
- Measure the distance between where you are standing and the base of the object and how high your eyes are above ground level.
- Add these two measurements together to get a fairly accurate measurement of the height of the object.

3 Use this diagram to explain why we want a 45° angle on the clinometer. Think about what you know about isosceles triangles.

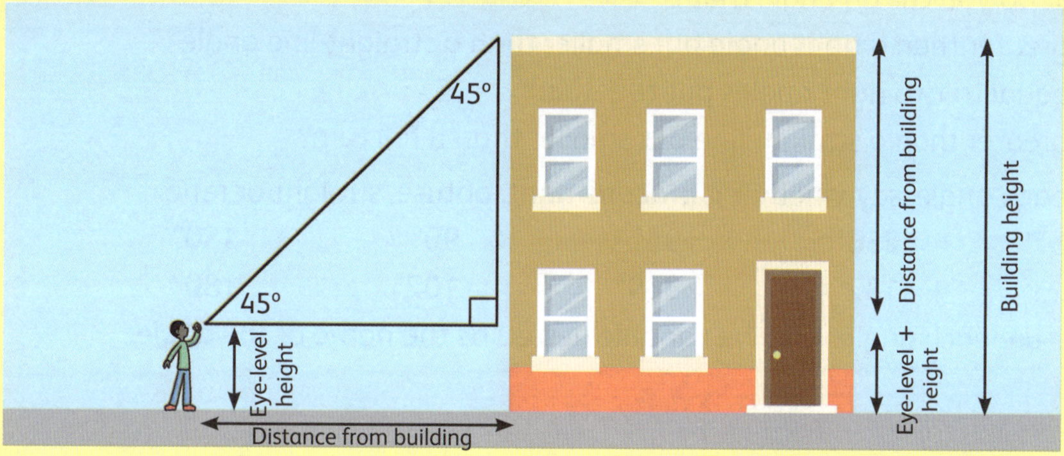

Case 2

1 Jermaine is 8.7 metres from a tree and his eyes are 0.85 metres above the ground level. How tall is the tree?

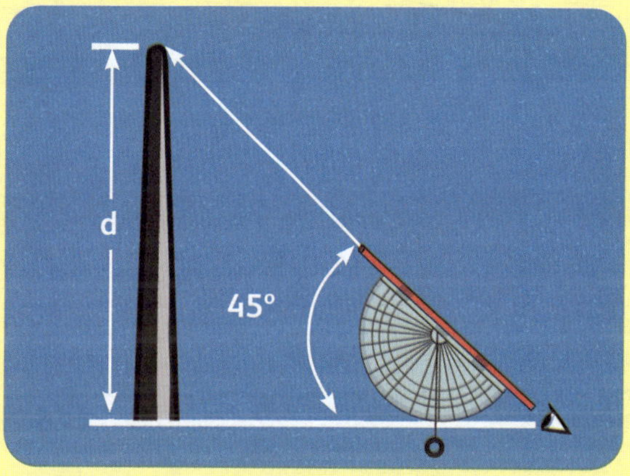

2 Sandra is 24.8 metres from the phone tower. The distance from her eyes to the ground is 1.02 metres. How high is the phone tower?

What I have learned

Copy the mind map. Draw diagrams and write short notes to summarise the main points under each topic.

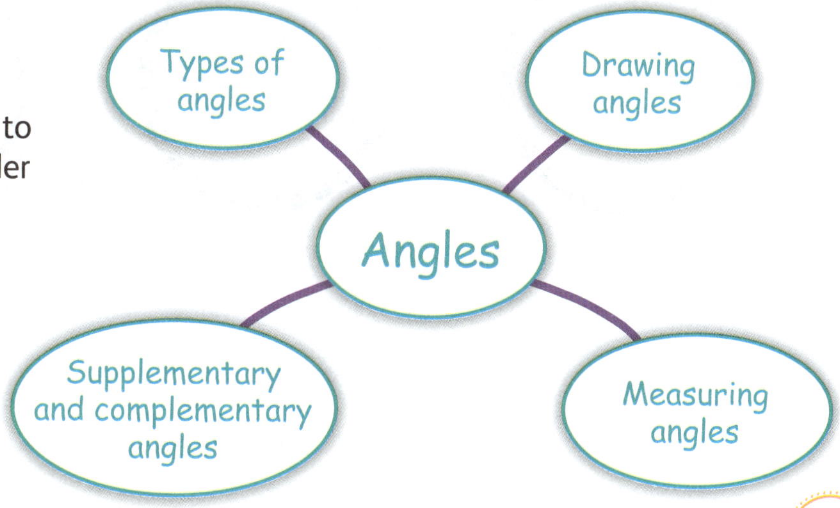

Practice questions

1 What do you call an angle that is:

a greater than a right angle but smaller than a straight-line angle

b equal to two right angles put together

c greater than a straight line but smaller than a full turn?

2 For each angle, say whether it is acute, right, obtuse, straight or reflex.

a 87°	b 256°	c 95°	d 90°	e 150°
f 23°	g 182°	h 359°	i 102°	j 180°

3 Measure each angle. Use the variable (letter) as the name of the angle.

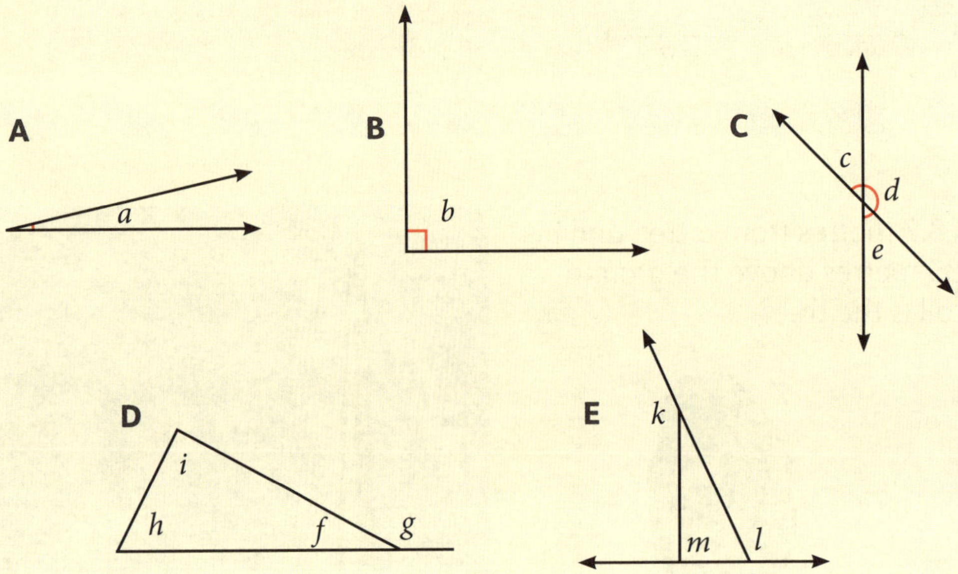

4 Draw these angles accurately.

a ∠XYZ = 35° b ∠KLM = 100° c ∠PQR = 200°

5 Draw two different right-angled triangles. Use a protractor to measure the size of the other two interior angles in degrees.

6 Classify two different angles formed by the hands on each clock.

22 Self check

	Check you can do this:	If you have difficulty, try this:
1	Estimate the size of angles.	Go back to page 339 and remind yourself how to judge whether an angle is acute, obtuse, reflex, right or straight. You can use this knowledge to help you estimate angle sizes.
2	Use a protractor to measure angles accurately.	Revise page 343. Make sure you know how to position the protractor with the cross-hairs over the vertex of the angle, and the baseline of the protractor on one arm.
3	Use a protractor to draw angles accurately.	Revise page 345 and practise drawing some angles of different sizes using a ruler and protractor.
4	Name the different types of angles.	Revise pages 339–341 and revise the different terms for describing angles.

Unit 4: Algebra

Chapter 23 Solving simple equations

In this chapter you will explore this focus question:

To answer this question, you will:

- use arithmetic operations to solve simple equations and word problems
- determine the solution to a simple equation with one variable
- solve word problems involving the four arithmetic operations with one- or two-digit numbers.

Starting point

1 Work with a partner. Work out the missing answer. Then explain to the class how you worked out the value of each shape.

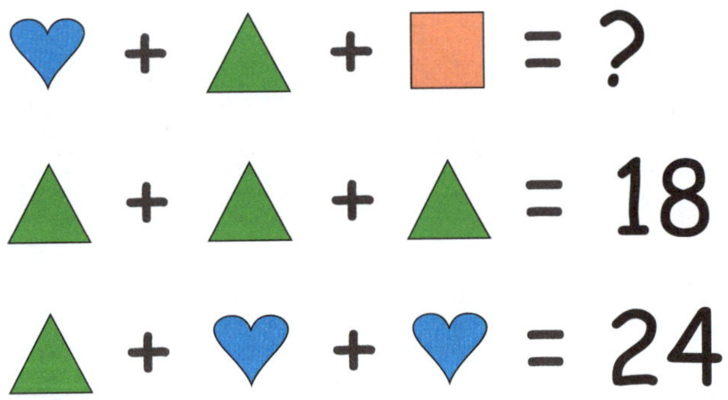

2 Write an expression to represent each of these statements. Use n to represent a number.

a four times a number

b half of a number

c 7 less than a number

d a number divided by 7

e 12 more than double a number

3 A rectangle has a perimeter of 32 cm. The length of the rectangle is 6 cm longer than its width. Write an equation and solve it to work out the lengths of the sides.

Simplifying and evaluating algebraic expressions

Key maths idea

You have worked with **algebraic expressions** before. These are mathematical expressions where we use letters to represent a value, amount or **quantity**.

$2y$	means $2 \times y$
$2 + n$	means 2 more than n, or n plus 2
$3q^2 + 1$	means $(3 \times q \times q) + 1$

Sometimes an expression has like terms, which we can **simplify** by grouping the like terms. For example:

$$n + n + p + p + p = 2n + 3p$$

Like terms have the same variable, and the same exponent.

Consider: $4k + 3k$. You can simplify the expression to $7k$.

Now consider: $4k + 3k^2$ You cannot simplify the expression because the expressions have different exponents.

If you are given a value for the letter in the expression, you can use substitution to evaluate the whole expression.

Key words

algebraic expressions
quantity
simplify

1 For each of these real-life descriptions, write a variable to represent each kind of fruit. Then write it as an algebraic expression and simplify it if you can. The first one has been done for you.

a 2 pineapples, 3 more pineapples and 2 mangoes $5p + 2m$

Can you see that p represents pineapples and m represents mangoes? So we can add the 2 and the 3 to make 5.

b 7 papayas and 4 bananas

c 10 coconuts, 3 jackfruit and another 4 coconuts

d 4 mangoes, 5 oranges, 6 more mangoes and 5 more oranges

2 Simplify these algebraic expressions.

a $j + 3j$

b $a + b + a + b$

c $5a + q + 3a + 2q + 6$

d $3n + 8 + 5n$

e $d \times d \times 2$

3 Now evaluate the expressions in Exercise 2 using the following values:

a $j = 7$

b $a = 10$ and $b = 9$

c $a = 11$ and $q = 25$

d $n = 100$

e $d = 4$

Writing equations to solve problems

Key maths idea

Think about a square. We can calculate the perimeter of a square using this **equation**:

$P = 4 \times s$,

where P represents the perimeter and s represents the length of one side.

- If one side is 3 cm long, $P = 4 \times 3$ cm
- If one side is 5 cm long, $P = 4 \times 5$ cm, and so on.

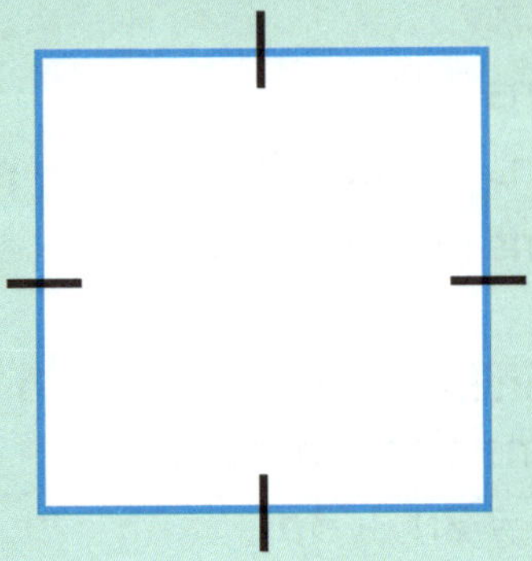

We say s is a **variable**, because it represents a quantity whose value can change.

In the expression $P = 4s$, we say 4 is a **constant**. Its value cannot change.

Key words

equation
variable
constant

1 Josh says: 'I think of a number, let's call it x. I add 3. The answer is 5. What was my number?'

a How can you work out the starting number?

b How can you check that your answer is correct?

2 Write your own 'Think of a number' instructions for a partner. Let them work out your starting number. Do this using:

a a starting number, followed by one operation ($+$, $-$, $\times$ or $\div$)

b a starting number, followed by two different operations

c a starting number, followed by three different operations.

3 Nicky earns half as much as Shannon earns. Let Nicky's earnings $= n$ and Shannon's earnings $= s$. Which equations are correct?

a $n = s \div 2$ **b** $s = 2n$ **c** $2n = s + 1$

4 When Carrie gets a bag of candies, she takes out 10 of the candies, and divides the rest evenly among three containers. Let the total number of candies in the bag $= c$ and the amount in one container $= n$. Which of the following equations are not true?

a $(c - 10) \div 3 = n$ **b** $n \div 3 = c$ **c** $(c - n) \div 3 = 10$

5 Nina is 8 years older than her sister Rose. Let Nina's age $= n$ and Rose's age $= r$.

a Write an equation that starts $n = \ldots$

b True or false: n and r are both variables. Give a reason for your answer.

Balance the weights

Key maths idea

On these scales, the square weights are each equal to 1 kg. The triangles represent an unknown weight. We can use an equation to describe the relationship between the two sides.

How can you work out the value of the triangles?

Look at how these students worked it out:

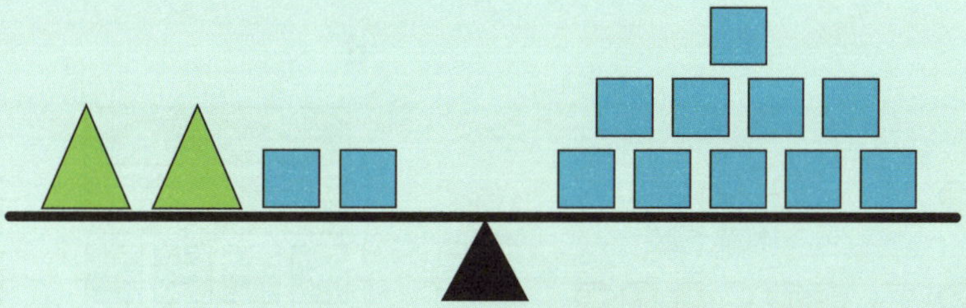

I imagined taking 2 square weights (2 kg) away from each side. Then I can see that 2 triangles is equal to 8 kg. I know that half of 8 kg is 4 kg, so the triangle must be equal to 4.

I used equations:

Let ▲ = n

$2n + 2 = 10$	In other words, 2 triangles plus 2 kg = 10 kg.
$2n = 8$	We take away 2 from both sides of the equation.
$2n \div 2 = 8 \div 2$	We divide both sides by 2.
$n = 4$	

Remember: whatever you do to one side of the equation, you need to do the other side. This keeps the equation balanced.

1 Work with a partner. Find the value of the unknown weights on each scale. In each diagram, the squares represent 1 kg each.

a

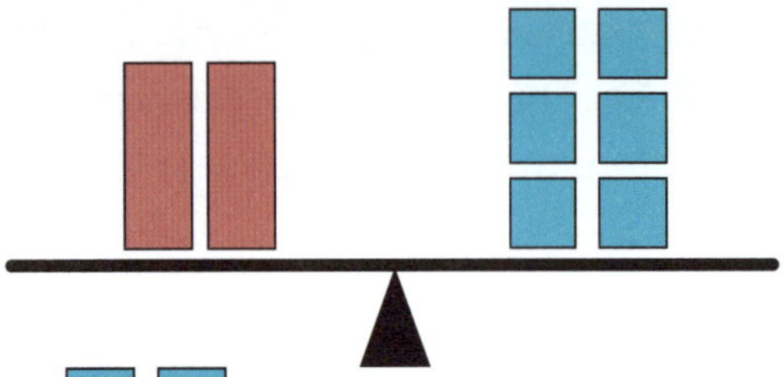

b

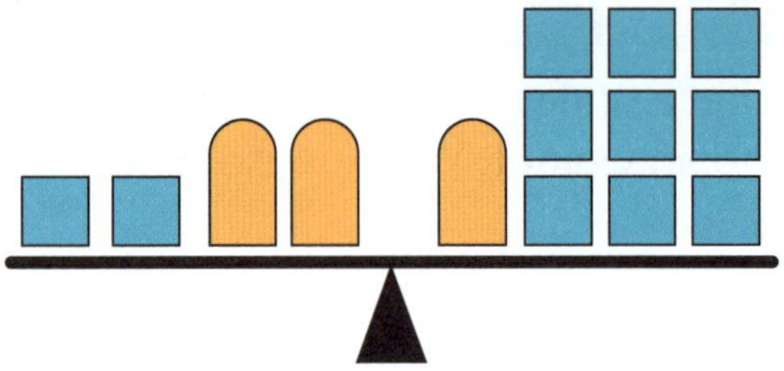

c

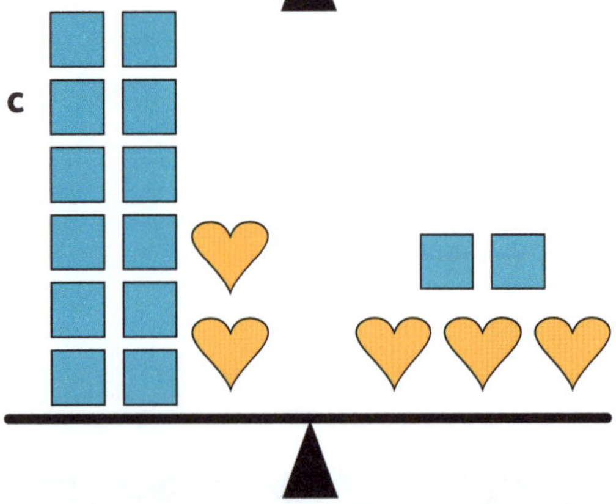

d

2 Make up your own balance weight problem for a partner to solve.

Solving equations

Key maths idea

In Exercise 1, you **solved** each puzzle with the weights. The **solution** to an equation tells you the value(s) of the variables.

Sometimes you will need to solve an equation simply by working mathematically with the expressions. It is important that you remember how to read algebraic expressions.

Remember:

- $5n$ This means $5 \times n$ or $n \times 5$. A constant always comes before an unknown. You can also group like variables in this way, so $n + n + n = 3n$.
- $5n + 5$ This means $(5 \times n) + 5$
- $4k^2$ This means $4 \times k \times k$
- $(4k)^2$ This means $4k \times 4k$. We can simplify it to $16k^2$.

Key words

solve

solution

When you solve an algebraic equation, you work out the possible values of the variables.

Examples

1 $3x - 6 = 9$

Add 6 to both sides: $3x = 9 + 6$

$3x = 15$

Divide by 3: $x = 15 \div 3$

Solution: $x = 5$

2 $8x + 3 = 35$

Subtract 3 from both sides:

$8x = 35 - 3$

$8x = 32$

Divide both sides by 8:

$x = 32 \div 8$

$x = 4$

1 Write an equation that describes each bar model. Then use the equation to work out the value of the variable.

a

x	x	x	x
16			

b

k	3
15	

c

5	x
21	

d

p	p	7
31		

2 Solve each equation.

a $3a = 12$ **b** $30 = 6y$ **c** $100 = 10n$

d $30 = 5q$ **e** $15k = 45$ **f** $12m = 144$

3 Solve each equation.

a $4y + 6 = 38$ **b** $5d - 2 = 48$ **c** $100 - 3x = 70$

d $10x + 20 = 140$ **e** $12x + 2x = 56$ **f** $11x + 4 = 125$

4 A number y is divided by 6. When 7 is added to the number, the result is 10. What is y?

5 Three times a number is added to four times the same number. The result is 63. What is the number?

6 When 11 times a number is added to 26, the sum is 48. What is the number?

7 Six times a number p is subtracted from 80. The difference is 38. What is the value of p?

8 Seven times a number q is added to 15. The sum is 260. What is the value of q?

Real-world maths: Using a graph and its equation

1 A Caricom task group is planning a meeting in Miami. They need to hire meeting rooms for different events. They get this information from the conference venue.

Hiring costs:

The total cost of a meeting room is $c = 15h + 40$, where c is the total cost and h is the number of hours the rooms is hired for.

Use this graph to calculate costs for different amounts of time.

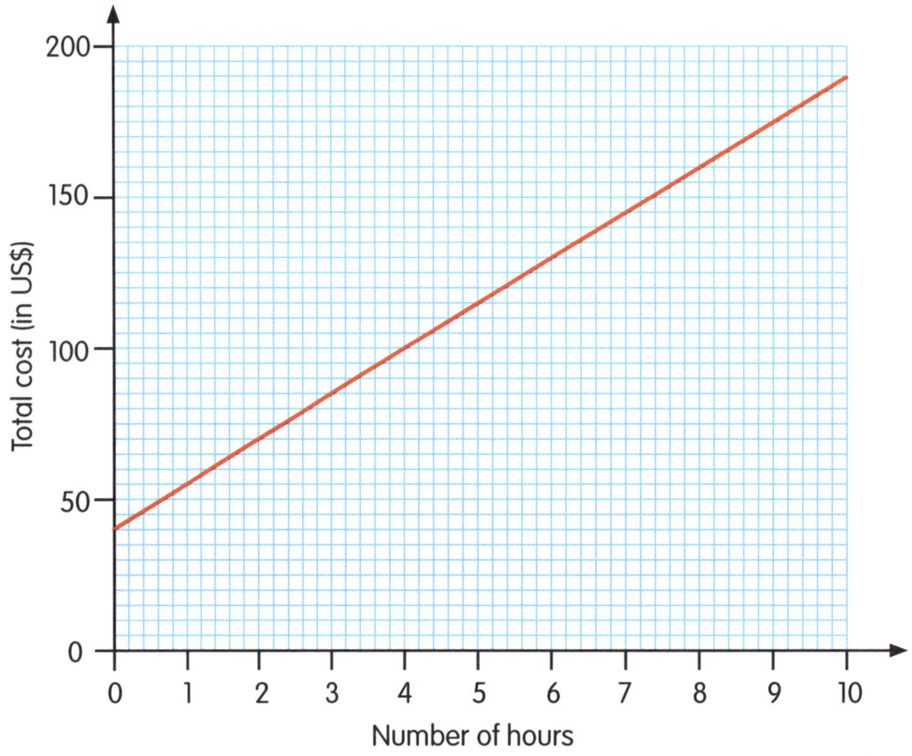

2 Answer these questions about the costs in US dollars.

- **a** What is the cost of hiring a meeting room for 2 hours?
- **b** If the hiring cost is $100, how many hours was the room hired for?
- **c** If $h = 5$, what is c?
- **d** The group needs to hire four rooms, for 5 hours each. What will the total cost be?

3 Jamaica, Barbados and Grenada share the flight costs in US dollars of sending one person to the meeting. Jamaica pays three times as much as Grenada and Barbados pays $800 more than Jamaica. If the total cost is $2 900, how much do they each pay?

Solving word problems with algebra

Key maths idea

Sometimes, we can use algebra to help us solve word problems.
First we need to turn the problem into an equation to be solved.

Example

Jayden and Lamar are brothers.
This formula can be used to calculate Lamar's age, compared to Jayden's age.

$j + 4 = A$

j stands for Jayden's age, A stands for Lamar's age.

When Lamar is 17, how old will Jayden be?

$j + 4 = 17$

So $j = 13$

Jayden will be 13 years old.

1 The distance from Kingston to Constant Springs is twice the distance from Kingston to Half-Way-Tree. If the sum of the two distances is 9 km, what is the distance from Kingston to Half-Way-Tree?

2 Anthony is three times his sister's age. If the total of their ages is 24 years, how old is Anthony?

Let his sister's age $= s$, then write the equation to solve for s.

3 Jordan and Alex divide \$600 between them. If Jordan has twice as much as Alex, how much does Alex have?

4 Caylin's age is three times Andre's age, and half of Iris' age. If Iris is 90 years old, how old is Andre?

5 Peta-Gaye has ten times as many cards as Sophia, and half as many as Alicia. The three girls have 372 cards altogether. Work out how many cards each girl has.

6 a A regular hexagon has a perimeter of 42 cm. If s = the length of one side, work out the value of s for this hexagon.

b A regular octagon has each side length equal to (n cm + 2 cm). If the perimeter is equal to 120 cm, what is the value of n?

Maths detective: Solving unknown quantities

Use what you have learned about equations and about working with money, shapes and measures to solve problems involving unknown quantities.

Case 1

For each problem, write an equation and solve it. Remember to say what any letters you use stand for.

1 On a visit to the Bahamas, a tour group paid B$60 for ferry tickets from Nassau to Paradise Island. The tickets cost B$2 for children and B$4 for adults. If there were 6 fewer children than adults, how many people were in the tour group?

2 Amie had saved $500 and her sister Marie had earned $1 500 at her part-time job. Their aunt gave them each the same amount of money as a gift. If Marie now has double the amount that Amie has, how much did their aunt give them?

3 a How many whole dollar amounts can you find for $a + b = \$6$?

b Which of those values work when $a + b = 6$ and $a - b = 4$?

Case 2

For each problem, draw a sketch if none is given and then write an equation and solve it. Remember to say what any letters you use stand for.

1 Each of these shapes has a perimeter of 48 cm. Work out the lengths of each unknown side.

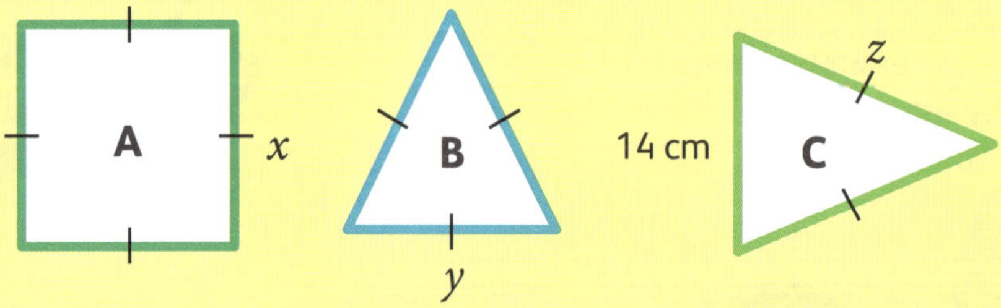

2 In a triangle, the biggest angle is four times the size of the smallest angle. The third angle is 24° bigger than the smallest angle. Work out the size of each angle. (Do you remember how many degrees the angles of a triangle add up to?)

3 A rectangle has a perimeter of 32 cm. The length is three times the width. What is the area of this rectangle?

Case 3

For each problem, write an equation and solve it. Remember to say what any letters you use stand for and include units in your answers.

1 Two students have a combined mass of 88 kg. If one student is 6 kg heavier than the other, what do they each weigh?

2 What is the mass of each yellow block on this balance scale?

3 A contractor uses the formula $n = 4m + 1$ to work out how many planks of wood (n) are needed to build a fence of different lengths in metres (m).

a How many planks of wood are needed for a fence that is 6 m long?

b If the fence is 12.5 m long, how many planks of wood did the contractor use?

c How long is the fence, if the contractor used 113 planks?

What I have learned

For each term below, write a definition in your own words and give an example to show what it means.

Expression | Equation | Substitute

Constant | Variable | Solution

Unknown quantity | Inverse operation | Write an equation

Practice questions

1 Write algebraic expressions to represent these instructions. Use n to represent the unknown number each time.

a Multiply a number by 7 and then subtract 10.

b Add 14 to a number and then divide the total by 2.

c Double a number and then subtract 25.

2 Simplify these expressions.

a $p + 2p + h + 7p + 10h$
b $n + 50 + 2n + 5n$
c $18q - 5q + 4z - z$
d $r \times r \times r \times 10$
e $n \times n \times 2$
f $t \times t \times 8 + 9$

3 When seven times a number m is added to 14, the sum is 84. What is the number?

4 Marcus put a bucket under a dripping tap to save the water. At the end of the first day, there were 3 litres of water in the bucket. Each day after that, he saved q more litres in the bucket. Write an expression to show how many litres he had saved:

a at the end of day 2
b at the end of day 5
c at the end of day 19
d at the end of day m?

5 Use the bar models to help you solve for x.

a

x	18
145	

b

1 000	
x	x

c

x	x	30
180		

d

x	x	x	x
8 400			

6 Solve these equations.

a $2m + 8 = 72$
b $r^2 = 81$
c $5p - 1 = 49$
d $7b + 5 = 40$
e $2n^2 = 200$

23 Self check

Check you can do this:	If you have difficulty, try this:
1 Write algebraic expressions.	Go back to page 351 and make sure you know how to write expressions using letters, numbers and symbols.
2 Construct equations.	Revise page 352.
3 Determine the solution to a simple equation with one variable.	Go over pages 353–355.
4 Solve word problems involving the four arithmetic operations with one- or two-digit numbers.	Go over the work you did on page 357.

Unit 5: Statistics and probability

Chapter 24 Probability

In this chapter you will explore this focus question:

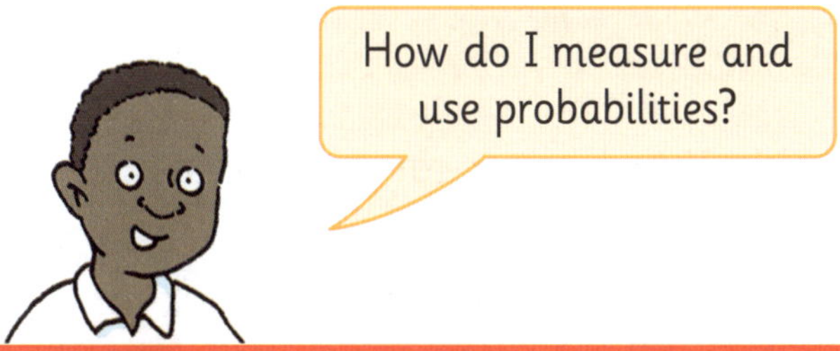

To answer this question, you will:

- understand what we mean by probability
- identify possible outcomes of an experiment such as tossing a coin or rolling a die
- state the probability of a simple event
- calculate a range of probability values
- perform and report on a variety of probability experiments
- draw conclusions from experiments and collected data.

Starting point

How to play:

Both players count to three, then make one of the signs: rock (fist), paper (flat hand), or scissors (two fingers, as shown in the middle here). Rock beats scissors; scissors beats paper; paper beats rock. If you both make the same sign, no one wins.

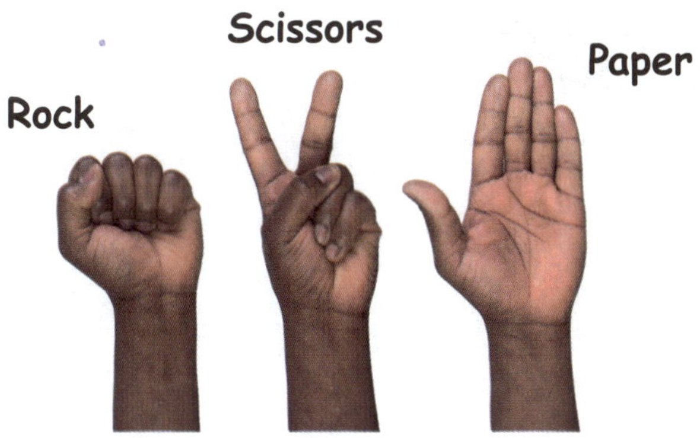

1 Play a few 'best out of 10' rounds of rock-paper-scissors. Keep a note of your results, like this.

Player 1	Player 2
R	P
P	R
R	S
S	P
S	R
R	S
S	P
P	R
R	S
R	P

2 Jacob draws this table to show the different possible outcomes for the game.

	R	P	S
R	RR	PR	SR
P	RP	PP	SP
S	RS	PS	SS

a How many ways are there for no one to win a point?

b How many ways has player 1 got for winning a point?

c How many ways has player 2 got for winning a point?

d Do you think the game is fair or unfair? Why?

3 Compare your results to the results in the table. Which result came up more often than others?

4 Which of these is the most likely result of ten rounds? Explain why you think so.

a Player 1 wins all the rounds.

b Player 2 wins all the rounds.

c Player 1 and Player 2 each win around half of the rounds.

Probability concepts

Key maths idea

Imagine that you pulled a marble at random from a bag containing mostly blue marbles and a few yellow marbles. It would be **possible** to pull out a blue marble or a yellow marble, but you would quite probably pull out a blue one. Now imagine that the bag contained only blue marbles. Then you would be **certain** to pull out a blue marble. It would be **impossible** to get a yellow. This is the language of **probability** – the area of mathematics that helps us to calculate how **likely** or **unlikely** it is that something will happen.

Key words

possible
certain
impossible
probability
likely
unlikely

1 Read each statement. Discuss with a partner whether each statement is impossible, possible or certain.

- **a** If I measure a right angle with my protractor, the measurement will be 45°.
- **b** I will catch a cold next winter.
- **c** Next week will contain 10 days.
- **d** Next January will have 31 days.
- **e** The sun will rise tomorrow.
- **f** I will hear dogs barking tonight at 7 o'clock.
- **g** I will get a pet dinosaur for my next birthday.
- **h** Your teacher will give you a problem-solving challenge in your next maths class.
- **i** It will snow in Jamaica next week.

2 **a** Write two of your own sentences using the words 'impossible' or 'certain'.

- **b** Share them with the class.
- **c** Discuss: Were there any sentences that you disagreed about? Did you need to use any other probability words in your discussion?

3 Write your own definitions of:

a possible **b** likely **c** unlikely **d** impossible

The probability scale

Key maths idea

A **probability scale** looks like a number line. The lowest value on the scale is 0, because when an event is impossible, there is 0 chance of it happening. The highest value on the scale is 1. When an event is certain, there is a 100% chance of it happening.

All other probabilities lie between 0 and 1. We express them as fractions or percentages.

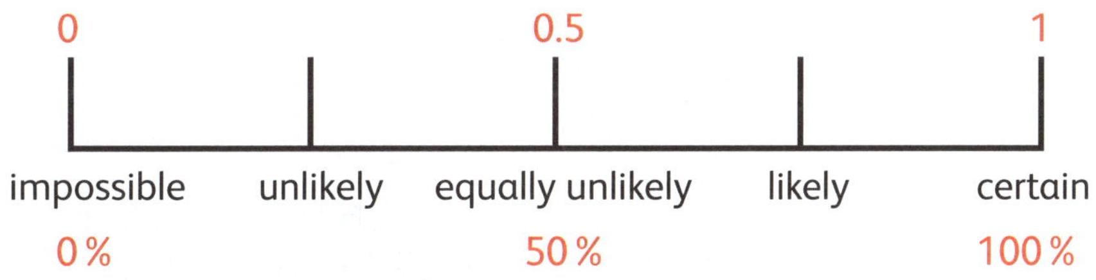

Key words

probability scale
equally likely
even chance

Remember: a percentage is a fraction out of 100. $\frac{100}{100} = 1$.

Think about what happens when you toss a coin. It could either land heads up, or tails up. We say there is an **equally likely** chance of these two possibilities.

We can express two equally likely possibilities in a few ways. We can say:

- **even chance**
- a chance of 1 in 2
- a 0.5 chance
- a 50 % chance

The closer to 1 a probability is, the more likely it is to happen.

1 For each sentence, describe the probability using the words likely, unlikely, certain, impossible, possible. Show a partner where you would put this event on the probability scale.

- **a** If today is Wednesday, tomorrow will be Thursday.
- **b** If today is Sunday, tomorrow will be Wednesday.
- **c** If I ran a sprint against Usain Bolt, he would win.
- **d** The sun will set at 11:00 a.m. tomorrow.
- **e** If I flip a coin, it will land on heads.
- **f** If today is rainy weather, tomorrow will also be rainy.

2 Draw a probability scale like the one on the previous page. Think about the outcome when you roll a fair six-sided die. Write the letter in the correct position on your probability scale.

- **a** an even number facing up
- **b** the number facing up is greater than 0 and less than 7
- **c** the number facing up is greater than 1
- **d** the number is greater than 6

Possible outcomes

Key maths idea

Some **events** have a fixed number of possible **outcomes**.
For example, flipping a coin is an event that has two possible outcomes – the coin could land on heads (H), or on tails (T).

Now think about the possible outcomes of flipping a coin twice.
We can show the possible outcomes in a two-way table, like the one below.

		Coin 1	
		Heads	**Tails**
Coin 2	**Heads**	HH	HT
	Tails	TH	TT

You can see that there are four possible outcomes in this situation: HH, HT, TH and TT. Because the outcomes on each coin are equally likely, the four combined outcomes are also equally likely.

Key words
- event
- outcome

1 Mikayla visits a restaurant with her friends. This is the menu.

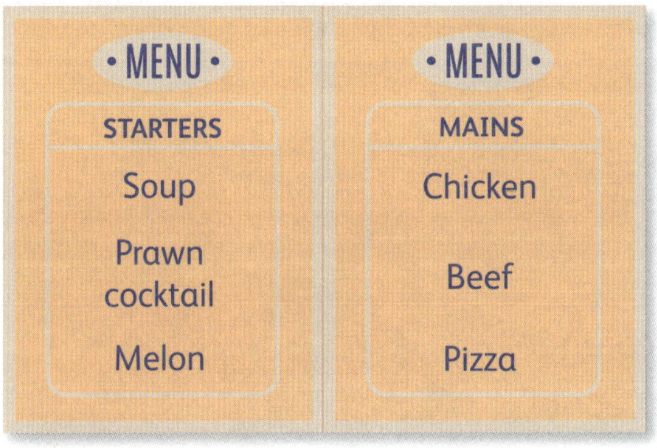

Mikayla chooses one starter and one main.
Work out all the possible combinations Mikayla could choose.

2 Anthony wants to choose one drink and one snack from a menu.

Work out all the possible combinations Anthony could choose.

3 Jayden spins two spinners numbered 1 to 5. He records the sum of the values each spinner lands on.

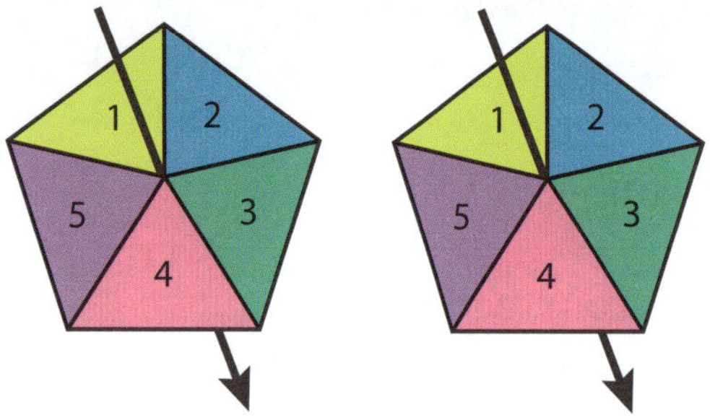

Copy the table. Then list all the possible outcomes that Jayden could have by completing the table. One outcome has been done for you.

		Spinner 1				
		1	**2**	**3**	**4**	**5**
Spinner 2	**1**					
	2		4			
	3					
	4					
	5					

Calculating probability

Key maths idea

If an event is possible but not certain, you can express the probability as a fraction – a number between 0 and 1. You calculate the probability as a fraction.

Example 1

What is the probability of throwing a 4 when you roll a die?

There are six possible outcomes when you roll a die. It could land on 1, 2, 3, 4, 5 or 6. There is an equal chance of each of the six outcomes.

Rolling a 4 is one of these six outcomes.
So the probability of throwing a 4 is 1 out of 6, or $\frac{1}{6}$.
We write this as $P(4) = \frac{1}{6}$.

Example 2

What is the probability of the spinner landing on orange?

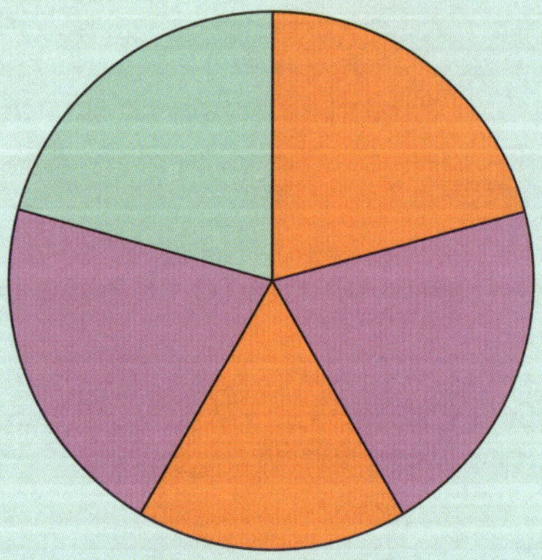

This spinner has five areas of equal size. It has an equal chance of landing on any of the five areas.
Orange takes up 2 of the 5 areas, so:
P(landing on orange) = $\frac{2}{5}$ or 40 %

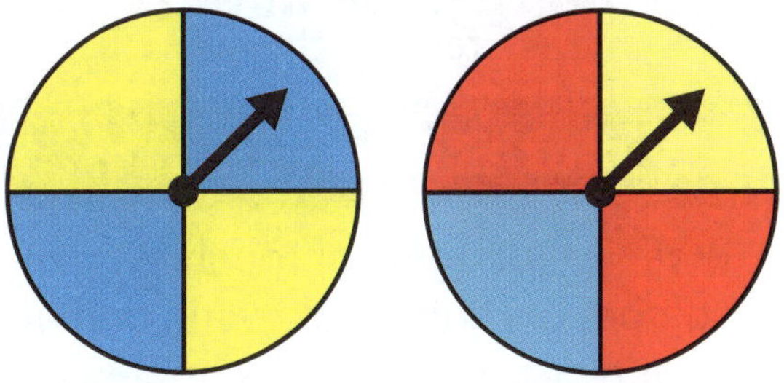

1 Look at spinner A.
Express as a fraction the probability of landing on:
- a blue
- b yellow
- c any other colour

2 Look at spinner B.
- a How many possible outcomes are there?
- b List the possible outcomes and express the probability of each as a fraction.

3 Together with a partner, use a real pack of 52 playing cards to help you answer these questions. Make sure to remove any jokers.
If someone is drawing a card at random from a full, shuffled pack, what is the probability of drawing:

- a a red card
- b a 7
- c a queen of hearts
- d a diamond
- e a number card with the number 2, 3, 4, 5, 6, 7, 8, 9 or 10
- f an ace
- g a picture card (jack, queen or king)?

4 Write out your first name and your surname. Count the number of letters in each name, and how many vowels and consonants. Express these events as probabilities:
- a If you pick a letter at random from your first name, what is the probability it will be a vowel?
- b If you pick a letter at random from your surname, what is the probability that it will be a consonant?

Express the probabilities as fractions, decimals and percentages.

In pairs or groups, check each other's calculations.

5 In small groups, develop a song or poem about probability. Use some of the key words you have learned in this chapter.

Probability experiments

Key maths idea

When we perform a probability experiment, we carry out a number of trials of an event, and measure how many times each result occurs. For example, on this spinner, there are four different colours. The probability of landing on each of the four colours is 1 in 4, or 25%. This is called the **theoretical probability**.

However, this table shows the results of 20 spins:

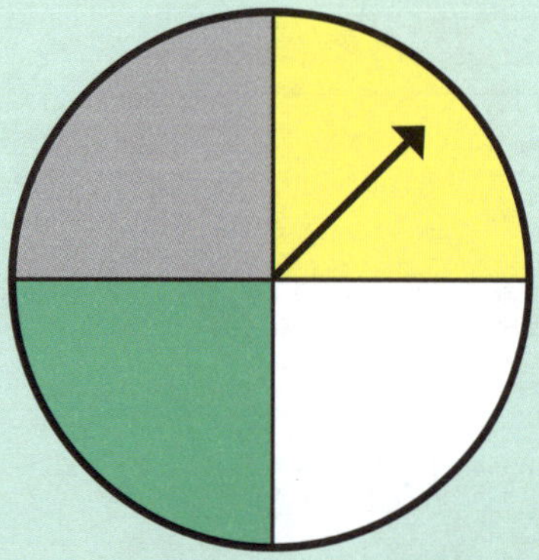

Green	////
White	𝍸 /
Grey	///
Yellow	𝍸 //

In this experiment, $\frac{7}{20}$ of the spins landed on yellow, $\frac{3}{20}$ on grey, $\frac{6}{20}$ on white and $\frac{4}{20}$ on green. These fractions show us the **experimental probability**, or what actually happened.

Key words
theoretical probability
experimental probability

Remember: the experimental probability is what actually happened.

1 **a** In pairs, design a probability experiment where you have an equal chance of each outcome. You can decide how many possible outcomes there are, but it should be more than 1 and fewer than 6.

b Carry out 20 trials. Record your results in a tally table.

c Write the experimental probability for each colour as a fraction, and as a percentage.

2 Danny flips a coin 4 times, and says afterwards that he had a 100% probability of landing on heads.

a What happened in his experiment?

b If he repeats the experiment, does he have 100% chance of landing on heads again? Why or why not?

For this activity, you will need a cloth bag that you cannot see through, and some marbles of different colours. You are going to explore the experimental probability of drawing a marble of each colour from the bag.

Real-world maths: Grocery store choices

A large grocery store kept track of popular items scanned by an express checkout cashier.

1 These are the results.

Item	Number sold
Bread (loaves)	750
Water (bottled)	280
Soda (cans)	150
Chocolate (milk)	100
Chocolate (dark)	120
Chips (packets)	250
Patties	200
Milk (litres)	75
Cookies (bag)	75

a What is the total number of items?

b What is the probability of choosing bread if you pick an item at random?

c Which items have the same probability of being picked?

d Is it correct to say that the probability of choosing chocolate is 11 %? Explain your answer.

e If you remove bread from the list, what is the probability of choosing chips if you pick an item at random? Give your answer as a decimal.

f Which items are you less likely to pick at random – chips or chocolate? Why?

2 Draw a probability scale 20 cm long in your book. Mark and label the probability of choosing each item on the list at random.

Maths detective: Exploring probability

Use your vocabulary skills and what you know about probability to solve these problems.

Case 1

Choose words from the box below to describe the probability that the next person to enter your classroom will be:

1 bald
2 wearing glasses
3 male
4 female
5 wearing shoes
6 someone you know
7 Jamaican
8 a teacher
9 a firefighter
10 the school principal
11 an alien from outer space

certain	likely	unlikely
very unlikely	impossible	
very likely	even chance	possible
impossible	probable	

Talk about

Compare your answers with others in your group.
Be prepared to explain your choices.

Case 2

How many possible outcomes are there in each of these situations?

1 Choosing a letter from the word JAMAICA
2 Tossing a coin
3 Choosing a day of the week at random
4 Tossing two coins at the same time
5 Rolling a die and tossing a coin
6 A student from your class is chosen to speak at assembly

Case 3

The quality control department of a large factory tested 500 packs of batteries and found that 4 packs were sub-standard while all the rest were perfect.
What is the probability that batteries produced in this factory would be:

1 substandard
2 perfect?

Case 4

Marsha read on a hotel website that May was a good time to visit Jamaica as rain was unlikely and sunny weather was highly likely.

She found the following weather information on the internet:

Average weather in May		
Sunny days	Cloudy days	Rainy days
13	10	8

Was the hotel website correct? Give reasons to support your answer.

What I have learned

Write short notes to answer these questions and summarise what you learned in this topic.

What words do we use to describe probability?

What are possible outcomes?

How many numerical ways can you represent a probability of 75%?

How do you calculate the probability of getting an even number when you roll a six-sided die?

What is a probability experiment?

Practice questions

1 Describe each statement using one of these terms: impossible, unlikely, equal chance, likely, certain.

a The sun will rise tomorrow.

b There will be snow outside the classroom tomorrow.

c It will rain sometime tomorrow.

d There will be six moons in the sky tonight.

e You will go to bed after midnight tonight.

f You will do some homework this afternoon.

g One hour after 5 o'clock will be 6 o'clock.

2 Complete each sentence in your own words.

a There is a 50 % chance that …

b There is 0 chance of …

c There is 100 % chance that …

3 Milla throws a standard six-sided die. What is the probability that it lands on a prime number?

4 Joshua has an 8-sided spinner.
Use the words impossible, unlikely, even chance, likely or certain to describe the probability of each outcome on a given spin:

a a number less than 10
b the number 11
c the same number three times in a row
d an odd number.

5 Draw a spinner of your own. Then draw a probability scale, and mark three different positions on the scale. Match each position to an outcome on your spinner.

24 Self check

Check you can do this:	If you have difficulty, try this:
1 Use and understand terms we use to describe probability.	Revise the words you learned on pages 362–364.
2 Identify the possible outcomes of an event.	Revise page 365.
3 Express probability as a fraction or percentage.	Revise page 366.
4 Formulate and carry out probability experiments, and draw conclusions from the results.	Go over page 368. Remember, an experiment is where you try out actually making an event happen and record the actual outcomes.
5 Work out the probability of something happening in real life.	Look back at the real-life problems you solved on pages 369–371.

Practice test

These questions will help you prepare for the multiple-choice items on your curriculum-based test.

Instructions

1 The test consists of 60 multiple-choice items. Each question has four possible answers (A, B, C and D). You have to choose the correct answer.

2 Your teacher may give you an answer sheet. If so, circle the letter of the correct answer for each item. If not, write the question number and the letter of the correct answer in your book like this:

1 B

2 D

3 You will need some paper for working.

4 Read all questions carefully to make sure you understand what is being asked.

Answer all questions.

1 What is the sum of 245, 765 and 90?

a 11 000 **b** 110 **c** 1 100 **d** 1 090

2 What is the difference between 896 and 498?

a 1 394 **b** 398 **c** 406 **d** 392

3 The product of 1 027 and 8 is:

a 1 035 **b** 8 216 **c** 128.4 **d** 7 196

4 16 499 ÷ 7 is

a 16 492 **b** 16 506 **c** 2 357 **d** 1 649.2

5 $\frac{1}{2} + \frac{5}{6} + \frac{2}{3} + \frac{1}{4} =$

a $\frac{9}{6}$ **b** $\frac{1\,521}{2\,634}$ **c** $\frac{9}{4}$ **d** $\frac{29}{12}$

6 Which fraction is the greatest?

a $\frac{13}{20}$ **b** $\frac{7}{10}$ **c** $\frac{1}{2}$ **d** $\frac{4}{5}$

7 What is $\frac{5}{8}$ as a percentage?

a 0.625 % **b** 62.5 % **c** 58 % **d** 6.25 %

8 0.035 as a percentage is:

a 35 % **b** 3.5 % **c** 0.35 % **d** 0.00035 %

9 A car travels at a speed of 60 km/h. What distance will it cover in 2.5 hours?

a 24 km **b** 150 km **c** 175 km **d** 1 500 km

10 Which of these ratios is **not** equivalent to the other three?

a 15 : 10 **b** 3 : 2 **c** 1 : 2 **d** 6 : 4

11 What is the missing number in the sequence 81, 64, _____, 36, 25?

a 48 **b** 47 **c** 49 **d** 45

Use this Venn diagram for questions 12, 13 and 14.

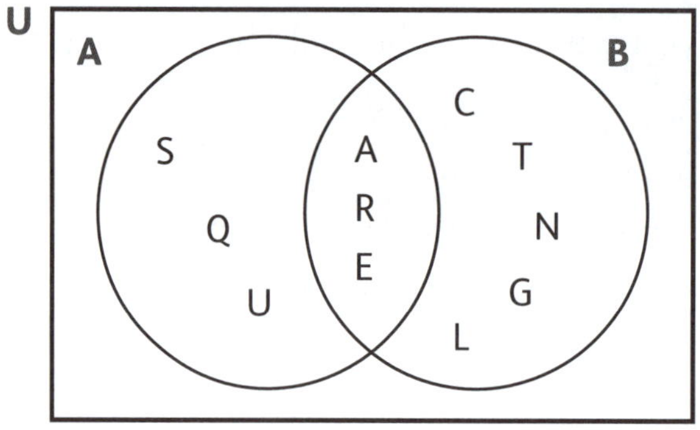

12 The elements of set A are:

a {S, Q, U}

b {C, T, N, G, L}

c {S, Q, U, A, R, E}

d {R, E, C, T, A, N, G, L, E}

13 $A \cap B =$

a {S, Q, U, A, R, E}

b {C, T, N, G, L}

c {S, Q, U, A, R, E, C, T, N, G, L}

d {A, R, E}

14 $A \cup B =$

a {S, Q, U, A, R, E}

b {S, Q, U, A, R, E, C, T, N, G, L}

c {S, Q, U, A, R, E, R, E, C, T, A, N, G, L, E}

d {A, R, E}

15 A flight left the airport at 07:55 and landed at its destination at 14:45. How long was the flight?

a 7 hours **b** 6 hours 50 minutes **c** $7\frac{1}{4}$ hours **d** 22 hours 40 minutes

Use this scale diagram of the floor of a community centre for questions 16, 17 and 18.

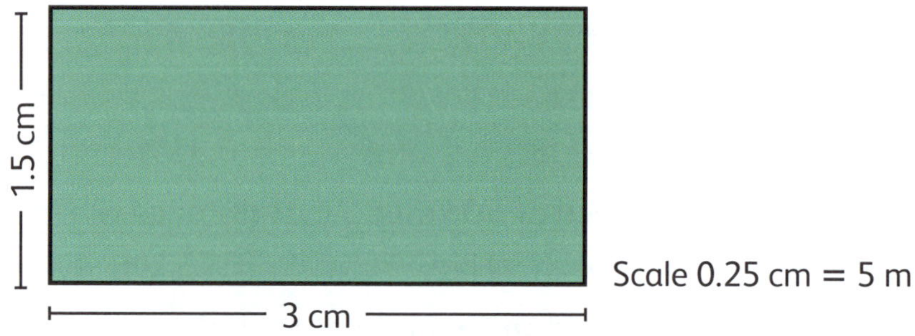

16 The width on the diagram is 1.5 cm. What is the width in real life?

a 15 m **b** 40 m **c** 30 m **d** 18 m

17 The perimeter of the scaled rectangle is:

a 4.5 cm **b** 4.5 cm **c** 9 cm **d** 4.5 cm^2

18 Expressed as a ratio in the form 1 : *x*, the scale would be:

a 1 : 2000 **b** 1 : 1250 **c** 1: 5250 **d** 1: 1000

19 Which is the most accurate description of a square?

a A quadrilateral with right angles

b A quadrilateral with equal sides

c A quadrilateral with equal sides and equal angles

d A quadrilateral with parallel and equal sides

20 Which of these solids is a regular polyhedron?

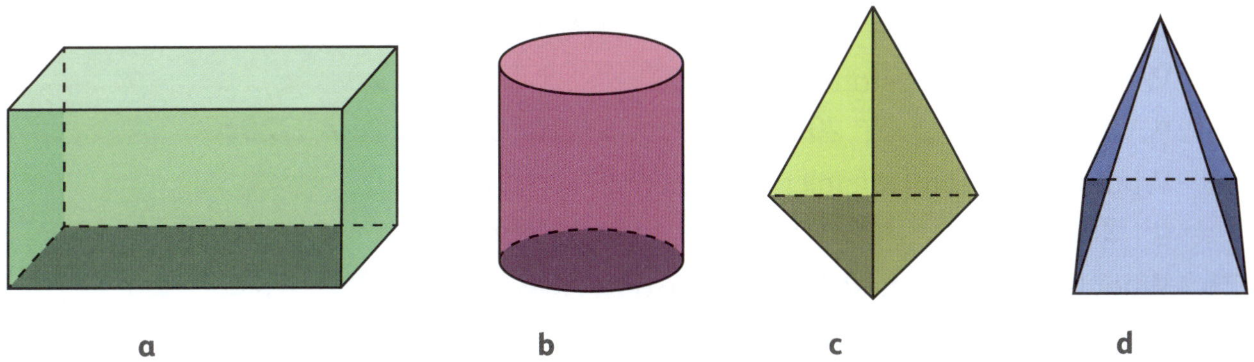

Use this diagram of a cube for questions 21, 22 and 23.

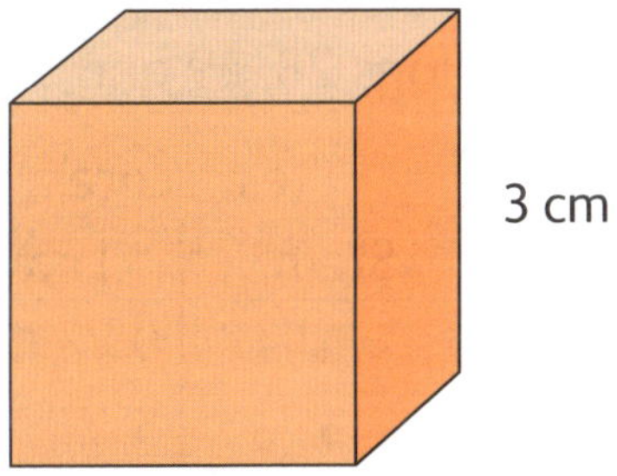

21 How many edges does a cube have?

a 8 **b** 9 **c** 12 **d** None

22 What is the volume of the cube shown in the diagram?

a 9 cm^3 **b** 24 cm^3 **c** 27 cm^3 **d** 30 cm^3

23 Which of these calculations will give you the total surface area of the cube?

a 9 cm × 3 cm

b 3 cm × 3 cm × 3 cm

c 3 cm × 24 cm

d 3 cm × 18 cm

Use this graph for questions 24, 25, 26 and 27.

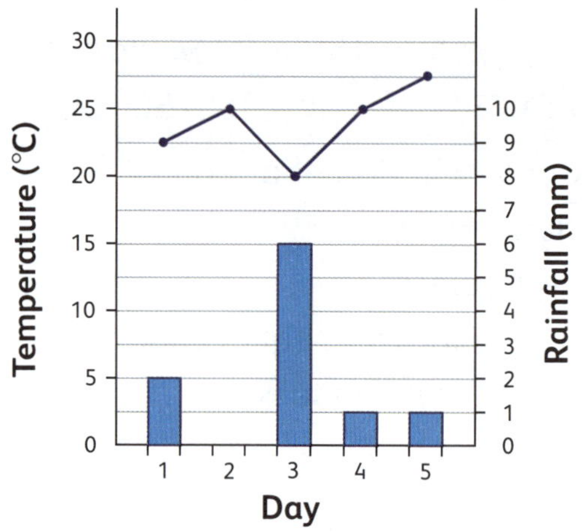

24 What was the temperature on day 2?

a 0 °C **b** 20 °C **c** 10 °C **d** 25 °C

25 What was the total rainfall for this period?

a 25 mm **b** 10 mm **c** 15 mm **d** 120 mm

26 What is the mean rainfall for the five days?

a 6 mm **b** 1 mm **c** 2 mm **d** 24 mm

27 What was the range of temperature in this time?

a 7.5 °C **b** 6 °C **c** 5.5 °C **d** 20 °C

Use this stem and leaf plot for questions 28, 29 and 30.

Test scores

Stem	Leaves
4	9
5	5, 7
6	6, 6, 8
7	2, 8, 8, 8
8	4, 5, 7, 8, 8
9	1, 5, 5
10	0, 0

Key: 4|9 = 49

28 How many students scored between 80 and 90 marks?

a 845 788 **b** 6 **c** 5 **d** 8

29 What was the lowest mark scored?

a 49 **b** 40 **c** 100 **d** 9

30 Which test score was the most common (the mode)?

a 100 **b** 66 **c** 78 **d** 88

31 What is the area of this plot of land?

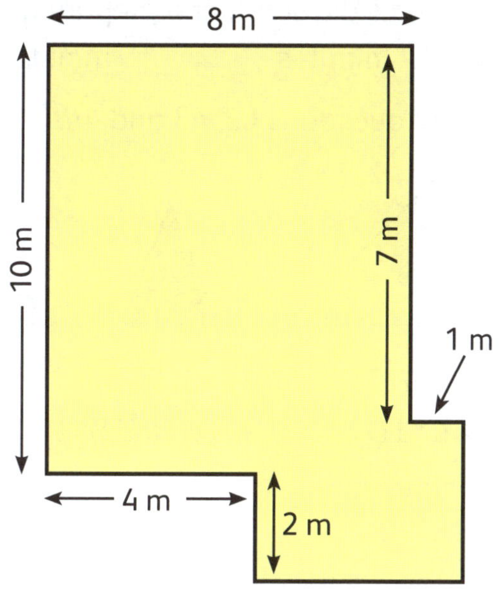

a $42\ m^2$ b $84\ m^2$ c $93\ m^2$ d Impossible to work out

32 A square has an area of $169\ m^2$. What is the length of each side?

a 14 m b 13 m c 42.25 m d 676 m

33 Which of these is **not** a prime number?

a 47 b 53 c 57 d 59

34 What is the HCF of 15, 20 and 30?

a 5 b 10 c 15 d 20

35 The LCM of 42 and 70 is:

a 2 946 b 7 c 210 d 112

36 The product of the prime factors of 72 is:

a $2^3 \times 3^2$ b 8×9 c $4 \times 2 \times 3 \times 3$ d $2^2 \times 3^3$

37 What is the sum of $\frac{3}{8} + \frac{3}{6} + \frac{3}{12}$?

a $9\frac{1}{8}$ b $\frac{9}{26}$ c $1\frac{1}{8}$ d $\frac{1}{8}$

38 Josh had $4\frac{3}{4}$ m of wire and he used $3\frac{2}{3}$ m to build a model. How much wire is left over?

a $1\frac{1}{3}$ m b $1\frac{1}{12}$ m c $\frac{1}{12}$ m d 1 m

39 A recipe calls for $1\frac{2}{3}$ cups of sugar for 12 brownies. If Junior only wants to bake 6 brownies, how much sugar will he need?

a less than $\frac{1}{2}$ cup b between $\frac{1}{2}$ cup and $\frac{3}{4}$ cup

c Between $\frac{3}{4}$ and 1 cup d more than 1 cup

40 What is 5 divided by $\frac{5}{7}$?

a 5 b 7 c 12 d 14

41 It took Marcia $\frac{3}{4}$ of an hour to make dinner. $\frac{1}{3}$ of that time was spent preparing the ingredients. How many minutes did she spend preparing the ingredients?

a $\frac{1}{4}$ of a minute **b** 12 minutes **c** 15 minutes **d** 20 minutes

Use this set of shapes to answer questions 42, 43 and 44.

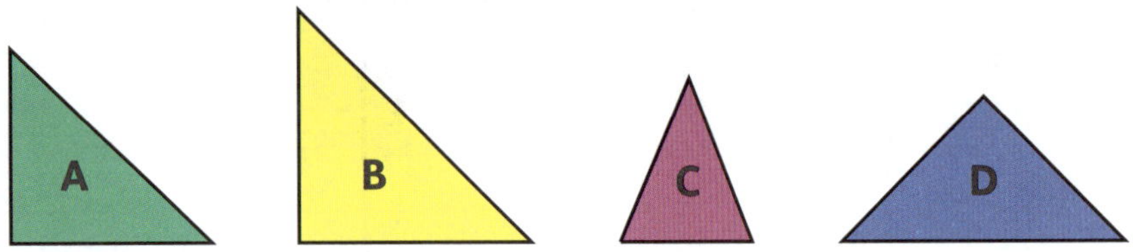

42 Which two shapes are similar?

a C and D **b** A and D **c** A and B **d** None of them

43 Which two shapes are congruent?

a A and B **b** A and C **c** A and D **d** None of them

44 Triangle C is an isosceles triangle. The sum of its angles can be expressed as $x + 2x + 2x$. What is the size of each of the equal angles?

a 30° **b** 45° **c** 60° **d** 72°

45 How many lines of symmetry does this polygon have?

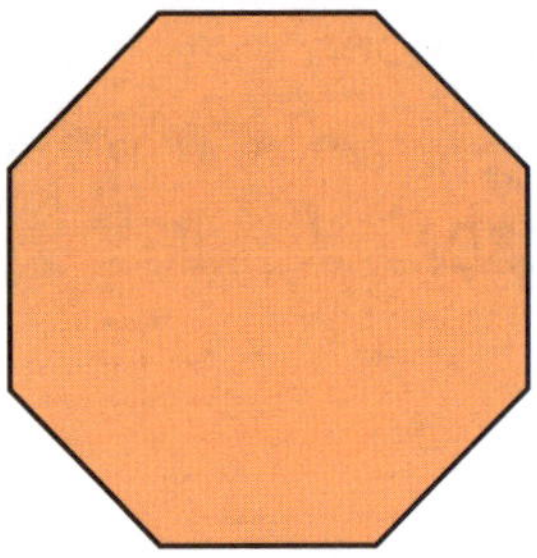

a 2 **b** 4 **c** 6 **d** 8

Use this diagram for questions 46, 47, 48 and 49.

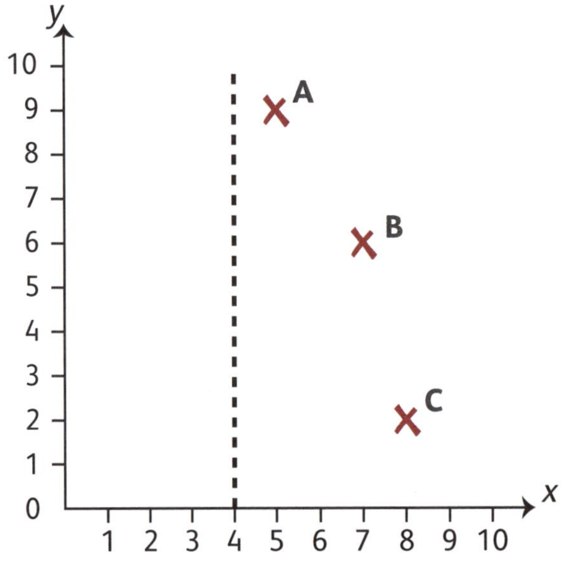

46 Which pair of coordinates represents point B?

a (6, 7) **b** (7, 6) **c** (3, 6) **d** (8, 2)

47 Point A is reflected about the dotted line. The coordinates of the image will be:

a (5, 9) **b** (9, 3) **c** (3, 9) **d** (4, 9)

48 Point D is plotted at (8, 9) and points A, C and D are joined to make a triangle. What type of triangle is formed?

a right-angled scalene **b** right-angled isosceles

c equilateral **d** none of these

49 Which are the correct coordinates for the reflection of point C in the mirror line?

a (8, 2) **b** (2, 8) **c** (0, 2) **d** (2, 0)

50 If $2x + 9 = 19$, x must be:

a 10 **b** 5 **c** 14 **d** 28

51 Paul had \$480 and Mirri had 25 % more. How much did Mirri have?

a \$605 **b** \$600 **c** \$60 **d** \$125

52 A school sold 300 tickets for a concert. Tickets cost \$400 or \$250. They sold half as many \$400 tickets as \$250 tickets. How much money did they make on the \$250 tickets?

a \$40 000 **b** \$90 000 **c** \$50 000 **d** \$80 000

53 If $x = 4$, $y = 5$ and $z = 8$, the value of $\frac{(2x+ 3z)}{2y}$ is:

a 320 **b** $3\frac{1}{5}$ **c** $\frac{32}{5}$ **d** $6\frac{1}{5}$

54 Femi can type 200 words in *t* minutes. Which expression shows how many words he can type in 1 minute?

a $\frac{200}{t}$ **b** $t - 200$ **c** $\frac{t}{200}$ **d** $200 + t$

55 Ms Thompson had a balance of \$45 800 dollars in her savings account. She withdrew half the money to invest in a new business. The business was so successful that she earned a dividend of 15 % on her investment. She deposited this amount into her savings account. What is her balance after this?

a \$22 900 **b** \$3 435 **c** \$26 335 **d** \$52 670

56 The diagram shows how \$120 000 was spent to renovate a library.

New shelving	Painting	Furniture	Doors
51 %	25 %	14 %	10 %

What was the amount spent on shelving?

a \$60 000 **b** \$58 800 **c** \$612 000 **d** \$61 200

57 A circle has a diameter of 14 cm. Which expression can you use to work out the circumference?

a $\pi \times 49$ cm **b** $\pi \times 14$ cm **c** $\pi \times 7$ cm **d** $\pi \times 3.14$ cm

58 What are the sizes of the angles in this diagram?

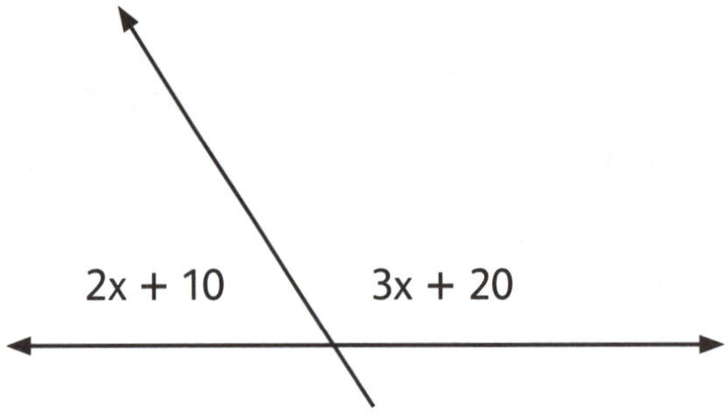

a 60° and 120° **b** 70° and 110° **c** 30° and 150° **d** 90° and 90°

59 Shania has four white T-shirts and six red T-shirts. If she picks a T-shirt without looking, what is the probability that it will be white?

a 10 % **b** 25 % **c** 40 % **d** 60 %

60 Which statement is true for this spinner?

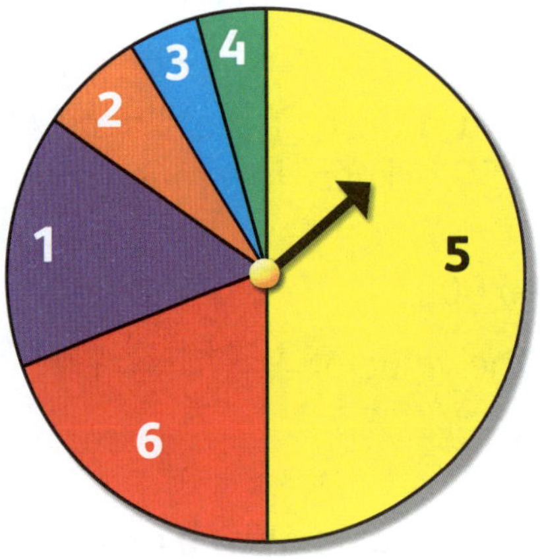

a It is certain that the arrow will land on 5.

b It is more likely that the arrow will land on 6 than on 5.

c It is impossible for the arrow to land on 3.

d It is more likely for the arrow to land on 1 than on 2.

Glossary and Index

12-hour notation: a time measurement system that divides the day into two 12-hour periods 56

24-hour notation: a time measurement system with which you can tell the time using all 24 hours of the day 56

2-D shape: 2-dimensional shape, a shape that has only two dimensions we can measure: length and width 66

3-D solid: 3-dimensional solid, a solid object that has three dimensions: length, width and height 79

A

a.m.: the hours from 12 o'clock at night until 12 o'clock in the day 56

abscissa: the distance from a point to the vertical axis measured along the horizontal axis 251

acute: an angle that sums up to less than 90° 339

adjacent: very near, next to or touching one another 78

adjacent angles: two angles that are set on the opposite sides of a common arm and share the same vertex 341

algebra: a type of mathematics where we use symbols and letters to represent unknown values, quantities or amounts 264

algebraic expression: a mathematical expression where a combination of symbols (letters and numbers) are used to represent unknown values, quantities or amounts 263, 299, 307, 351

alternate (angles): two angles at opposite ends and on opposite sides of a transversal that crosses two lines 343

angle: the space between two intersecting lines or surfaces to the point where they join, measured in degrees 103, 339

apex: the top or the highest part of an object or shape 79

arc: the shape of a part of the line which forms the outside of a circle or other curved line 314

area: the space covered by a 2-dimensional shape on a flat surface 211, 309

arm (of an angle): either of the two lines that form the angle 339

average speed (rate of travel): how fast someone or something is moving during a specific amount of time, calculated by dividing the overall distance travelled and the overall time spent travelling 327

axes (sing: axis): fixed lines that show the position of a point on a graph 95

B

Bank of Jamaica: the central bank of Jamaica, located in Kingston 283

bar graph: a type of diagram where different amounts are represented by vertical or horizontal rectangles of different heights but equal widths 95

bar model: an image that represents a mathematical problem or concept in which bars or boxes stand for known and unknown quantities 302

base: the number that is multiplied by an exponent 31, 273

blueprint: an architectural drawing or plan of a building or a space, containing dimensions and the information you need in order to make or build something 48

BOMDAS: an acronym which helps you remember the order of operations in mathematics, that is, Brackets, Of, Multiplication and Division, Addition and Subtraction 265

building society: a business or organisation that lends money to people in order to buy a house or pays interest to people who invest their money with them 283

C

Cartesian plane: an imaginary surface, formed by a horizontal and a vertical axis which intersect at a point and divide an area into four quadrants 251

central bank: the national bank of a country that offers services to a government, issues and controls the amount of money and cost of credit in a national economy 283

centre: the middle point of an object, shape or area 314

certain: something that is sure to happen .. 362

chord: a straight line that divides a circle into two distinct segments 314

circle: a closed plane shape with its points always having the same distance from its centre 103, 314

circle graph: a circular chart divided into different segments that represent different quantities or percentages of a total 95, 103

circumference: the distance around the edge of a circle, the perimeter of the circle 103, 314

class interval: the range of values grouped together in a category 100

closed sentence: a mathematical statement or equation that is always true or always false 268

coefficient: a value that multiplies another value 264

collect data: to gather information for a specific research purpose 107

commercial bank: a bank with different branches that provides services to the general public and businesses 283

common denominator: a number that is divided exactly by all the denominators, that is, the numbers below the line, in a set of fractions 134

common difference: in an arithmetic sequence, it is the constant difference between consecutive terms 272

common factors: whole numbers that when multiplied together they produce another whole number 134

common multiple: a whole number that can be divided exactly by two or more different whole numbers 25

compare: look for and identify differences between two or more things 152, 175

complementary angles: two angles that when added together their sum is 90° 341

composite number: a whole number that has more than two factors, that is, it can be divided by numbers other than itself and number 1 .... 25, 122

compound shape: a shape that is formed by combining two or more basic shapes 244

congruent (shapes): two or more shapes having corresponding sides of equal length and corresponding angles of equal size 241

consecutive terms: numbers that follow one another without an interruption 272

constant: a value that multiplies another value 264, 352

continuous data: numerical information can take any value within a range 100

convention: an agreed upon rule that is followed by all people 265

convert: to change one thing into a different form or properties 42, 152, 181, 302

coordinates: pair of numbers and/or letters that are used to show the exact location on a map 251

corresponding: connected with or equivalent to something; two angles in matching corners of a transversal crossing two other lines 241, 343

credit union: a financial organisation that offers low-interest loans to its members 83

cube: an object which has six square surfaces of equal size 79, 229

cubic unit: a unit that measures volume, that is, the space that is occupied by a solid 227

cuboid: an object which has six rectangular surfaces 79, 229

curly brackets (braces): braces that are used to enclose words, numbers or lines so that are considered together 9

curved surface: the rounded surface of a shape or object 79

D

data: facts or information 92

data handling: the process of gathering, recording and presenting information 92

decimal: a number that is expressed with a counting system that is based on number ten 302

decimal fraction: a number that is expressed with a counting system that is based on number ten 152

decimal places: the position of a number following a decimal point 160

decreasing: becoming less or fewer 272

degree: represented by the sign °, it is the 360th part of a circle's circumference 339

denominator: in fractions, it is the number below the line 138

deposit (slip): a record of how much money you are paying into your account 286

design: draw or make a plan for something.. 82

diameter: the length of a straight line that cuts the circle exactly in half and goes through the centre 314

digit: any number from 0 to 9 27

direct observation: watching and recording an event that you see or how people behave 93

discrete data: information that takes certain values, usually whole numbers 100

disjoint sets: two sets that do not have common members 18

distance: the space between two places or things; in mathematics, the area that is measured in a straight line 41, 327

distribution: the manner in which something is spread in a place or to a group of people; in mathematics, the times each value of a variable is found in a data set 105

divide: split something up into two or more parts; splitting a number up into equal parts 152

division: one of the four basic operations in mathematics where an amount is divided into equal-sized parts 165

divisor: a number by which another number is divided 167

dodecahedron: a solid shape or object that has twelve flat sides 81

E

edge: the outer point of a solid where its faces or surfaces meet 79, 228

element: any of the distinct objects that comprise one set 9

environment: the natural world where we live along with animals and plants 82

equal sets: two or more sets that have the exact same members or elements 11

equally likely: having the same possibilities of happening 364

equation: a mathematical statement in which you show that two amounts are equal using mathematical symbols....... 263, 352

equilateral (triangle): a triangle with all three sides equal in length 68

equivalent: having the same amount, value or quantities 167

equivalent fractions: fractions that have different numerators and denominators but they have the same value 134, 153, 191

equivalent ratios: two or more ratios that are a multiple of one another 177, 179

equivalent sets: sets that have the same number of members 11

estimate (v): make an educated guess that is close to the real answer 24, 208

estimate (n): a rough calculation of the approximate value, number or quantity of something 160

even chance: having an equal probability of happening 364

event: an action or situation 365

expanded form: a way to write a number as a sum of the value of the digit in each place 28

experimental probability: a probable result that is calculated by conducting a series of experiments and observing the outcomes ...368

exponent: a number showing how many times the base is multiplied by itself 31, 124, 273

exponential form: a way to write a number that is multiplied more than one times by itself 31, 124

expression: a sentence of letters and/or numbers linked by an operations symbol...... 270

exterior angles: the angles on the outside of a shape 341

F

face: the flat side of a solid shape or object 79, 228

factor: a whole number that divides another number an exact number of times without leaving a remainder........... 25, 124

finance (v): provide or lend money for something 288

finances: the financial conditions and money available to a person 287

finite set: a set with limited members 12

flip: place a shape across a mirror line........... 254

formula: a mathematical rule or principle, expressed in algebraic symbols 69, 211

line segment: a line that forms one side (arm) of an angle 339

loan: an amount of money a person borrows and is expected to pay back at an agreed upon time 283

loss: the money you lose if you sell a product or service for less that its cost 291

lowest common denominator (LCD): the smallest number that can be a multiple of the denominator of two or more fractions ... 134

lowest common multiple (LCM): the smallest number that is exactly divisible by each number in a specific set of numbers 25, 127, 134

M

map: a drawing or a plan of the earth's surface or a building, showing the position and shape of different features as you see them from above 44

map scale: a ratio that shows the relationship of the distance between two points on the map and the same two points in real life 43

measurement (unit of): a standard quantity that expresses a physical quantity 326

member: element; any of the distinct objects that belong to one set 9

mirror line: a line of symmetry; an imaginary straight line that divides a plane shape into two congruent shapes 238

mixed number: a whole number combined together with a fraction 134

mortgage: a large loan used to buy a home 288

multiple: a multiple of a number is the result of multiplying this number by another.... 25, 127

N

net: a 2-dimensional shape or pattern that can be folded to make a 3-dimensional solid 83

notation: the way you write sometimes – the symbols or abbreviations you use 56

***n*-th term:** the position of a term (n) in an arithmetic sequence 275

null set (empty set): a set that has no members 10

numerator: the number above the line in a fraction 138

O

object: a shape 254

obtuse: an angle that measures between 90° and 180° 339

octagon: a 2-dimensional shape that has eight straight sides 69

octahedron: a 3-dimensional solid that has eight triangular faces of equal size 81

open sentence: a sentence that is not known if it is true or false 268

operation: a mathematical process such as addition (+), subtraction (-), multiplication (x) and division (÷) 265

operator: a symbol, such as +, that tell us which operations to perform on numbers and variables 264

order: listing numbers from least to greatest or greatest to least 152

order of operations: a convention that tells you the sequence you have to follow when you solve a mathematical expression with two or more operations 265

ordered pair: a pair of coordinates written in this order (horizontal distance first, then vertical distance) 251

ordinate: the place of a point on a vertical axis 251

outcome: the result or the effect of an event 365

P

p.m.: the hours from 12 o'clock in the day until 12 o'clock at night 56

pair of compasses: a V-shaped instrument used for drawing circles 316, 344

parallel: two or more lines or surfaces that have the exact same distance from one another and they never cross 78

parallelogram: a 2-dimensional shape with four sides and with its opposite sides having equal length and being parallel 78

pattern: a repeated shape, design or arrangement 272, 307

pedestrian: a person moving around a town or city on foot 332

pentagon: a 2-dimensional shape that has five sides and five angles 69

per cent: out of 100 190

percentage: a proportion of the whole 190

percentage decrease: a proportion of the whole becoming less 200

symmetrical (parts): two parts that mirror each other and can be folded perfectly.........238

T

table: a simple set of rows and columns......... 96

tally chart: a useful way to collect information by drawing tally marks for each item you count.......... 96

tally marks: a set of lines, four vertical and one diagonal, that represent five items........... 96

tax: an amount of money we pay to the government so they can provide us with public services..........286

term: the position of the separate elements in a sequence..........298, 307

tessellation: when a surface is covered with a pattern of two-dimensional (2-D) shapes so that there are no overlaps or gaps.............208

tetrahedron: a solid shape that has four triangular faces.......... 81

theoretical probability: the probability of an outcome, calculated before doing an experiment..........368

three dimensional (3-D): a solid object that has three dimensions: length, width and height..........227

tiling: when a surface is covered with a pattern of two-dimensional (2-D) shapes so that there are no overlaps or gaps..........208

time (unit of): a standard amount of time used in measurement, for example, a minute..........327

time interval: how much time has passed between events, or how long something takes.......... 59

traffic (rules): all rules that tell you how to act as a pedestrian or a driver..........332

transformation: a change in position or orientation of a 2-dimensional shape or image..........254

trapezium: a 2-dimensional shape that has two parallel sides.......... 78

two-dimensional (2-D): a shape that has only two dimensions we can measure: length and width..........208

tyre footprint: the amount of tyre that touches the ground..........310

U

union: the combined members of two sets....16

unit: a standard measure used to express amounts..........41

unit of measure: a standard quantity that expresses a physical quantity..........213, 302

unit solid: the three dimensions of a solid shape that are all equal..........227

universal set: a set that contains all the possible members you may use in one mathematical problem..........10

unlikely: something that is not probable to happen or be true..........362

V

variable: a letter that represents an unknown value or a value that can change...........264, 352

Venn diagram: a graph that uses overlapping circles in a frame to show sets and the relationships between them..........14

vertex (pl: vertices): the point of a solid shape where three or more faces meet or the point where the two rays of an angle meet..........81, 228, 339

vertical: a straight up line that forms a 90° angle with a horizontal line or surface...........251

vertically opposite (angles): the angles that are formed when two straight lines cross one another and the angles are opposite one another..........343

volume: the amount of space a solid takes up..........227, 229

W

weight: how heavy someone or something is; the force of gravity on someone or something..........309

whole: the complete quantity of something..........190

withdrawal (slip): a record of how much money you take out from your bank account..........287

X

x-coordinate: the first element in an ordered pair, with its value resulting from measuring parallel to a horizontal axis (x-axis)..........251

Y

y-coordinate: the second element in an ordered pair, with its value resulting from measuring parallel to a vertical axis (y-axis)..........251